Sea of Sand

Public Lands History

GENERAL EDITORS

Ruth Alexander
Mark Fiege
Adrian Howkins
Janet Ore
Jared Orsi
Sarah Payne

MICHAEL M. GEARY

Sea of Sand
A History of Great Sand Dunes National Park and Preserve

UNIVERSITY OF OKLAHOMA PRESS : NORMAN

Also by Michael M. Geary:
A Quick History of Grand Lake (1999)

Library of Congress Cataloging-in-Publication Data

Name: Geary, Michael M.
Title: Sea of sand : a history of Great Sand Dunes National Park and Preserve / Michael M. Geary.
Other titles: History of Great Sand Dunes National Park and Preserve
Description: Norman, OK : University of Oklahoma Press, [2016] | Series: Public lands history series ; volume 2 | Includes bibliographical references and index.
Identifiers: LCCN 2015031829 | ISBN 978-0-8061-5210-3 (hardcover) ISBN 978-0-8061-6914-9 (paper) Subjects: LCSH: Great Sand Dunes National Park and Preserve (Colo.)—History. Classification: LCC F782.G78 G43 2015 | DDC 978.8/49—dc23
LC record available at http://lccn.loc.gov/2015031829

Sea of Sand: A History of Great Sand Dunes National Park and Preserve is Volume 2 in the Public Lands History series.

The paper in this book meets the guidelines for permanence and durability of the Committee on Production Guidelines for Book Longevity of the Council on Library Resources, Inc. ∞

Copyright © 2016 Michael M. Geary. Published by the University of Oklahoma Press, Norman, Publishing Division of the University. Paperback published 2021. Manufactured in the U.S.A.

All rights reserved. No part of this publication may be reproduced, stored in a retrieval system, or transmitted, in any form or by any means, electronic, mechanical, photocopying, recording, or otherwise—except as permitted under Section 107 or 108 of the United States Copyright Act—without the prior written permission of the University of Oklahoma Press. To request permission to reproduce selections from this book, write to Permissions, University of Oklahoma Press, 2800 Venture Drive, Norman, OK 73069, or email rights.oupress@ou.edu.

For Kerri and Evan

CONTENTS

	List of Illustrations ix
	Acknowledgments xi
PROLOGUE	Reflections in the Sand 3
ONE	Prehistoric Dunes 10
TWO	Outskirts of Empire 32
THREE	Dunes in the Great Desert 50
FOUR	Arid Eden 79
FIVE	Monumental Dunes 102
SIX	Sand and Water 131
SEVEN	Racing for a National Park 157
EIGHT	Unimpaired for Generations 176
EPILOGUE	Lessons in the Sand 206
	Notes 213
	Bibliography 249
	Index 271

ILLUSTRATIONS

0.1 The San Luis Valley   7

1.1 Hayden's map of the San Luis Valley   12
1.2 Great Sand Dunes from the International Space Station   15
1.3 The Ghost Forest   17
1.4 The Great Sand Dunes Tiger Beetle   18
1.5 Clovis and Folsom stone points   22
1.6 Depiction of Paleoindians in the San Luis Valley   23
1.7 San Luis Lake   27

2.1 Don Juan de Oñate   33
2.2 Depiction of a Spanish expedition into the San Luis Valley   38
2.3 Detail of mural in Luther Bean Museum, Adams State College   41
2.4 Pirate George Lowther from 1959 comic book *The World around Us*   43

3.1 Zebulon Montgomery Pike   51
3.2 Depiction of Pike at the Great Sand Dunes   54
3.3 Long expedition map   57
3.4 Sketch of the San Luis Valley   61
3.5 Drawing of the Great Sand Dunes   62
3.6 Map of the "San Luis Parc"   71
3.7 Charles Richardson's mysterious lake   73
3.8 Earliest known photograph of the Great Sand Dunes   75
3.9 Photograph of San Luis Lake   76
3.10 Booster-era pamphlet promoting the San Luis Valley   78

4.1 Ulysses Virgil Herard, circa 1930   85

4.2 Crestone, Colorado 87
4.3 Hotel Liberty 89
4.4 "Master of Them All" steam tractor 93
4.5 Artesian well in the San Luis Valley 99

5.1 The Great Sand Dunes in 1927 107
5.2 Medano Creek, the "Lost River" 108
5.3 Ruins of the Volcanic Mining Company's gold-processing mill on the banks of Medano Creek 116
5.4 Remains of a gold dredge, Medano Creek 117
5.5 The Great Sand Dunes in 1932 118
5.6 Access road to Great Sand Dunes National Monument 120
5.7 Main entrance to Great Sand Dunes National Monument 122
5.8 U.S. Army troops training at the Great Sand Dunes in 1942 123
5.9 Monument headquarters/entrance station 124
5.10 1948 map of Great Sand Dunes National Monument 125
5.11 Monument entry road and sign 127
5.12 The new Great Sand Dunes National Monument Visitor Center 128
5.13 Pinyon Flats campground 128

6.1 Center-pivot sprinkler patent application 134
6.2 Satellite image of the San Luis Valley 135
6.3 Closed Basin Project Main Conveyance Channel 137
6.4 Closed Basin Project solar-powered monitoring well 139
6.5 Various dune types 148

7.1 Scott McInnis and Bill Wellman racing 158
7.2 Entry gate at the Medano Ranch 166
7.3 Outbuildings at the Medano Ranch 167
7.4 The Summit in the Sand 173

8.1 Cover of *Colorado Central Magazine* 179
8.2 Official designation ceremony 196
8.3 North gate of the Baca National Wildlife Refuge 197
8.4 New entrance sign at Great Sand Dunes National Park and Preserve 198
8.5 Updated map of Great Sand Dunes National Park and Preserve 200

9.1 Bison grazing on the banks of Big Spring Creek 207
9.2 Indian Spring 209

ACKNOWLEDGMENTS

THIS BOOK OWES ITS EXISTENCE to a wide variety of individuals, institutions, and organizations other than the author. Chief among them is the National Park Service (NPS), for its leading role in the preservation and resource management of the Great Sand Dunes, as well as its continuing commitment to safeguarding wilderness landscapes across the United States "by such means as will leave them unimpaired for the enjoyment of future generations." The NPS generously provided funding for the research and compilation of *Sea of Sand* through a cooperative agreement with the Rocky Mountains Cooperative Ecosystem Studies Unit (RM-CESU) at the University of Montana–Missoula. Thanks to Kathy Tonnesson and Pei-Lin Yu of the RM-CESU for their guidance and patience with this project.

Speaking of patience, I owe a tremendous debt of gratitude to Maren Bzdek, program manager for the Public Lands History Center at Colorado State University. Maren surely ranks among the most patient people I've ever had the pleasure of meeting, and she played a pivotal role not only in helping to secure funding for this project, but also in providing an inexhaustible supply of support in acquiring research materials and interlibrary loans. More importantly, she continually offered me kind words of encouragement when unforeseen circumstances repeatedly threatened to derail my progress. For this and a thousand other kindnesses large and small, I am eternally grateful.

NPS regional historian Robert Spude reviewed the initial draft of the manuscript and offered incisive commentary, encouragement, and suggestions for improvement, for which I am much obliged. The superb staff at Great Sand Dunes National Park and Preserve (GRSA) also deserves special recognition for, among other things, their impressive professionalism, collective expertise, and pervasive good humor. Sincere thanks are due to former Great Sand Dunes superintendents Bill Wellman, Steve Chaney, and Art Hutchinson for offering their timely input and keen insights about the dunes. Thanks also go to their respective support personnel past and present, including Patrick Myers, Kris Illenberger, Tom Morin, Sue Judis, Rocket, Barb Irwin, and at least a dozen other staffers whose names I've forgotten but whose generosity I have not. For assistance above and beyond the call of normal duty, my gratitude to GRSA park biologist Phyllis Pineda Bovin, who kindly allowed me free rein to rummage through the park archives and museum in search of obscure photographs, specimens, and documents; to GRSA park geologist Andrew Valdez, who never once failed to provide expert answers to my countless (and often ignorant) questions, and whose offer to drive my brother and me up Medano Pass one summer long ago began an adventure that has lasted nearly twenty years; and finally, to my dear friend and GRSA chief of resource management Fred Bunch, a heartfelt and very humble thank-you for your wit, your wisdom, your enthusiasm, and your guidance. Fred has already forgotten far more than I will ever know about the Great Sand Dunes, and his willingness to share his prodigious knowledge of this remarkable landscape is as inspiring as it is enlightening. Thanks, Fred—the next round is on me.

Nearly two decades ago, my education about the Great Sand Dunes began at Colorado State University (CSU), where I had the tremendous good fortune to learn from (among others) Dr. Janet Ore, Dr. Ruth Alexander, Dr. Daniel Tyler, Dr. Liston Leyendecker, Dr. Jared Orsi, and especially Emeritus Professor John Albright, who effortlessly combines a piercing intellect and razor-sharp wit with a contagious enthusiasm for history. Not only was he instrumental in my decision to begin researching the history of the Great Sand Dunes, he continually pushed me to excel and offered kind words of encouragement and advice throughout my education at CSU. To all of them, as well as to the superb faculty and staff of the CSU Department of History and Morgan Library, most notably Doug Ernest and Naomi Lederer, I offer my grateful thanks.

My deepest appreciation and gratitude is reserved for Dr. Mark Fiege, principal investigator on this project and my graduate adviser at CSU, who guided, inspired, advised, encouraged, cajoled, berated, and befriended me as I navigated my way through the minefields of graduate school. As a founder of CSU's Public Lands History Center, Mark was instrumental in securing funding for this project and convincing me to revisit the Great Sand Dunes long after I had graduated and immersed myself in the vicissitudes of earning a living as an historian. Through countless hours of reviewing my drafts, refining my arguments, and polishing my prose, his penetrating insights and incisive commentary have proven invaluable, and he never let me lose sight of the importance of history in understanding the human condition. I'm convinced that I've become a better writer and historian (and human!) thanks to his efforts. I sincerely hope he realizes how much I admire and respect his intellect, his dedication to history, and his abilities as both a teacher and a mentor.

I am deeply indebted to former U.S. representative Scott McInnis (R-Colo.), along with his staffers Mike Hesse and Christopher Hatcher, and former U.S.

senator Wayne Allard (R-Colo.) for their learned opinions, factual input, and kind assistance during my research for this book. Mr. McInnis cordially agreed to my interview request over coffee at a truck stop in Denver and answered every one of my questions with humor and insight, while Mr. Allard graciously granted me access to his personal papers in the archives of the Western History/Genealogy Collection at the Denver Public Library. Mr. Hesse and Mr. Hatcher both allowed me to contact them by phone and email and offered their unique perspectives on the legislative battle to secure national park status for Great Sand Dunes National Monument. Former Alamosa County commissioner Bob Zimmerman also consented to be interviewed for my research, as did Erin Smith, former reporter for the *Pueblo Chieftain*; Mark Burget and Paul Robertson of The Nature Conservancy; Ralph Curtis of the Rio Grande Water Conservation District; and Dr. Hobie Dixon of Adams State University. To all of these fine folks, I offer my genuine thanks for their kindness and generosity.

I am also indebted to Denver attorney David W. Robbins, who initially drafted the bill to create Great Sand Dunes National Park and Preserve and whose knowledge of San Luis Valley water issues is unsurpassed; and to Christine Canaly, director of the San Luis Valley Ecosystem Council and tireless advocate for the Great Sand Dunes and San Luis Valley water resources. Both Mr. Robbins and Ms. Canaly kindly took time out from their busy schedules to educate me about the intricacies of water issues in the San Luis Valley, as well as those of the Baca Ranch and the entire Great Sand Dunes ecosystem. Their respective work on behalf on the landscapes and inhabitants of the San Luis Valley is much appreciated, as is their willingness to share their expertise.

For their generous assistance with researching and securing rights for the assorted images, artwork, and maps included in *Sea of Sand*, I'd like to thank Peter Blodgett of The Huntington Library for his help with the Kern prints; Coi Drummand-Gehrig, Ellen Zazzarino, and Abby Hoverstock of the Western History/Genealogy digital collections at the Denver Public Library; Kat Olance of the Luther Bean Museum at Adams State University in Alamosa; Katie Lopez of The Tom Lea Institute in El Paso, Texas; Andrea Ashby of the Independence National Historic Park in St. Louis, Missouri; Jody Russell of NASA-Johnson Space Center in Houston, Texas; David Rumsey at the David Rumsey Historical Map Collection; Sarah Gilmore at History Colorado in Denver; Eric Bittner, archivist for the National Archives, Denver branch; Jenny M. Stevens of the U.S. Geological Survey (USGS), Denver branch; artist Monica Griesenbeck; and Mike Rosso, editor of *Colorado Central Magazine*.

To the ladies of the PEO Sisterhood–Monte Vista Chapter, especially Carla Clutter and Pam Self, and Peg Schall and Peggy Barr of the Monte Vista Historical Society, my sincere gratitude for providing copies of letters and documents describing the pivotal role played by the PEO in establishing Great Sand Dunes National Monument in 1932. Thanks also to Laura Gomez and Bill Martin of the Colorado State Land Board for answering my questions about public lands in the San Luis Valley. For their assistance and input during the early days of my research, my thanks to David Hammond of CSU/NPS; Peter Rowlands and Cliff Martinka of the USGS; John Koshak at San Luis Lakes State Park; Rob Sontag, Tom Wylie, Glen Bean, and Greg Kendrick of the NPS; Al Ossinger of the Colorado Mountain Club; the staff of the Stephen H. Hart Library at the Colorado Historical Society (now History Colorado); and especially all of the members past and present of the Friends of the Dunes.

To Paul Brunswig and my former colleagues at Historical Armory in Fort Collins, including Ayn, Ben, Kim, Paula, Brooke, Heather, Julie, Kristen, Matt, Louie, and assorted others over the years, I offer my sincere thanks, and I wish you continued success in all of your aspirations. It sure was fun while it lasted, and I honestly appreciate your dedication over the last decade, as well as the inspiring and innovative work we did together. Special thanks to artist Heather Paul, truly one of the most creative people I've ever met, for her enduring kindness and incredible artwork depicting historical scenes of the Great Sand Dunes, as well as her uncanny ability to transform my ridiculous, arm-waving descriptions into useful, original art. Finally, a huge, extra-special-sauce, double-secret-probation thank-you to graphic artist/designer/engineer/jack-of-all-trades extraordinaire Tim (the Wizard) Deal, an amazingly talented, creative, and gracious individual, possessed with skills beyond measure, who has helped me in ways that I cannot even begin to fathom, and whose friendship has been a blessing to me over the years.

At the University of Oklahoma Press, my deepest gratitude to Charles Rankin, Rowan Steineker, Stephanie Evans, Brittney Berling, Amy Hernandez, and Jay Dew for their collective expertise and limitless patience. Highest of praise also goes to copyeditor Christi Stanforth for her incomparable and insightful editing skills.

On a personal note, I'd like to express my sincere appreciation for the various friends and family who have supported and encouraged me on this long, strange journey, including the entire extended James and Brownlee families, who welcomed me from the moment I met them and who long ago made Colorado my home away from home; the cherished friends who comprise my Seattle connection, most especially Chris O'Hara, Bill Cruikshank, and Joe Lowry, for filling my life with irreverent humor and for always keeping me humble; my late father, David Patrick Geary Jr., and my late brother, Timothy Patrick Geary, for a lifetime of lessons taught and lessons learned; my dear mother, Lee Geary, for her unwavering love, incredible kindness, and boundless compassion, forever offered without judgment or condition; and to my brother Daniel David Geary, for his formidable intellect, profound wisdom, and for half a century of side-splitting laughter. As the mountain men of the American West used to say, "He tends a warm fire."

Finally, this book is dedicated to my remarkable son, Evan Daniel, who continues to astonish me daily with his sharp mind, contagious curiosity, and genuine kindness, and who has made me proud to be a father from the moment he entered this world; and to the love of my life, Kerri Lynn, who has somehow always managed to catch me when I stumble, and whose relentless optimism, quiet dignity, and unbelievable inner strength in the face of extraordinary adversity has taught me more about the resiliency of the human spirit than I ever thought possible. This book simply would not exist without her encouragement and support. I am indeed blessed to have both Kerri and Evan in my life. The adventure continues. Onward!

Sea of Sand

PROLOGUE | Reflections in the Sand

When I use the term "environmental history," I mean specifically the history of the consequences of human actions on the environment and the reciprocal consequences of an altered nature for human society.

Richard White, *Trails: Toward a New Western History*

THE HEADWATERS OF MEDANO CREEK gather in the lofty peaks of the rugged Sangre de Cristo Mountains of south-central Colorado, swelling with spring water and seasonal snowmelt before cascading down the rocky western side of Medano Pass. The upper portion of the creek resembles countless other watercourses in the Colorado high country, but once freed from the relative confines of the Sangres, Medano Creek takes on a decidedly unique appearance. Its sparkling waters rush headlong into the vast inland sea of sand known as the Great Sand Dunes, where the pervasive forces of sand, wind, water, and time have combined to create one of the world's most distinctive landscapes.

The Ute Indians who once roamed this region called the towering dunes Sowapophe-uvehe, "Land That Moves Back and Forth." In 1807, explorer Zebulon Pike and members of his expedition crossed Medano Pass and encountered the vast collection of dunes, which Pike described in his journals as appearing "exactly that of a sea in a storm." Contemporary visitors recognize them as the primary geologic feature of Great Sand Dunes National Park and Preserve, home to the tallest sand dunes in North America. My education about the Great Sand Dunes began on the banks of Medano Creek near the remains of an old gold-processing mill, long since abandoned. As part of an ambitious environmental history project studying what was at the time Great Sand Dunes National Monument, the National Park Service (NPS) had arranged a guided tour to familiarize project participants with the stunning landscape of the dunes.

Our tour commenced in a popular parking area that normally offered visitors close, convenient access to the main dunefield. On this particular day, however, the trail from the parking lot to the dunes had vanished, submerged beneath the swirling waters of Medano Creek.

Judging by the faces of our NPS guides, we were indeed fortunate to witness such high water. Normally the creek coursed along the perimeter of the dunes in a braided network of shallow, sandy channels, but the deep snows of the previous winter had turned the reluctant trickle into a raging torrent. As we stood transfixed by the rushing water, our guides made sure that we noticed the rare and curious hydrological phenomenon known as "surge flow," which is peculiar to only a handful of streams in the world, most notably Medano Creek. Every twenty seconds or so, a clearly defined wave rippled past like a tidal current and then dissipated into the hazy distance. The phenomenon can be traced to the swift flow of shallow water over the sandy creek bed, which causes the formation of small sand ridges known as *antidunes*; these antidunes confine the water until sufficient weight and volume cause the ridge to collapse, resulting in a rush of water that resembles nothing so much as an ocean wave heading for an appointment with a distant beach.

Our tour continued with a visit to the Ghost Forest on the east side of Medano Creek, where drifting sand had buried several large stands of ponderosa pine, leaving only ghostly skeletal remnants of a once-thriving forest. We wandered the ruins of the old Herard homestead, settled in 1875 near the western approach to Medano Pass, where horses belonging to pioneer settler Ulysses Herard may have inspired the enduring legend of the "web-footed mustangs," a herd of wild horses with hooves flattened and webbed from generations of traversing the soft sand. Later, as the deep shadows of the gathering twilight softened the stark contours of the bronze-colored dunes, we stopped to examine the park's numerous culturally peeled ponderosa pine trees, their massive trunks scarred by hungry Ute and Apache in search of food and medicine sometime in the not-too-distant past.

After the tour, in the quiet warmth of a conference room at park headquarters, NPS personnel revealed the goals and rationale for the environmental history project. In 1986, an audacious plan by American Water Development Inc. (AWDI) to export groundwater from the San Luis Valley had sparked an extended legal battle over water rights in the valley. AWDI lost that particular battle in 1991, but the threat to the valley's water resources remained, as did the potential threat to the Great Sand Dunes, which are hydrologically connected to the valley's surface and groundwater. As the NPS sought data to prove those hydrological connections and strengthen its case against AWDI in court, resource management personnel at the park concluded that only by fully understanding the intricacies of the entire Great Sand Dunes ecosystem, along with its historic connections to the broader landscapes of the San Luis Valley, could they protect the sand dunes from future threats. The environmental history project had therefore been initiated as part of an ambitious effort to gather as much information as possible about the myriad geological, hydrological, and cultural forces that had shaped the Great Sand Dunes over time.[1]

Specifically, NPS personnel sought detailed information about the various climatic and biological changes that had occurred at the sand dunes throughout history, an important step in attempting to distinguish between natural and anthropogenic (human-caused) ecological changes at the dunes. In addition, they wanted to identify the key indicators of environmental change in the region (an indicator represents the state of certain environmental conditions over a given area and a specified period of time), as well as explore the disparate land-use practices that had occurred in the long history of human habitation in the San Luis Valley, the largest of Colorado's four legendary mountain "parks" (North, Middle, and South Park are the other three). Finally, the NPS expressed concern over the shrinkage and disappearance of a series of interdunal ponds on the western edge of the main dunefield, and wondered if their disappearance could be attributed to climatic change, groundwater pumping, or some other unknown variable. By accumulating this information, park resource managers sought a more detailed scientific and historical understanding of the Great Sand Dunes. Their ultimate goal was the creation of appropriate resource management strategies for protecting one of the most unusual and incongruous landscapes in North America.

The innovative project consisted of three distinct fields of inquiry: biological, geographical/geological, and cultural/historical. My portion of the project focused on the cultural history of the Great Sand Dunes and the San Luis Valley, which seemed to offer an intriguing array of research possibilities. The valley is a veritable crossroads of human history, a cultural corridor spanning eons of time, from Paleolithic hunters to pre-Columbian Native American cultures, from Spanish conquistadors to fur trappers, from transcontinental explorers to modern tourists in motor homes. The region also serves as a microcosm of the broader history of the American West, with most of the familiar historical elements in place: intrepid adventurers, frontier forts, ambitious settlers, farmers and ranchers, mineral strikes, railroads, land grabs and water wars, massive resource extraction and exploitation, and the continual ebb and flow of the "boom and bust" cycles that typified America's economic development during the nineteenth and early twentieth centuries.

My initial research uncovered an abundance of written historic references to the San Luis Valley, references that tended to reflect the cultural biases of their authors. Generally, early Spanish explorers wrote about the region's rivers, geography, and prospects for plunder, while later American explorers focused on the valley's climate, geology, and suitability for railroads. The admittedly few records left behind by fur trappers described the valley's mountain passes and abundant streams teeming with beaver and other fur-bearing critters, while the "boosters," those great champions of America's Manifest Destiny, heralded the region's fertile soils, copious water, and unrivaled prospects for settlement. Taken together, these written references formed an intricate and detailed mosaic of the cultural forces that combined to create the valley's human history since the first Spanish incursions of the late sixteenth century. Finally, when I discovered a remarkable sketch of the dunes in the archives of the Denver Public Library Western History Collection, drawn in 1872 by Charles S. Richardson and never before published, my slowly simmering interest in the Great Sand Dunes exploded into full-blown excitement.

This generous helping of written and visual history promised to make for a stimulating research project, yet initially I struggled with the abundant data, seeking a unifying theme for the historical narrative I had been formulating in my mind. In subsequent meetings with NPS personnel, I proposed a number of what I thought were novel approaches, including focusing exclusively on extractive activities in the valley throughout history and how they affected the dunes, or exploring the concept of landscapes being altered and exploited as a result of specific differences in cultural values and mores. These ideas met with generally positive yet vaguely unenthusiastic approval from park staff, which left me both frustrated and curious about how exactly I was going to proceed.

The breakthrough finally came during a one-on-one meeting with park superintendent Bill Wellman, who summed up his aspirations for the environmental history project with one simple phrase: "Follow the water." Intuitively aware that water was the key to understanding the history of the Great Sand Dunes, and by extension the entire San Luis Valley, Wellman patiently explained his desire for the cultural component of the project to focus on the role that water has played in determining the extent of human activity in the valley over time, with an emphasis on the history of water at the dunes and how they were connected hydrologically to the greater San Luis Valley. Wellman's guidance proved invaluable, for in the very midst of my quest for a unifying narrative theme, I had casually overlooked the fact that water is undoubtedly the crucial thread that binds together the rich cultural tapestry of San Luis Valley history. The availability of water is what first attracted prehistoric animals and Paleolithic hunters to the valley, just as it attracted later Native Americans, Spanish and American explorers, Hispanic and Anglo-American settlers, huge agribusinesses, and aspiring water developers. To be sure, an in-depth examination of water as a fundamental force in human history was far from a revolutionary concept. The history of much of the arid American West is intimately, inextricably linked with water, and countless authors from John Wesley Powell to Wallace Stegner to Marc Reisner have explored the intricacies of that relationship. To my knowledge, however, a history of the Great Sand Dunes and the San Luis Valley that specifically emphasized water remained largely unexplored territory.

Ironically, the San Luis Valley is the driest part of Colorado, receiving an average of only seven to ten inches of rain per year, yet its roughly 8,000 square miles sit atop a series of enormous freshwater aquifers that some scientists estimate

Figure 0.1 The San Luis Valley of south-central Colorado, a paradoxical landscape defined by a distinctive combination of aridity and moisture. Original artwork by Heather Paul.

hold as much as 2 billion acre-feet of water (an acre-foot equals approximately 326,000 gallons, enough to cover an acre of land with water one foot deep, roughly what a typical suburban American family of four uses annually).[2] The region also contains a plethora of creeks and streams brimming with seasonal runoff from the surrounding mountains, and the fabled Rio Grande del Norte, the Great River of the North, courses through the southern portion of the valley. Water is at once scarce and abundant, and in the midst of this paradox is the quintessential embodiment of aridity, the drifting, towering heaps of sand that comprise the Great Sand Dunes. Their very existence depends on water, and yet they seem the ultimate representation of a desert. This curious incongruity of aridity and moisture continues across the San Luis Valley floor, where sprawling sheets of sand coexist with enormous fields of fertile soil and irrigated crops, where plentiful groundwater and artesian wells are countered by scant rainfall, parching winds, fierce blizzards, and bone-cracking cold. The environment is both hostile and accommodating, and throughout centuries of history, humans have managed to adapt to this unusual dichotomy only by utilizing the valley's one truly indispensable resource: water.

Superintendent Wellman's suggestion to focus on water also proved to be remarkably prescient for the near future of the Great Sand Dunes. In the mid-1990s, close on the heels of AWDI's unsuccessful attempt to export groundwater from the valley, yet another ambitious water-mining proposal emerged, this time from a group called Stockman's Water Company. Once again, local opposition was fierce, and the subsequent battle over the San Luis Valley's water resources ended up galvanizing an unprecedented coalition of politicians, ranchers, scientists, farmers, governmental agencies, water managers, business owners, environmentalists, conservancy groups, and just plain old regular folks who united in an effort to end the war over water in the San Luis Valley, once and for all. The vigorous defense of the valley's water took several years, but in the process, in what surely must be one of the most remarkable success stories in the long history of the National Park Service, enormous tracts of land that had been held in private hands for well over a century became part of the public domain, and the towering sand dunes that had been protected as a national monument since 1932 finally achieved official recognition as a national park in 2004. In essence, the threat to export the valley's precious water resources became the very catalyst for the process that ultimately led to the entire Great Sand Dunes ecosystem being safeguarded in perpetuity.

More than a decade after I had finished my portion of the original environmental history project, an unlikely and unforeseen series of coincidences that included a cooperative agreement between the National Park Service, the Rocky Mountains Cooperative Ecosystem Studies Unit, and Colorado State University led to an opportunity to update my work and to recount the fascinating story of how Great Sand Dunes National Monument finally became a national park. I had long since moved on to other adventures and the exigencies of earning a living, but the chance to revisit the dunes intrigued me immediately, and before long I began digging through old boxes looking for my research files from the original project. Fittingly, the first file I opened contained notes from the last formal field trip I had taken at the dunes, a memorable journey to the interdunal ponds on the extreme western edge of the main dunefield. For years, NPS personnel had been monitoring the disappearance of these ponds and were actively seeking to explain why the ponds were slowly vanishing.

Having never visited the remote western reaches of the monument, I had eagerly accepted an invitation from Resource Management Specialist Fred Bunch and joined his crew in an NPS four-wheel-drive jeep. After a bone-jarring ride across the stark and undulating landscape immediately south of the park boundary, my fellow researchers and I set off on foot across the wind-swept dunes, stopping occasionally to examine the various hardy plants that had gained a foothold against the inexorable tide of sand. The wind howled incessantly, stinging any exposed skin and infiltrating every crease and fold of our clothing. Finally, cresting the windward summit of a steep dune, we came upon what appeared to be a large rock garden, with broken pieces of stone scattered around the remains of a dried-up pond.

Curious about this seemingly incongruous gathering of rocks, I poked around the sand and eventually discovered two small stone points, one a deep black color and the other an iron-oxide red, both of them intact and obviously worked by human hands. On closer examination, the arrow-shaped stones exhibited a number of characteristics typical of Folsom points: fine razor-sharp tips, delicately flaked edges, and distinct narrow flutes incised down either side. Clearly, these were prehistoric artifacts brought to the site by Paleolithic hunters perhaps as early as 10,000 years ago. More significant, at least from my perspective, was the fact that I had discovered the points at the edge of a long-vanished pond, the remnants of an oasis in the midst of the forbidding dunefield, where the hunter or hunters had gathered long before the dawn of recorded history.

Unearthing the points was a thrilling experience, and after a few photographs and a gentle reminder to return the points to the sand where I had found them, we continued our journey to the interdunal ponds, where to our surprise we discovered a large beaver dam heaped in the midst of the largest pond. Again, the importance of water in this arid, seemingly desolate place became starkly evident. The water that had sustained Paleolithic hunters and their prey thousands of years ago, that had attracted successive waves of Spanish and American explorers, and that had inspired extensive Hispanic and Anglo-American settlement and irrigation efforts was still supporting wildlife in the very midst of the enormous dunefield, continuing the cycle of survival in apparent defiance of the region's persistent aridity.

Reading the notes from that field trip, I clearly remembered how profoundly I had been affected by the discovery of those Folsom points so many years before, and in the weeks that followed, I found my thoughts continually drifting back to them. The human history of the Great Sand Dunes clearly began with those points, and so, believing it best to begin at the beginning, I once again commenced my journey into the past in the howling wind near the water's edge, provisioned with curiosity and compelled by stone points in the shifting sand.

ONE | **Prehistoric Dunes**

It is clear that the single most important factor governing the human habitation of the San Luis Valley is the availability of water.

Dennis Stanford, "History of Archaeological Research"

SCULPTED INTO THEIR PRESENT graceful contours by countless centuries of wind and water, the Great Sand Dunes sprawl along the eastern fringes of the vast San Luis Valley in south-central Colorado, covering an area of nearly thirty square miles. Hard against the rugged slopes of the Sangre de Cristo Mountains, they are the tallest aeolian (wind-produced, also spelled *Aeolian* or *eolian*) dunes in North America, heaping mounds of sand that tower up to 750 feet above the flat immensity of the valley floor. Despite their impressive height, the dunes are dwarfed beneath the lofty summits and thickly forested foothills of the Sangres. Strangely incongruous and otherworldly, seemingly isolated and remote at the base of the majestic mountains, the dunes are nonetheless intimately connected with the landscape that surrounds them, a massive intermontane basin that is often described as the largest alpine valley in the world.[1]

Among the most enduring questions asked by generations of visitors to the massive dunefield is some variation of "How did all of this sand accumulate in the middle of the Rocky Mountains?" The answer to that query is constantly evolving, but most modern geologists agree that the origins of the Great Sand Dunes lie in the deepest recesses of geologic time, when large surface plates in the earth's crust beneath what is now the San Luis Valley began splitting apart in a process known as *rifting*. Crustal uplift accompanied this rifting, along with extreme volcanic activity that eventually unleashed the cataclysmic eruption of the La Garita Caldera about 27 million years ago, which many geologists believe was among the largest volcanic explosions in earth's history.[2] In the aftermath, the San Juan

Mountains began forming on the western fringes of the valley, born from the hot ashes and lava flows caused by the violent fracturing of the planetary shell. The Rio Grande Rift, a long, deep chasm that stretches from central Colorado all the way to northern Mexico, also appeared around this time, another product of the continual tectonic rending of the earth's crust. Around 18 to 19 million years ago, rotation of a large crustal plate forced the relatively rapid (geologically speaking) uplift of the Sangre de Cristo Mountains on the eastern edge of the valley, which further widened the rift and caused it to subside, deepening the chasm. The Rio Grande Rift has been collecting eroded sand, silt, gravel, and assorted detritus known as *rift fill* from the surrounding mountains ever since. In some places, this rift fill measures up to 17,000 feet thick, and it may go even deeper, with the very top layer visible as the floor of the modern San Luis Valley.[3]

Geologic evidence further indicates that along with this rift fill, water has been accumulating in the valley for millions of years, creating ephemeral lakes and wetlands that appeared and disappeared as the climate changed over countless millennia. Most notable was an enormous body of water known to modern science as Lake Alamosa, which covered a considerable portion of the San Luis Valley beginning about 3.5 million years ago, expanding and contracting and depositing sediment in response to the fluctuating glacial and interglacial climates of the Pleistocene Epoch. In 2002, geologists confirmed the existence of Lake Alamosa with the discovery of lakebed deposits on low hills in the southern portion of the valley. At its fullest expanse, the ancient lake extended over sixty miles from north to south and nearly thirty miles east to west, making it one of the largest high-altitude lakes in North America. Lake Alamosa existed for roughly 3 million years, until around 440,000 years ago, when its waters overtopped a low point in the San Luis Hills, cut a deep gorge, and began draining southward, eventually joining the ancestral Rio Grande River.[4]

Through untold eons of rifting and uplift, volcanic eruptions, sedimentary deposition, erosion, and the appearance and disappearance of transitory lakes both large and small, the material necessary for the creation of the Great Sand Dunes slowly accumulated on the floor of the San Luis Valley. Exactly how old this material is, as well as exactly where it originated, has long been debated by geologists, beginning in 1869 with Ferdinand V. Hayden's official United States Geological Survey report on the geology of Colorado and New Mexico. Hayden theorized that the sand came from "loose materials" of the Santa Fe Formation, the principal sediment in the Rio Grande Rift. These materials began accumulating sometime during the Miocene Epoch, which lasted from roughly 23 million to 5 million years ago. Curiously, Hayden, a renowned geologist who had traveled extensively throughout the American West, evidently found the prospect of sand dunes piled up to 750 feet high at the base of a range of mountains that included several 14,000-foot peaks to be so unremarkable that he devoted but a single sentence to them in his otherwise lengthy report on the region. Nonetheless, his map of the San Luis Valley, which first appeared in his

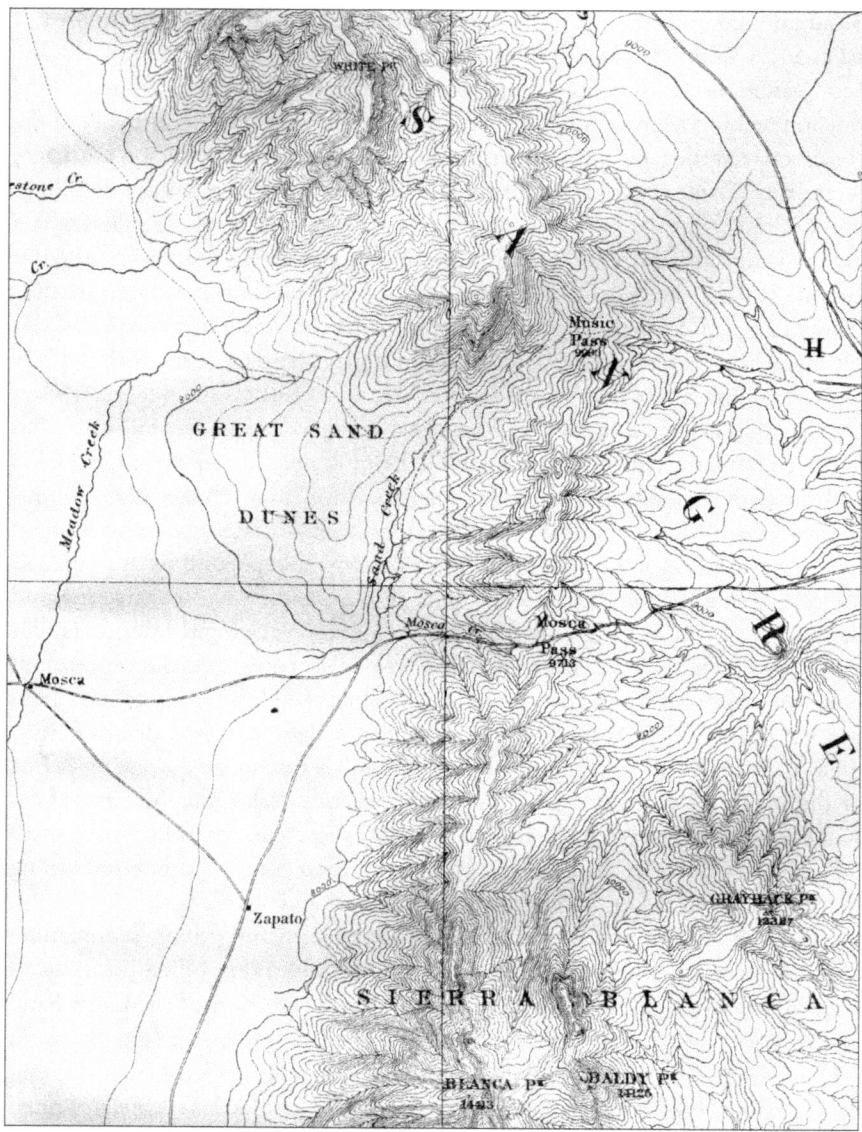

Figure 1.1 Ferdinand Hayden's map of the San Luis Valley, compiled from surveys undertaken in 1874–75. Perhaps the first time the term "Great Sand Dunes" appeared in print. U.S. Geological Survey.

Geological and Geographical Atlas of Colorado (originally published in 1877) formally labeled the vast dunefield the "Great Sand Dunes," which may be among the earliest uses in print of this now widely recognized term.[5]

Geologist F. M. Endlich reached a different conclusion than Hayden, writing in 1877 that the "dunes seem to be of comparatively recent date, geologically speaking, and belong to the Post-Glacial age," which indicates that he thought the dunes began forming roughly 12,000 years ago, near the end of the last Ice Age. In addition, Endlich advanced the theory that the sand in the dunes came from eroded volcanic rocks in the San Juan Mountains. He declared, "The sand came across the valley from the mountains on the southwest, being driven to its present locality by the prevailing southwest winds . . . especially collected in the reentrant angle in the mountain front near Mosca Pass as an effect of eddying currents in the winds, caused by the low gaps in the mountains near this point."[6] Recent studies have confirmed Endlich's theory, indicating that roughly 70 percent of the sand in the Great Sand Dunes originated in the San Juans, and the remainder came from the Sangres.[7] In 1910 geologist Claude Siebenthal produced a comprehensive report on the San Luis Valley and concluded, like Hayden, that the sand dunes were a remnant of the Miocene Santa Fe Formation that had been "broken down by the winds."[8] He also theorized about the existence of an ancient lake in the middle of the valley, where in "the depression caused by the down warping in the center of the valley there was formed an extensive fresh-water lake, in which was deposited the series of sands and clays that make up the Alamosa formation."[9] Siebenthal emphasized that he was not the first to propose the existence of such a lake; instead he conferred that honor upon an itinerant fur trapper named Jacob Fowler, who first visited the San Luis Valley in 1822 as he made his way up the Rio Grande. Siebenthal wrote of how "this observant trapper grasped the character and later geologic history of the valley"[10] and quoted Fowler's observation that "I Have no doubt but the river from the Head of those Rocks up for about one Hundred miles Has once been a lake of about from forty to fifty miles wide and about two Hundred feet deep—and that the running and dashing of the Watter Has Woren a Way the Rocks So as to form the present Chanel [sic]."[11] Nearly two centuries later, geologists confirmed the existence of the lake described by Fowler and Siebenthal and christened it Lake Alamosa. No doubt the fur trapper and the geologist would have been pleasantly surprised at how prominently their lake figured in future geologic studies and written descriptions of the San Luis Valley. Indeed, the idea of a large lake in the middle of the valley later became a recurring theme in the writings and advertisements of the so-called boosters who were trying to attract ranchers, farmers, and settlers to the valley during the latter half of the nineteenth century.

Between 1910 and 1962, several short papers were published in academic journals that essentially echoed or combined the ideas expressed by Hayden, Endlich, and Siebenthal regarding the Great Sand Dunes, and at least one author concluded that the age of the dunes was still open to question, given the conflicting opinions of previous geologists.[12] Then, in 1967, geologist Ross Johnson proposed an entirely new interpretation for the origin and age of the sand dunes that became widely quoted and accepted for nearly thirty years. His theory, similar to Endlich's, suggested that the dune-building process began near the end of the last Ice Age, roughly 12,000 to 15,000 years ago, in an age known as the Late Pleistocene.[13] Johnson surmised that during this period, as the last of the great continental ice sheets were retreating northward, the waters of what is now the Rio Grande River exited the San Juan Mountains near Del Norte and flowed directly eastward toward the vicinity of the present-day San Luis Lakes, where the river abruptly turned south. Over thousands of years, as melting glaciers poured their waters into the valley, the course of the Rio Grande gradually shifted to the southwest. In the process, the river deposited enormous quantities of eroded rock and sand from the San Juans in great alluvial fans across the center of the valley floor. According to Johnson's theory, this mineral debris provided the raw material for the eventual formation of the Great Sand Dunes.[14]

As the understanding of aeolian geology and dune systems evolved throughout the late twentieth and early twenty-first centuries, a more rigorous scientific analysis of the geologic processes at work in the San Luis Valley presented alternatives to Johnson's hypothesis. For instance, a 2007 study that received considerable support in the scientific community, authored by U.S. Geological Survey (USGS) geologist Richard Madole and his colleagues, proposed that the sand dunes might actually be much older than the Late Pleistocene. Moreover, the geologists asserted that the sand in the dunes may not have originated in the alluvial-fan deposits of an ancestral Rio Grande River, but instead came from ancient dry lakebeds on the floor of the Closed Basin, a portion of the San Luis Valley known locally as the "sump" that has no natural drainage aside from downward seepage. The geologic study concluded that the dunes were the product of multiple episodes of sand deposition and transport that were influenced by climatically driven fluctuations of the underlying water table over tens, or even hundreds, of thousands of years. In other words, as the climate of the San Luis Valley continually changed over time, the peculiar hydrology of the Closed Basin caused a series of shallow lakes to repeatedly form and then disappear over a time span of unknown duration, eventually creating the extensive sand sheet that became the primary source for the sand in the Great Sand Dunes. West of the sand sheet, the mineralized hardpan of the *sabkha* (an Arabic word roughly translated as "salt flat") bears further witness to this process, as fluctuations in the water table repeatedly brought alkaline minerals to the surface and created a hardened crust of sand cemented together by the evaporative concentration of salts.[15]

Since the dunes could only have begun forming when sufficient sand was available on the valley floor, and since the presence of an enormous prehistoric Lake Alamosa in the middle of the San Luis Valley would have necessarily precluded the movement of sand across the valley, Madole and his colleagues concluded that the unique geologic and climatic conditions most conducive to the initial formation of the sand sheet and the Great Sand Dunes probably began coalescing sometime *after* Lake Alamosa began draining, perhaps around 440,000 years ago. Thus the dunes could in fact be considerably older than the previously accepted estimate of approximately 12,000 years old. Using a process called *optically stimulated luminescence* (OSL), which measures how long sand grains have been in complete darkness deep under the surface, recent analysis of sand cores drilled at various locations in and around the main dunefield seems to confirm this conclusion, indicating sand ages between 18,000 and 67,000 years old. Testing of drill cores taken from gravel deposits elsewhere in the valley revealed sand up to 130,000 years old, placing the age of the dunes well into the Middle Pleistocene.[16]

Despite over a century of shifting theories regarding the age and origin of the Great Sand Dunes, geologists unanimously agree that the valley's strong prevailing southwesterly winds bear primary responsibility for the initial formation and continual replenishment of the main dunefield. Loose sand on the floor of the San Luis Valley is continually swept up by the prevailing winds and transported eastward, either as airborne particles or bounced along the ground in a process known as *saltation*, toward a series of three low passes clustered together in a distinct pocket or indentation (known in geological terms as a *reentrant*) in the Sangre de Cristo Mountains: Music Pass to the north, Medano Pass in the middle, and Mosca Pass to the south. Funneled into these low passes by the broad flanks of the Sierra Blanca massif (also known as Blanca Peak and Mount Blanca) and by the formidable barrier of the high Sangres, the sand-laden wind loses its momentum to friction and turbulence as it rises. No longer capable of carrying its load, the wind continually deposits sand at the foot of the Sangres, resulting in the towering accumulation that modern visitors recognize as the Great Sand Dunes.[17]

Figure 1.2 Photograph of the Great Sand Dunes taken from the International Space Station on October 26, 2007, depicting such distinctive geological features as the main dunefield, a portion of the sand sheet and sabkha to the west (*left*), and the Sangre de Cristo Mountains, including the Sierra Blanca massif at lower right. National Aeronautics and Space Administration and Johnson Space Center, Houston, Texas, ISS016-E-066986.

PREHISTORIC DUNES | 15

Far from static, the dunes are in a constant state of flux, their contours forever changing as the pervasive forces of erosion and deposition perpetually shape and reshape the shifting sands. Yet the dunes also exhibit a remarkable degree of long-term stability. Geological and photographic studies indicate that the main dune mass itself has moved very little over the last century, and quite possibly has remained that way for considerably longer. A number of factors account for this stability, including the fierce northeasterly "reversing winds" that occasionally gust through the low passes in the Sangres whenever low-pressure weather systems are located east of the San Luis Valley. These winds blow in the opposite direction of the valley's prevailing southwesterly winds, which temporarily halts the relentless eastward migration of sand and contributes to the dunes' tremendous height.[18]

Intermittent clumps of vegetation (blowout grasses, scurfpea, Indian ricegrass) further anchor the sand in the active dunefield, although the role played by plant life in stabilizing the dunes is relatively minor. Far more crucial are the thin ribbons of water that course along the edges of the Great Sand Dunes, where centuries of natural geologic processes have produced a rather exceptional sand "recycling" system that partially relies on the waters of Medano and Sand Creeks. These waterways, so incongruous in the very midst of the heaping mountains of sand, serve both as barriers to the shifting dunes and as conduits for transporting sand. In a process that has continued ever since the dunes began forming, drifting sand constantly threatens to bury Medano and Sand Creeks, which respond by carving away the eastern and northern perimeters of the massive dunefield. The creeks then wash this eroded material out to the sand sheet to the west, where the valley's pervasive winds continually lift the sand and deposit it back on the dunes. Coupled with the reversing winds that blow opposite the prevailing southwesterlies and the vegetative life that struggles to survive in the depressions between the dunes, Medano and Sand Creeks are critical elements in the dynamic and elegant process of sand erosion, transport, and deposition that ensures the long-term development of large dune forms at the Great Sand Dunes. However, the presence of the "escape dunes" on the far eastern fringes of the main dunefield confirms that even this formidable combination of factors cannot hold back the entire, inexorable tide of shifting sand. The escape dunes have crossed Medano Creek and climbed the foothills of the Sangres, where they have buried, killed, then reexposed entire groves of ponderosa pine trees, creating the eerie Ghost Forest, which stands as stark testimony to the continual transformation of this remarkable landscape.

Another factor in the relative stability of the Great Sand Dunes is somewhat more controversial. Geologists have long suspected a connection between the dunes and the prodigious store of groundwater that lies just below the surface of the San Luis Valley, but the exact dynamics of that relationship were unknown until relatively recently. One hypothesis that emerged during the contentious battles over the valley's groundwater in the late 1980s and early 1990s theorized

Figure 1.3 A portion of the Ghost Forest as it appeared in 1951. National Park Service, Great Sand Dunes National Park and Preserve, GRSA-2257.

that water from the subterranean aquifer beneath the dunes continuously flowed upward in a wicking process known as *capillary rise*, saturating the sand from below to anchor the main dunefield. Although this theory was not actually subject to rigorous testing at the time, its sensational nature made it popular with the local press and led to fears that groundwater pumping would irreversibly lower the water table, causing the dunes to literally dry up and blow away. Modern geologic studies have failed to prove definitively that the dunes are in fact substantially stabilized by groundwater, and currently the most accepted explanation is that the moisture found in the dunes is meteoric in origin. In other words, eons of rain and snow have soaked the dunes from above and filtered downward, thereby contributing to sand cohesion. Still, geologists and National Park Service personnel generally acknowledge that groundwater does play an integral role in the Great Sand Dunes sand transport and recycling system, primarily through its proven hydrologic connections to Sand and Medano Creeks.[19]

Save for the isolated pockets of vegetation that punctuate the sand in sporadic patterns, the dunes appear desolate, barren of life. Located in a vast, arid valley whose average elevation exceeds 7,500 feet, scorched by searing summer heat and frozen by bitter winter temperatures, the relentlessly windy dunes seem at best an extremely forbidding environment. Yet a surprising array of animal life thrives in or near the Great Sand Dunes, including at least seven species of endemic insects that are found nowhere else in the world, most notably the predatory Great Sand Dunes Tiger Beetle.[20]

Other species have likewise found their own unique ecological niches at the Great Sand Dunes. Kangaroo rats, masters of water conservation, thrive in the vegetated areas among the shifting sands, nourished by plant seeds gathered during nocturnal forays into the dunes. Raccoons, porcupines, squirrels, and cottontail and jackrabbits browse the grasslands and piñon-juniper forests that surround the dunes. Ravens, swallows, nighthawks, and golden eagles ride the area's persistent thermal winds, while one of nature's most sublime spectacles occurs each spring and autumn when tens of thousands of sandhill cranes return to the verdant wetlands of the San Luis Valley, including those in the vicinity of the dunes. Elk, mule deer, bison, and pronghorn graze amid the shrubs and grasses that dot the broad valley floor. Native predators such as black bears, mountain lions, bobcats, and coyotes prowl the woodlands and foothills adjacent to the dunes. Higher up, bighorn sheep thrive on the rugged flanks of the lofty Sangres. Each of these species has adapted to the oftentimes harsh, arid conditions of the region, and each contributes a distinctive facet to the compelling ecological diversity and complexity of the Great Sand Dunes.[21]

Created in a swirling, symbiotic dance of wind, water, sand, and time, the Great Sand Dunes are ultimately a striking paradox of aridity and moisture, of abundant life thriving in the midst of apparent desolation, of constant fluctuation and geologic forces coexisting in a state of dynamic equilibrium. Countless eons of erosion and deposition produced the graceful contours and stark beauty of the dunes, and generations of evolution and adaptation filled the region's

Figure 1.4 The Great Sand Dunes Tiger Beetle (*Cicindela theatina*), a predatory beetle found nowhere else on earth. National Park Service, Great Sand Dunes National Park and Preserve.

ecosystems with an impressive variety of vegetative and animal life. At some point during this long prehistoric continuum of geologic time, a moment that can be approximated but never precisely pinpointed, the unique and complex landscapes of the Great Sand Dunes and the sprawling San Luis Valley finally witnessed their first human visitors.

First Footprints

Mystery surrounds the first humans who visited the San Luis Valley. Modern science refers to them as "Clovis," named for the town in eastern New Mexico where archaeologists first excavated artifacts associated with these prehistoric hunter-gatherers in the early 1930s. The more generic term *Paleoindian* is also widely used to designate the earliest humans in North America. Archaeologists have determined that significant numbers of Clovis peoples were definitely present in North America by about 13,000 years ago, but their exact route (or routes) and the precise timing of their arrival remains open to debate. The prevailing theory during the twentieth century suggested that the ancestors of the Clovis first began entering the North American continent across the Bering Land Bridge sometime during the last great Ice Age, perhaps 16,000 years ago, and then spread southward as melting glaciers created an ice-free corridor between the Arctic and the rest of the Americas.[22]

More recent archaeological research has offered alternatives to this theory, including the so-called Solutrean hypothesis, first proposed in 1998 and based on striking similarities between ancient stone tools associated with the Solutrean culture of southwestern Europe (specifically the Iberian Peninsula) and later Clovis tools found throughout North America. The controversial hypothesis, which continues to inspire lively debate, theorized that sometime around 20,000 years ago, during a period known as the Last Glacial Maximum, when continental ice sheets were at their greatest extent, small groups of prehistoric hunters used primitive boats to follow the edge of the continental ice shelf northward up the European coast and across the North Atlantic to North America, where their tool-making methods influenced the later Clovis stone tool technology.[23] Another theory suggests a trans-Pacific origin for the Clovis, while excavations at a Paleoindian site in southern Chile called Monte Verde uncovered evidence indicating human occupation that dates at least 1,000 years earlier than previous sites, pushing human habitation of the New World back to some 14,000 years ago, possibly much earlier.[24]

Newly found archaeological evidence challenging long-established theories seems to surface almost yearly, but regardless of the route they took or when they first arrived, very little evidence remains that would tell us how the Clovis perceived themselves or the world around them, or how intensively they altered that world. The scant traces of their presence in the San Luis Valley consist mostly of stone tools and weapons, often discovered in conjunction with fossilized animal bones, as well as a few instances of petroglyphs and pictographs (rock art) discovered at various locations, mostly on the western fringes of the valley. This scarcity of evidence renders any attempt to reconstruct Clovis lifeways a tenuous combination of hard science and supposition, of artifacts coexisting with inference. Radiocarbon dating of projectile points, bone scrapers, and other stone tools discovered near the Great Sand Dunes indicates that humans associated with Clovis projectile points first entered the San Luis Valley sometime during the Late Pleistocene, perhaps as early as 11,500 years ago. One Clovis site near the dunes, situated at the edge of what appeared to be a shallow Ice Age pond, contained small chert flakes and a bone scraper alongside some badly decomposed Columbian mammoth bones. Unfortunately, local artifact collectors removed several Clovis points from the site in the 1950s, seriously impairing subsequent attempts to analyze the Clovis artifacts in their original context.[25]

Why the dunes? What did the immense sea of sand offer to the Clovis? Precise reconstruction of prehistoric climates is a tricky business, but scientists generally agree that the planet experienced intense climatic fluctuations toward the end of the Pleistocene Epoch, roughly 12,000 years ago. Climatic conditions conducive to the growth of continental glaciers gradually changed, stalling the expansion of ice across North America. As global temperatures increased, continental glaciers retreated both northward and to higher elevations. This glacial meltdown, combined with a dramatic change in global weather patterns, triggered a variety of variations in regional climates and vegetational distributions across the planet.[26]

In the prehistoric San Luis Valley, these climatic fluctuations produced a moist landscape of interconnected lakes, ponds, and marshes, which inspired the growth of nutritious grasses and assorted plant life that attracted migrating herds of mammoth and bison to the region, which in turn attracted the Clovis. Judging from the available archaeological evidence, these large mammals were the favored prey of the Clovis hunters, primarily because a single mammoth or a few bison could provide meat for weeks on end, while hides, tusks, and bones furnished material for weapons, shelter, and clothing. Lacking the relative size, speed, and strength to overcome their prey individually, the Clovis instead most likely hunted in cooperative groups, relying on cunning and guile to overcome their comparative physical deficiencies. Spears and darts made with razor-sharp projectile points attached to wooden shafts augmented their considerable tracking and ambush skills and turned the Clovis into formidable hunters.[27]

It would be a mistake, however, to assume that the Clovis only hunted larger game animals. Archaeological finds at several locations, including the Jake Bluff site in Oklahoma, indicate that the Clovis had an eclectic diet that included not only bison and mammoth but also smaller animals such as deer and rabbits. One Clovis site in Texas even included alligator bones, while excavations at another site near the Delaware River turned up bones from various species of fish. Dennis Stanford, a renowned paleoarchaeologist at the Smithsonian National Museum of Natural History who is one of the world's foremost authorities on prehistoric cultures and well acquainted with Clovis sites in the San Luis Valley, claims the Clovis basically "hunted and ate anything that moved." Wild plants, birds, and various aquatic animals probably rounded out their diet.[28]

The fossil record from the Late Pleistocene also indicates a rather sudden and widespread disappearance of several Ice Age big-game species known collectively as *megafauna*, including the Columbian mammoth, the mastodon, and the giant sloth. A great deal of research and literature has been devoted to explaining what exactly caused the sudden extinction of so many large mammals. During the 1960s, Paul Martin of the University of Arizona proposed a provocative hypothesis that blamed human hunters for the demise of the megafauna in what was later described as "this dreadful syncopation—humans arrive, animals disappear."[29] In essence, Martin suggested that the Pleistocene extinctions might have been the first instance of human-induced biological and ecological catastrophe in North America.[30]

Martin's hypothesis, sometimes referred to as the "overkill" or "blitzkrieg" theory, initially garnered much support and even more publicity, although at least some in the scientific community later disputed his conclusions. One critic of the theory noted, somewhat hyperbolically, that it was simply too much to believe that "a few thousand Indian men with pointed sticks could run around a continent and bring to extinction 135 species in maybe 400 years."[31] Another theory blamed the rapidly changing climate and successive series of droughts that followed the end of the Ice Age for the megafaunal extinctions. Given that the Clovis and their prehistoric prey lived in a period of severe climatic fluctuation, this argument has considerable merit.[32] Perhaps the increasingly arid environment of the postglacial Holocene Epoch so reduced grazing land and reliable water sources that the largest Pleistocene mammals simply could not survive.[33] Yet another theory emerged in the late 1990s that implicated rampant disease for the extinction of the megafauna, which presumably had no natural immunity to the lethal pathogens introduced by the first waves of humans in North America, while a more recent hypothesis blames asteroid or cometary impacts for contributing to the demise of the megafauna. In this scenario, which resembles the devastating extraterrestrial impact event that caused the extinction of the dinosaurs, one or more "ET objects" exploded over northern North America around 12,900 years ago, and the ensuing shockwave and thermal pulse ignited extreme wildfires across the continent. These fires presumably wiped out the megafaunal food supply and produced massive clouds of toxic ash and soot that blocked out the sun and triggered an abrupt period of intense global cooling known as the Younger Dryas (named for a particular layer of pollen left by a species of flowering plant that thrives in cold conditions) that lasted for roughly 1,500 years.[34]

These various "overkill" or "overchill" hypotheses have proven to be contentious, and consensus among scientists who support or dispute them has been elusive, so the rather sudden extinction of the megafauna remains somewhat of a mystery. A more recent (2011) article published in the respected scientific

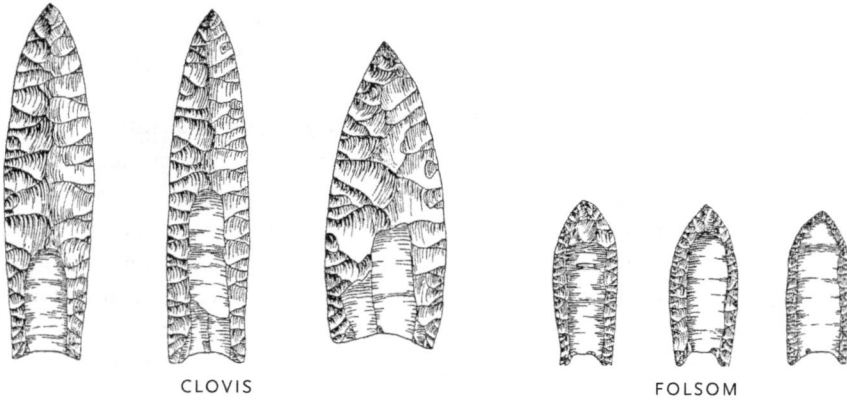

Figure 1.5 Examples of Clovis and Folsom stone points, similar to artifacts discovered at Great Sand Dunes National Park and Preserve. Department of Anthropology, Colorado State University, Fort Collins.

journal *Nature* presented evidence gleaned from studies of over 800 fossil DNA samples, nearly 3,000 fossil bones, and over 6,000 archaeological-site records tracing populations of several Pleistocene-era mammal species and concluded that these populations continually ebbed and flowed with fluctuating warm and cold periods, suggesting that climate has been a major driver of species population change over the past 50,000 years. However, because each species responded differently to the effects of climate change, habitat redistribution, and pressure from human hunters, climate alone cannot be blamed for all of the megafaunal extinctions. The study concluded that the disappearance of the Ice Age mammals is most likely attributable to some combination of climatic change and anthropogenic (human-influenced) effects on animal populations over a long period of time.[35] Indisputable evidence for an extraterrestrial impact event or rampant disease may yet emerge to perpetuate or definitively end the debate, but whatever the precise cause for the Pleistocene extinctions, the demise of the Ice Age megafauna also signaled the decline of the Clovis culture, which had lasted for a little over half a millennium, from about 11,500 to 10,800 years ago. In its wake emerged another group of Paleoindians known as the Folsom, who, like the Clovis, were named for a town in New Mexico where archaeologists first excavated examples of their smaller and more advanced fluted projectile points in 1926.[36]

In the San Luis Valley, archaeologists as well as casual collectors discovered numerous Folsom sites dating from roughly 10,500 years ago (ca. 8500 B.C.E., or Before the Common Era) near the Great Sand Dunes, usually in conjunction with bones from a now-extinct species of bison called *Bison antiquus* that survived the Pleistocene extinctions. One site in particular, located just south of the dunes and known as Stewart's Cattle Guard, yielded an astonishing wealth of information in the early 1980s after archaeologists excavated the remains of

Figure 1.6 Depiction of Paleoindians hunting *Bison antiquus* in the San Luis Valley. Original artwork by Heather Paul.

dozens of *Bison antiquus* that had been killed and butchered by Folsom hunters. *Bison antiquus* evidently managed to thrive where other large Ice Age mammals failed by adapting its diet to take advantage of the gradual expansion of drought-resistant, short-grass grasslands in the postglacial period. In fact, the continual climatic fluctuations of this period resulted in an abundance of prime bison habitat in the San Luis Valley. Plant fossils and pollen specimens taken from San Luis and Como Lakes indicate that the postglacial valley teemed with hardy grasses able to withstand the increasingly arid climate. The availability of such grasses apparently induced the bison population in the valley to multiply, a development that most likely proved irresistible to Folsom hunters.[37]

Like the Clovis before them, the Folsom enjoyed only a few centuries of hunting in the San Luis Valley before climatic changes drove them from the region. Sometime after 10,500 years ago, the gradual increase in arid conditions caused ponds and marshes to dry up and grasslands to wither, circumstances that probably compelled grazing animals like *Bison antiquus* to seek greener pastures. In all likelihood, this migration forced the Folsom to follow. Such vegetative changes were not unusual in the San Luis Valley. Plant communities near the dunes and on the floor of the valley have fluctuated dramatically over the millennia, at times consisting of sagebrush grasslands, then shifting to montane forest vegetation as local climatic conditions changed. Indeed, at some point in the distant past, the floor of the valley may have been thick with forests of Douglas fir.[38]

Scant evidence exists indicating human activity in the San Luis Valley for the subsequent 1,500 years or so. Paleoindian artifacts characteristic of the post-Folsom period commonly found on the Great Plains and Front Range of Colorado are virtually absent from the valley.[39] Then, around 9,000 years ago, climatic changes again resulted in an abundance of water in the San Luis Valley, and evidence for human activity reappears. Artifacts recovered from various sites in the valley indicate the presence of a series of post-Folsom cultures known collectively as the Cody Complex, distinguished by long, slender, unfluted projectile points. The Cody hunted bison and smaller game animals in the valley until they, too, were driven from the region by a return of the arid conditions similar to those that had driven out the Folsom. A gradual warming trend characterized this dry spell, which lasted from roughly 8,500 to 4,000 years ago (6500 to 2000 B.C.E.), an era often referred to as the Altithermal. Sporadic wet and dry climates alternated in the San Luis Valley during this period, and each wet period attracted a succession of Paleoindian cultures to the region. Specific distinctions between these post-Cody cultures (known variously as the Yuma, Scottsbluff, Eden, etc.) are not especially dramatic; the names given to them do not designate distinctly separate and different types of humans, but rather refer to differences in weapon and tool-making technologies.[40]

Yet, like the Clovis and Folsom, these cultures probably visited the San Luis Valley on a seasonal basis only, hunting wild game during the warmer months and fleeing when the cold winds of winter began to twist and howl across the valley floor. In fact, none of the cultures that frequented the valley during the prehistoric era left any evidence of permanent occupation, perhaps due to the valley's isolated location and severe climate. Protected by rugged mountain ranges, the San Luis Valley is by no means "on the beaten path," requiring a concerted effort to enter its vast interior, especially in winter, when mountain passes are frequently clogged with deep snow. Furthermore, the valley can get terribly cold during the winter months, both because of its elevation (roughly between 7,000 and 8,000 feet) and its tendency for thermal inversions, which occur when layers of warm air trap cold air close to the ground. Compounding this cold is the bitter winter wind that rakes incessantly across the exposed valley floor, often creating wind-chill factors well below zero. Lacking adequate shelter and resources to contend with these conditions, nomadic prehistoric peoples of the past probably found it easier to follow the migrating game animals to warmer climes during the long months of winter.[41]

The successive cultural groups that visited the San Luis Valley across thousands of years did share a number of survival traits. Although primarily big-game hunters subject to the vagaries of seasonal wanderings, all of these cultures probably supplemented their diets with small game and wild plants. In fact, prehistoric humans may have been part-time agriculturalists. Tending small plots of desirable plants along their customary migration routes provided insurance against drought, floods, or an unproductive hunt, although primitive agriculture was not the dominant cultural activity of these prehistoric people; they likely focused most of their time and energy engaged in hunting and processing their prey.[42]

Other cultural activities of the Paleolithic visitors to the San Luis Valley remain a mystery. Beyond the possible role of the Clovis in the extinction of the megafauna, little is known of the influence these prehistoric cultures had on the landscapes of the region. A lack of technology and small populations likely blunted their impact, although it is possible that primitive hunters did employ fire as a tool. Evidence of deliberate burning by prehistoric humans to alter a particular environment occurs throughout the world, and a growing body of literature on the subject suggests that the early human use of fire was both prudent and thoughtfully grounded in a clear understanding of the effects of fire on various types of vegetation. Paleoindians intentionally set fires to drive game animals into an ambush, or to convert forest to grasslands in an effort to attract desirable animals to a specific location. Such environmental manipulation may have occurred in the San Luis Valley, although any evidence to support this possibility has long since vanished, or has yet to be discovered.[43]

The Paleoindians of the San Luis Valley left behind more questions than answers. Information concerning their cultures, cosmology, and land-use practices remains incomplete, limited to inference and supposition gleaned from projectile points and bone scrapers. In August 2000 archaeologists did discover the remains of a prehistoric pit-house dwelling dug into a hillside near Indian Spring, on the western edge of the Great Sand Dunes. Dating from roughly 6,000 to 4,000 years ago, the site included a hard-packed floor and charcoaled hearth, along with a stone-drilled pendant, hundreds of pottery shards, and several stone tools and grinding stones. The lack of any roof structure suggests that the dwelling had once been covered with brush and was probably used on a seasonal basis only. Other artifacts and dwellings that might offer a keener insight into the lifeways of the Paleoindians are either still buried somewhere in the vast reaches of the valley or have vanished altogether, obliterated by centuries of wind and water.[44]

Throughout the Paleolithic Stage (ca. 10,000–5500 B.C.E.) and continuing during the Archaic Stage (ca. 5500 B.C.E. to A.D. 500) and the later Historic Aboriginal Stage (ca. 500–1881), the prehistoric cultures that entered the immense San Luis Valley and gazed upon the Great Sand Dunes undoubtedly influenced the various ecosystems and species they encountered in profound and lasting ways, yet modern science can only speculate as to the extent of that influence. One thing is certain: more than anything else, the availability of water dictated human interactions with the landscape of the San Luis Valley and the Great Sand Dunes.[45] The correlation is constant: when arid climatic conditions prevailed, evidence for human presence in the valley and at the dunes is decidedly scarce; when climatic fluctuations produced an abundance of water, evidence for human activity increases markedly. Water defined the historic limits of humanity in the San Luis Valley, so perhaps it comes as no surprise that water played an essential role in the first recorded human references to the Great Sand Dunes, found in the oral traditions of a later Native American culture that inhabited the valley long after the Clovis and their successors had disappeared. For the Tewa Pueblo Indians, the presence of water in the very midst of the Great Sand Dunes designated the precise spot where human existence began.

Ancient Origins

According to the origin myth of the Tewa Pueblo people, the first humans and animals were born in the underworld, a place where the sun shone at night like a pale moon. At some undefined point in the distant past, these primeval humans sought a way out of the underworld and climbed a tall fir tree to a lake called Sip'ophe (variously translated as "Sandy Place Lake" or "Lake of the Dead"). From this lake the first humans and animals emerged onto the earth, becoming the ancestors of future generations of Tewa and the creatures that inhabited their world. The Tewa also believed that the spirits of the dead reentered the underworld through this lake, thus rendering the site a particularly potent neighborhood for spiritual forces. Legend has it that Sip'ophe is a definite place, a small, dark, brackish lake in the midst of the Great Sand Dunes.[46]

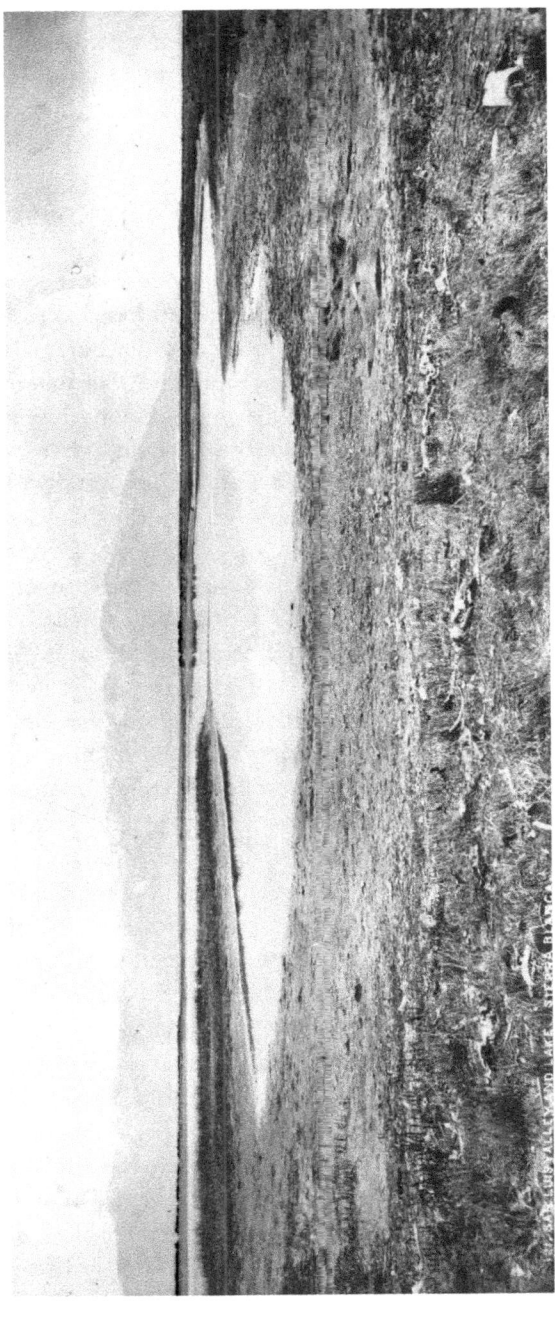

Figure 1.7 Historic photograph of San Luis Lake, taken in late September 1874 by famed frontier photographer William Henry Jackson. This region, roughly eight miles west of the Great Sand Dunes, contains San Luis, Dollar, and Head Lakes, as well as a number of interconnected ponds and marshes, one of which may be the possible location of Sip'ophe, the legendary Lake of the Dead of Tewa Pueblo mythology. The high peaks of the Sierra Blanca massif loom in the hazy distance. U.S. Geological Survey.

In 1892 Edgar L. Hewett, noted anthropologist and author of *The Pueblo Indian World*, visited the dunes in the course of his research and discovered a "small lake of very black, forbidding-looking water," roughly one hundred yards in diameter. Rotting on the shore around the perimeter of the lake was a continuous line of dead cattle. An old man who lived on the slopes of the nearby Sierra Blanca told Hewett that the lake never dried up and that many cattle died every season from drinking its water. The old man failed to mention whether the lake's forbidding appearance and legions of dead cattle could be attributed to bad water or malevolent spirits, but Hewett left the San Luis Valley satisfied that he had found the true location of Sip'ophe.[47]

Present-day San Luis Lake may be a likely candidate for the location of Sip'ophe, although when swollen with spring runoff, San Luis Lake is considerably larger than the lake described by Hewett. The Taos Pueblo origin myth, which is similar to that of the Tewa, places the location of Sip'ophe near the summit of the Sierra Blanca massif. Dollar Lake, a small pond just north of San Luis Lake, is another potential location for Sip'ophe, and given its nearly perfect circular shape, it seems an appropriate locus for supernatural forces.[48] Yet Hewett claimed that he had found Sip'ophe in the very midst of the Great Sand Dunes, and both San Luis and Dollar lakes are roughly eight miles to the west of the dunes. One possible explanation is that Hewett stumbled upon Indian Spring (also known historically as Big Spring) or one of the numerous interdunal ponds that punctuate the western fringes of the main dunefield and convinced himself that he had found Sip'ophe.

To the ancient Tewa, Sip'ophe marked both the exit and the entrance to the underworld, the wellspring of life coexisting with the realm of the dead. The Tewa recognized the astonishing paradox of water existing in the midst of the immense sea of sand, a natural anomaly so enchanting and mysterious that they conferred cosmological significance on it in their cultural origin myths and legends. Such mythology is often interpreted as humanity's attempt to reconcile the fundamental contradictions and dualities of existence (e.g., male/female, day/night, sun/moon, life/death).[49] If so, perhaps the legend of Sip'ophe is best explained as the Tewa attempt to reconcile the stunning discrepancy of sand and water existing together at the Great Sand Dunes. Whatever the explanation, the Tewa passed their myth of "Sandy Place Lake" from generation to generation for centuries. Their oral traditions provide the first known historic reference to the landscape of the Great Sand Dunes and thus define the beginning of the long and compelling relationship between humanity and water in the San Luis Valley.

Anthropologists and archaeologists suspect that the actual origins of the Tewa Pueblo are lost somewhere in the mists of the Ancestral Puebloan diaspora. Evolving from what is termed the Basket Maker period (ca. A.D. 1 to A.D. 450) to the Classic Pueblo period (ca. 1100 to 1300), the Ancestral Puebloans established themselves as a definitive cultural entity perhaps 2,000 years ago in the Four Corners region of the American Southwest, where they tended crops, hunted game animals, and built impressive villages, culminating with the architectural wonders of the cliff dwellings at Mesa Verde, Chaco Canyon, and elsewhere.[50] The cliff dwellers evidently abandoned their homes and fled the region sometime around the year 1300, and scholars continue to debate the exact cause for their flight. Dendrochronological (tree ring) data suggest that a persistent, widespread drought impacted the Southwest at roughly the same time as the disappearance of the Ancestral Puebloans. A combination of other factors associated with these drought conditions, including increasing pressure from neighboring tribes (who also may have been suffering the lingering effects of drought), food shortages caused by overhunting, and widespread depletion of crucial local resources such as wood for fuel and construction, likely contributed to their exodus from the region.[51] The displaced Ancestral Puebloans either drifted southward and were assimilated by other tribes, or wandered elsewhere, perhaps following desirable game animals away from the parched mesas of the Southwest toward the high country of Colorado and New Mexico, where rainfall tended to be both more abundant and reliable. Many apparently settled along the Rio Grande in north-central New Mexico, thus laying the foundation for later Pueblo cultures, including the Tewa and Taos.

The Tewa legend of Sip'ophe includes references to ancient Pueblo Indians living in the San Luis Valley, and Taos Pueblo legends describe bird-hunting expeditions to the Great Sand Dunes. The multitude of small arrowheads found near the sand dunes and at the San Luis Lakes lends credence to these stories. Other Pueblo tribes traveled to the valley searching for turquoise for use in religious ceremonies, for personal adornment, or for trade, yet no evidence exists for permanent human occupation in the San Luis Valley during either the Ancestral Puebloan or post–Ancestral Puebloan periods (ca. A.D. 1 to ca. 1300, and ca. 1300 to European contact). Presumably the valley's harsh winter climate continued to discourage permanent occupation, just as it had for the Clovis and Folsom.[52]

In addition to the Tewa and Taos Pueblo Indians and their ancestors, the San Luis Valley witnessed a variety of Native American visitors throughout the pre-Columbian era. At least seven distinct Native American tribes seasonally occupied or frequently visited the valley during the years following the disappearance of the ancestral pueblo, including the Apache, Arapaho, Cheyenne, Comanche, Kiowa, Navajo, and Ute The Jicarilla Apache, who eventually settled in northern New Mexico and reportedly still collect sand from the dunes for ceremonial purposes, called the dunes Seianyedi, meaning "it goes up and down," while Blanca Peak, just south of the dunes, has long been considered one of the four sacred mountains of the Navajo, who referred to the peak as Sisnaajinií, variously translated as "white shell mountain" and "horizontal black belt."[53]

Known variously as the Yutta, Uticahs, or "Blue Sky People," the Ute eventually emerged as the dominant Native American culture in the central Rocky Mountains.[54] Like the Tewa Pueblo, the exact origins of the Ute are unclear, although they too may have some connection to the Ancestral Puebloans. History identifies the seven generally recognized bands (or clans) of the Ute primarily by the territory they traditionally occupied. One band of Southern Ute, specifically the Capote (also spelled Kapota), vigorously defended the San Luis Valley, which they called Tela We-a-gat (meaning "Big Valley," and also spelled Tavi-we-a-gat) against neighboring tribes and claimed it as their exclusive hunting grounds. Similar to the Clovis and Folsom, the Ute hunted in the valley only when suitable climatic conditions attracted desirable game animals. What distinguishes the Ute from their predecessors and contemporaries is their place in the recorded history of the San Luis Valley: when Europeans first entered the valley in the late sixteenth century, they most likely found the Ute in control of the region.[55]

Prehistoric Legacy

How extensively did these pre-Columbian Native Americans alter the environment of the San Luis Valley and the Great Sand Dunes? A definitive answer may never be known, both because physical evidence is scant and because a considerable degree of conjecture and misinterpretation obscures the understanding of this subject. Thanks in part to the environmental movement of the 1960s and 1970s, a tendency has long existed in the United States to view Native Americans as the "original conservationists" or "primal ecologists," so intimately connected with the natural world that they left virtually no mark on the landscapes they inhabited. In reality, Native American interactions with the environment profoundly influenced the historical landscapes of North America. Evidence suggests that their subsistence activities caused significant changes to the continent's forests, prairies, and deserts, whether by carefully manipulating vegetation growth to encourage or discourage the spread of animal populations, or by actively altering and creating habitats best suited to human settlement. Some techniques, such as irrigation, tended to be localized; others, such as burning forests and grasslands, appear to have been employed across the continent.[56] The environmental impacts caused by the subsistence activities of prehistoric and Native American cultures may seem mild in comparison with later European and Anglo-American ecological manipulation and resource extraction, but this does not necessarily indicate that the Clovis, Folsom, Pueblo, Ute, and other indigenous people were somehow more "natural" than their Old World counterparts who flooded into North America after 1492. It only means that, relative to the Europeans, indigenous cultures had less impact on the ecosystems they inhabited, primarily due to their lower populations, less advanced technologies, and subsistence-level extractive activities.[57]

If the Native Americans did indeed alter their environments, is it possible to determine the effects of their activities on the Great Sand Dunes and the San Luis Valley? The probable role of Paleolithic hunters in the extinction of prehistoric megafauna across North America has been acknowledged, and given the similar hunter-gatherer cultural milieu of later Native Americans, the

Pueblo, Ute, and other tribes likely impacted the game animals that frequented the dunes and the valley. Unfortunately, stone points and fossil bones are all that remains to define the extent of that impact.[58] Similarly, proof that Native Americans purposefully burned the forests and grasslands adjacent to the dunes, or attempted to irrigate the surrounding landscape, has long since been obliterated by centuries of geological, biological, and climatological change. This does not mean that such activities did not occur; on the contrary, given the available historical evidence discovered in other locations, a strong possibility exists that these early inhabitants did alter the landscapes of the dunes and the greater San Luis Valley to some indefinable degree.[59]

Considerable evidence indicates that the Jicarilla Apache and Ute in particular had a profound effect on the ponderosa pine trees that grow in the rugged foothills of the Sangres just east of the Great Sand Dunes. The area contains at least one hundred culturally peeled ponderosa, scarred by generations of Ute and Apache Indians who gathered the bark for a variety of uses. Analysis of dendrochronological data indicates that the majority of the surviving ponderosa were peeled sometime in the early nineteenth century, long after European contact in the region. However, since ponderosa have a life span of up to six hundred years, it is probably safe to assume that the Ute and Apache peeled the ponderosa for centuries before that time.[60] The gathering of pine tree bark by Native Americans is well documented throughout western North America, for the trees provide a multitude of products. Outer bark typically served as building material for dwellings and fiber for baskets, while resin and pitch supplied adhesives and waterproofing agents, and inner bark and sap furnished poultices and medicine for a multitude of ailments, including stomach upset, infections, heart problems, and gonorrhea. Native Americans also harvested pine tree bark for food, and this appears to have been the case with the Ute and Apache who frequented the San Luis Valley and utilized pine bark as a thickening agent for meat soups and as a base for tea. Various Ute tribes also reportedly tied thin strips of pine bark into bundles and ate them with salt.[61]

Beyond the ponderosa scars, scattered arrowheads, and suspected role of human hunters in the decline of the megafauna, firm evidence to define what ecological impacts the Ute and their predecessors had on the Great Sand Dunes and the San Luis Valley is lacking in the historical record. Far more tangible are the cumulative effects inflicted on the San Luis Valley in the five centuries since European contact. On the cusp of the cultural and ecological collision that began in 1492, the indigenous peoples of the San Luis Valley and the rest of the Western Hemisphere simply had no indication of or warning about how radically their world was about to change.

TWO | Outskirts of Empire

In the name of the most Christian king, Don Philip . . . I take and seize tenancy and possession, real and actual, civil and natural . . . at this said Rio del Norte, without excepting anything and without limitations, including the mountains, rivers, valleys, meadows, pastures, and waters.

Don Juan de Oñate, 1598

They rode from one place to another on animals that looked like large dogs. As they rode, their iron weapons rattled noisily and sunlight glinted off their armor. Like a lost war party, they clanked and clanged their way across the American Southwest, stirring up great clouds of dust. We called them Naakaii, or Those-Who-Wander-Around. They were the Spaniards.

  Lawrence D. Sundberg, *Dinétah: An Early History of the Navajo People*

ACTING ON BEHALF OF A Spanish Crown concerned about the security and mineral wealth of the lands that comprised the northern frontier of Spain's New World possessions, Don Juan de Oñate officially took possession of Nuevo México in the name of Lord King Philip II of Spain on April 30, 1598, a little over one hundred years after Columbus first made landfall in the West Indies. All of the territory "from the leaves of the trees in the forest to the stones and sands of the river," which included the Great Sand Dunes, the San Luis Valley, and the entire upper Rio Grande watershed, now ostensibly belonged to the Spanish Crown. Oñate had signed a contract with King Philip granting him the right to explore, exploit, and settle the region on the condition that he covered his own expenses. Such a daunting financial burden did little to dissuade Oñate;

Figure 2.1 Tom Lea, *Conquistador, 1598—Juan de Oñate*, 1946. Pen and ink. The Tom Lea Institute, El Paso, Texas.

he had inherited a considerable fortune from his father and enjoyed the additional wealth and prestige of his wife, who was the granddaughter of the famous Spanish conquistador Cortés and whose lineage allegedly connected her with the Aztec emperor Moctezuma.[1]

Oñate set out in 1598 with two of his nephews, Juan and Vicente de Zaldivar, along with four hundred settlers, a handful of Franciscan friars, eighty-three baggage wagons, and 7,000 head of stock, including cattle, sheep, and goats, to conquer and settle the vast area to the north of Spanish colonial possessions in Old Mexico.[2] Leading his party to the confluence of the Chama and Rio Grande Rivers, about twenty miles north of present-day Santa Fe, New Mexico, Oñate reached the pueblo of Caypa, promptly rechristened it San Juan, and began setting up his temporary headquarters. Shortly thereafter he gave his nephews command over a series of expeditions to explore the area, including at least one with orders to follow the Rio Grande north to its headwaters. Although there is some confusion on the subject, historians generally agree that either Juan de Zaldivar or his brother Vicente thus became the first European to enter the immense interior of the San Luis Valley.[3]

Beginning with Columbus in 1492, Spanish and later European explorers introduced an astonishing array of biological organisms that set into motion a profound ecological transformation of the New World, a process that historian Alfred Crosby termed "ecological imperialism."[4] In essence, the newest characters on the North American stage had brought with them the means to change the very stage itself. Native Americans, too, had altered that stage, but the Europeans introduced a veritable cornucopia of flora and fauna, not to mention distinctive attitudes about using land and animals, that induced environmental changes on a profound scale. Old World sheep, pigs, cattle, and horses moved effortlessly into the various ecological niches of the Americas. As their numbers increased, so too did the damage inflicted on the region's ecosystems. Multitudes of sharp hooves and voracious appetites compacted fertile soils and consumed native grasses. Along well-worn trails, water followed the paths of migrating animals, carving deep gullies, or arroyos, that rapidly carried the water away rather than allowing it to soak into the soil. Areas that had once supported lush grasslands, dotted with trees and brimming with wildlife, began to diminish and, in some places, turned to desert. A menagerie of European weeds, grasses, and other biota further contributed to an ecological invasion that completely rearranged the environmental tapestry of the entire Western Hemisphere.[5]

The introduction of European species also had a tremendous cultural impact on the New World's indigenous human populations. For example, Native Americans had domesticated only one four-legged animal at the time of European contact—the dog. Whatever their virtues as man's best friend, dogs clearly are inferior sources of leather and food, and less effective beasts of burden, than the two major domestic European quadrupeds, horses and cattle. Historical records indicate that the Ute had acquired Spanish horses by at least 1637, possibly earlier. Access to horses, which some Native Americans initially believed were "magic dogs," fundamentally altered indigenous cultures, enabling them to build more mobile and militarily powerful societies, along with altering their trading and subsistence activities. Horses transformed Native Americans into more effective hunters and warriors, which in turn influenced their impact on both game populations and neighboring tribes. In addition to horses, pigs, sheep, and cattle provided dependable food and clothing supplies and thus became irresistible targets for newly mobile Indian raiding parties.[6]

A more ominous biological impact came from the deadly Old World viruses and bacteria that accompanied the Europeans across the Atlantic. Invisible, virulent microbes wreaked havoc on the indigenous humans, whose immune systems had no defense against the onslaught. Of course, disease was not unknown to the original inhabitants of the New World. Native Americans had suffered the ravages of maladies such as tuberculosis and parasitic infection for centuries. They were wholly unprepared, however, for the contagious diseases of densely populated Europe. Smallpox, chicken pox, bubonic plague, measles, typhoid fever, influenza, and assorted other pathogens were the least visible of Europe's transatlantic baggage, but their stunning ability to cripple and kill with ruthless efficiency absolutely devastated native populations.[7]

Europeans also introduced a sophisticated array of technology into the New World, including gunpowder, steel weapons and tools, and wheeled vehicles. The overwhelming superiority of this technology enabled the Europeans to dominate an indigenous population still toiling in the earliest stages of metallurgy. Emboldened by this superior technology and compelled by a heady mixture of religious, imperial, and personal motives, Europeans, and the Spanish especially, had tremendous success in the Western Hemisphere, conducting a rapid series of discoveries and conquests in lands where natives vastly outnumbered them. In the process, they began an ecological and cultural transformation the likes of which the New World had never seen. Yet even mighty Spain had serious difficulties extending its imperial reach into at least one portion of its newly conquered territories: the immense alpine valley at the upper reaches of the Rio Grande del Norte.

Spaniards in the Valley

Spain's conquest and subsequent cultural and biological appropriation of its New World possessions initially had limited adverse impact on the San Luis Valley. Isolated far inland, hundreds of miles from any seaports or navigable rivers, protected by rugged mountain ranges and vast stretches of open prairie and arid deserts, the San Luis Valley's remote location mostly insulated it from the radical transformations occurring elsewhere in the region. The valley had no permanent Native American population to be wiped out by contagious diseases; the Pueblo villages of the Rio Grande Valley were located far to the south, and the nomadic bands of Ute, Navajo, Apache, and Comanche that hunted in the valley came and went far too infrequently to ensure regular, consistent contact with the Spanish. This initial failure of Spanish imperialism in the San Luis Valley is illustrated by an ill-conceived attempt to domesticate a herd of bison in the region. In an official dispatch to the Spanish viceroy in Mexico in March 1599, Juan de Oñate dutifully informed his superiors that he had "acquired a possession so good that none other of His Majesty in these Indies excels it." His report included a description of an effort to corral a large herd of bison in the San Luis Valley, an encounter that at least one historian contends occurred "in the vicinity of Alamosa at the foot of the Sangre de Cristo Mountains."[8] Oñate noted that "more than three hundred buffalo were seen in some pools" and that the region was "continuously covered with an infinite number of cattle [bison], and the end of them was not reached."[9] Such a tempting target evidently proved irresistible to Oñate's men, who sought to assimilate the bison into their own herds of cattle. The attempt proved not only dangerous but also completely futile, the American bison being an animal that does not take kindly to domesticity. "For several days we tried a thousand ways of shutting them in or of surrounding them, but in no manner was it possible to do so," Oñate wrote. "This was not due to fear, for they are remarkably savage and ferocious, so much so that they killed three of our horses and badly wounded forty. They attack from the side, putting the head far down, so that whatever they seize they tear very badly."[10] Oñate's men had risked their lives in an attempt to transform the San Luis Valley into a profoundly Spanish landscape, complete with domesticated bison in place of more traditional cattle, but the valley largely maintained its peculiar resistance to Spain's imperial reach. The region's remote location, its bitter winter climate, and the limitations imposed by a short growing season continually frustrated Spanish ambitions to appropriate the San Luis Valley.

Extensive exploration by the Spanish did produce the first written descriptions of the valley, complete with Spanish names for its natural features, which led to maps that paved the way for further Spanish incursions, including those that sought to exploit the region's rumored mineral wealth. Spanish influence is clearly seen in the names of the San Luis Valley's rivers and mountains, some of which reflected Spanish conceptions of how these natural features resembled familiar elements in their lives, such as Culebra (snake), Cuchara (spoon), and Costilla (rib). Other names were purely descriptive, such as Sierra Blanca (white mountain) and Trinchera (trench), or referred to animals that frequented a particular area, such as the Rio de los Conejos (River of the Rabbits). This christening of the distinctive geologic and topographic features of the San Luis Valley reinforced Spanish beliefs that they were appropriating the region in the name of God and the Spanish king.

Early Spanish descriptions of the San Luis Valley also reveal the role that Spain's climate played in influencing Spanish perceptions of the New World. Since many Spanish explorers were familiar with the high, semi-arid plateaus of the Iberian Peninsula, they did not find the aridity of Old and New Mexico unusual, or even worth noting. On the contrary, Spanish descriptions of New Mexico frequently emphasized the region's numerous streams and apparent fertility, as in one report that declared, "The greater part of the country consists of immense plains and delightful valleys, clothed with very abundant pasturage."[11] Not surprisingly, Spanish descriptions of the San Luis Valley contain virtually no mention of the Great Sand Dunes. By contrast, a number of later Anglo-American explorers ventured into the Southwest from the humid climes of North America's Eastern Seaboard, where moisture and verdant foliage are common. The arid climate of the Great Plains and New Mexico had a tremendous impact on these explorers, which resulted in their general characterization of the area as barren and desolate. One Anglo-American explorer decried the "universal barrenness which pervades this country." Others described the region as "barren and uninteresting in the extreme," and later Anglo reports are filled with detailed references to the region's ubiquitous sand.[12]

Among the most telling examples of the San Luis Valley's resistance to Spanish imperial designs concerns the absence of permanent settlements. For over two hundred years, despite firmly establishing themselves on its southern threshold, the Spaniards never successfully colonized the San Luis Valley. Resistance from Native Americans, coupled with the relentlessly frigid winter climate, thwarted every attempt. Juan de Oñate's description of an encounter with a band of (presumably) Ute Indians clearly indicates their proficiency with weapons and tactics, which undoubtedly came in handy during the two centuries of efforts by native populations to keep the valley free from European settlement. "The Indians are numerous in all that land," noted Oñate. "Their weapons consist of flint and very large bows, after the manner of the Turks. We saw some arrows with long thick points, although few, for the flint is better than spears to kill cattle. They kill them at the first shot with the greatest skill."[13]

Oñate, like countless other Spaniards after him, failed to establish a permanent settlement in the San Luis Valley, although he did continue his colonization efforts elsewhere in New Mexico well into the 1600s, often with mixed results. Faced with resistance from indigenous people, Oñate could be brutal in his oppression. In one instance, after a surprise attack by Acoma Pueblos had killed his nephew, Juan de Zaldivar, Oñate retaliated by killing some five hundred men and three hundred women and children. Those who survived his wrath were either sentenced to a life of slavery or had their arms or legs amputated. Sadly, Oñate's treatment of the Pueblo Indians was neither unusual nor uncommon.[14]

Initiated as an entrepreneurial venture to enrich the Spanish Crown with mineral wealth and convert the souls of Native Americans, by the middle of the seventeenth century the effort to colonize northern New Mexico had degenerated into a military occupation, with the unwelcome Spanish invaders maintaining their control through terror, coercion, intimidation, and superior firepower.[15] Tensions with the Spanish increased until local Pueblo Indians began killing the

Figure 2.2 Depiction of a Spanish expedition into the San Luis Valley, circa 1700. Original artwork by Heather Paul.

foreign missionaries stationed in their midst. Spanish retaliation was invariably swift and violent, mostly consisting of imprisonment, flogging, amputation of limbs, or hanging. The situation exploded in the Pueblo Revolt of 1680, when a number of pueblos, in a rare instance of cooperation, united in rebellion against their Spanish overlords and drove them out of New Mexico, killing 21 missionaries along with over 350 colonists.[16]

The cooperation that had briefly unified the Pueblos in their revolt soon evaporated, and by 1693 the Spanish had once again moved into New Mexico, this time under the leadership of Don Diego de Vargas, who initiated a campaign of reconquest. In July 1694, after looting the pueblo at Taos, Vargas headed north to avoid potential Pueblo reprisals on his way back to Santa Fe, a route that included passage through the San Luis Valley:

> And so seeking a way out, even though it should be a dilatorious [*sic*] route . . . the older natives of the land . . . said that if I was willing to take a long and round-about route, they would direct me and the expedition through a region whereby the Villa of Santa Fe would be succored. . . . And so it was decided that the departure should be by way of the Yuttas (Utes), a nation which is very friendly towards the Spaniards.[17]

Vargas and his forces reached as far north as the Culebra River in the southern San Luis Valley, then turned west toward the Rio Grande. After fording the river, Vargas and his forces were "suddenly attacked by Utes armed with bows and arrows and war clubs." The Spaniards successfully repulsed the ambush, after which the Utes entered Vargas's camp to apologize, explaining that their Pueblo enemies to the south often infiltrated their territory to hunt buffalo "disguised as Spaniards, mounted, and with leather jackets, leather hats, firearms, and even a bugle." Vargas eventually reached Santa Fe after a journey of roughly 350 miles, with "manifestations of friendship on both sides, and the Utes were invited to Santa Fe to trade with the Spaniards." Although Diego de Vargas's diary contains scant descriptions of the San Luis Valley, he does mention several of the valley's rivers and mountains by their Spanish names, further indication of Spain's attempted cultural appropriation of the region, despite its relative isolation on the far northern fringes of the Spain's New World empire.[18]

Sangre de Cristo (Blood of Christ)

The years following the Vargas expedition witnessed a number of Spanish incursions into the San Luis Valley and surrounding mountains. Concerned about the security of its northern frontier, the Spanish colonial government sought to increase its presence in the region in order to provide protection for Spanish expeditions involved in the relentless quest for gold. Indeed, legend has it that both the Sangre de Cristo Mountains and the San Luis Valley received their names during one such expedition that allegedly began sometime around the year 1700, when Father Francisco Torres, a Spanish nobleman who had entered a monastery after the untimely death of his bride-to-be, arrived in the New World. During his mission to spread the Gospel, he joined a party of Spanish explorers on their way to investigate rumors of gold in northern New Mexico. After a brief stop in Santa Fe, where they enlisted a few local Pueblo Indians to serve as guides, Torres and the rest of the Spaniards headed north into the rugged mountains of the frontier.[19]

A grueling journey brought them through the mountains to the northern reaches of the Rio Grande, which they followed until they came to an immense valley stretching away toward a range of high, jagged mountains to the east. Impressed with the sublime beauty of the valley, Father Torres named it after Santo Luis, the patron saint of his hometown of Seville. Crossing the valley, the travelers came to a large lake (perhaps one of the present-day San Luis Lakes) teeming with birds, antelope, and deer. After resupplying, Torres and the explorers headed east, past the Great Sand Dunes and into the foothills of the Sangres, where good fortune blessed them with the discovery of rich veins of gold.[20]

Despite the protestations of Father Torres, the Spaniards forced their Pueblo guides to dig mine shafts and begin hauling out the precious ore. The Pueblos endured the brutal conditions forced upon them, but also began surreptitiously plotting their revenge with a small band of local Ute who had also been pressed into service by the Spaniards. Shortly thereafter, at some preplanned moment, the combined force of Indians attacked with shovels and picks, mortally wounding Father Torres and driving their Spanish oppressors down the mountains, past the Great Sand Dunes, and into the vast reaches of the valley floor. In full retreat, hauling the bleeding Torres, the beleaguered Spanish headed for the lake they had passed on their journey across the valley, where they hastily constructed a raft and escaped just as the sun began to set. As he lay dying on the raft in the center of the lake, Father Torres asked his compadres to raise him up so that he might behold the crimson alpenglow on the nearby mountains. Safe for the moment from attack, the surviving Spaniards lifted his head (in some versions they help him to his feet) and held him in the soft twilight. "Sangre de Cristo, Sangre de Cristo [Blood of Christ]!" gasped Torres with his last breath, pointing to the jagged summits of the blood-red mountains, which forever after bore the name that the dying Father Torres had given them.[21]

Presumably, at least one of the hapless Spaniards escaped with his life to relate the stirring tale of Father Torres, unless of course the story is apocryphal, which seems likely, given that several details in the story simply do not add up. For instance, the construction of the raft has always puzzled historians, primarily because the region around the San Luis Lakes is conspicuously devoid of trees. Perhaps the Spanish made the raft out of wood from their wagons, or hauled the wood along with them as they fled from the mountains, but these explanations seem dubious, given that the Pueblo and Ute were presumably in hot pursuit, which would have left insufficient time to construct a raft. Moreover, it seems strange that the attackers did not simply wait along the shore for the Spaniards to starve, but maybe they figured their freedom from slavery was revenge enough and chose to leave the Spaniards adrift on their raft in the middle of the lake. Regardless, the legend survived to become a permanent and much beloved feature of San Luis Valley lore, eventually inspiring a beautiful mural in the Luther Bean Museum at Adams State College in Alamosa.

Other legends variously attribute the naming of the Sangres to assorted Spanish explorers such as Valverde or de Anza, or to a group of anonymous Spaniards who saw the sunrise reflecting off the crimson-colored scrub oak that

Figure 2.3 Detail of a mural in the Luther Bean Museum at Adams State College in Alamosa, Colorado, depicting Father Torres and the legendary naming of the Sangre de Cristo (Blood of Christ) Mountains. Kat Olance, Luther Bean Museum, Adams State College.

covers the eastern foothills of the range. Yet another legend attributes the name to a mortally wounded priest who was fleeing the Pueblo Revolt in 1680, and some historical sources place the Torres legend as occurring at the suspiciously late date of 1830. Whatever its origins, the name stuck, as did the persistent legends of gold in the Sangre de Cristos.[22]

The Spanish established a number of mines in the Sangres during the late seventeenth and early eighteenth centuries, including the legendary Spanish Cave, or La Caverna del Oro (the Cave of Gold), where a Maltese cross allegedly marked the entrance. In 1869 an early settler in the Wet Mountain Valley, Captain Elisha P. Horn, discovered a Maltese cross at the entrance to a cave on Marble Mountain, on the eastern slope of the Sangres. Horn explored the cave and found human remains and rusted Spanish armor, but none of the cave's legendary gold. In 1920 two U.S. Forest Service rangers, Paul Gilbert and Arthur Carhart, interviewed Apollina Apodaca, a 102-year-old woman who claimed to be a direct descendant of the early Spanish explorers and settlers of the San Luis Valley. The woman told tales of Spanish miners and enslaved Native Americans and recalled that when she was a child (about 1830) the mine had been known as La Mina de los Tres Pasos, the Three Steps Mine. The rangers later located the mine and discovered a rusted shovel, several lengths of rotted hemp rope, and a deep internal shaft, but no sign of precious metals. Subsequent expeditions, including one by the Colorado Mountain Club in the 1930s, reportedly found rotted timbers, a hoisting apparatus and ladders, a seventeenth-century hammer, even a human skeleton chained by the neck to the cave wall, but not one ounce of Spanish gold. Rumor also told of the existence of two huge oaken doors seven hundred feet down the shaft that guard a large storehouse, but to date no gold has ever been found in the cave, despite numerous attempts by modern spelunkers to unlock the mysteries of La Caverna del Oro.[23]

In contrast to the failure of modern explorers to find gold in La Caverna del Oro, Spanish explorers and prospectors evidently did locate some traces of gold in the Sangres, as indicated by other historical evidence found in the region. Early Anglo-American settlers in the San Luis Valley discovered the remains of an old Spanish *arrastra* (a circular stone and timber device used to crush ore) in North Arrastra Canyon immediately south of the Great Sand Dunes, just outside the park boundary. Unfortunately, local resident Howard Shockey later removed the arrastra from the canyon for display purposes, thus limiting analysis and dating of the artifact in its original archaeological context.[24] Legends persist concerning a "Spanish oxcart" laden with gold bars that Spanish miners had apparently hauled down from their mines in the Sangres, only to have it swallowed by the shifting sands of the Great Sand Dunes. From time to time, the relentless winds are said to uncover the oxcart with its fortune in gold, usually at last light, but the cart is invariably covered again by the next sunrise. Other legends concern the tales told by early American trappers of the "Sangry de Christy [*sic*]" gold mine, stories that always seemed to end long before the trappers revealed the true location of the mine.[25]

Figure 2.4 Page from the March 1959 edition of the comic book *The World around Us* depicting pirate George Lowther and his crew sailing up the Rio Grande in 1724 to hide their gold in La Caverna del Oro. Historical records have so far failed to verify that such a journey ever took place. Note the Maltese Cross marking the entrance to what is now known as Marble Cave, located on 13,262-foot Marble Mountain on the eastern slope of the Sangre de Cristos. *The World around Us 7—the Illustrated Story of Pirates* (New York: Gilberton, 1959). Fred Bunch, Great Sand Dunes National Park and Preserve.

Despite Spanish mining operations in the Sangres, the San Luis Valley remained free of permanent settlement throughout the eighteenth century. Still, Spanish explorers continued to frequent the region on a regular basis. In 1765 a gold-searching expedition led by Don Juan Maria de Rivera forged its way into the mountains of southwestern Colorado along the trade route from Santa Fe to the Spanish missions in Alta California (present-day Los Angeles) that later became known as the Old Spanish Trail, part of which passed through the San Luis Valley.[26] Yet even the most heavily traveled Spanish trails offered little protection from the roving bands of Native Americans who hunted in the region. Although the Ute claimed the vast San Luis Valley as their hunting grounds during this period, numerous other tribes including the Comanche, Navajo, and Apache continued to harass Spanish settlements and expeditions in northern New Mexico with disturbing regularity. In 1771 alone, attacks by various Native Americans claimed the lives of more than one hundred Spanish New Mexicans, not to mention the countless horses and cattle stolen or slaughtered by the raiding parties. Local Spanish citizens blamed most of the attacks on the Comanche, who had earned a reputation for their ferocity and skill with horses and who by the early eighteenth century had formed a mutually beneficial coalition with the Ute to dominate the northern borderlands of New Mexico. With the situation fast becoming intolerable, the Spanish provincial government responded to these provocations by authorizing what became quite possibly the last large-scale extension of Spanish imperial power into the region that included the San Luis Valley.[27]

In August 1779 newly appointed governor of New Mexico, Don Juan Bautista de Anza, set out from Santa Fe in hot pursuit of Cuerno Verde (Green Horn), the feared Comanche warrior believed responsible for attacking Spanish settlements on the northern frontier. Anza led his army of 573 mounted soldiers into the San Luis Valley, where he stopped briefly at the Conejos River to augment his forces with 200 Ute and Apache warriors seeking to join the campaign against Cuerno Verde. The expedition then continued north through the valley, eventually reaching "a pleasant pond named San Luis," perhaps the very same pond that had witnessed the demise of the legendary Father Torres some eighty years earlier.[28] Leaving the pond, Anza and his forces exited the valley through Poncha Pass, then turned eastward along the Arkansas River in pursuit of the Comanches. After a brief skirmish near Pikes Peak, Anza and his forces finally caught Cuerno Verde and his warriors at the base of Greenhorn Mountain, just north of present-day Walsenburg, Colorado. The ensuing battle claimed the lives of Cuerno Verde and his eldest son, along with four of his leading captains.[29]

Anza's victory over the Comanche forces temporarily restored Spanish hegemony in the region, but the glory days of the empire at last began to fade. The end of the eighteenth century witnessed a gradual erosion of Spanish imperial power in the Americas. Consumed by the political and military intrigues of continental Europe, including the ominous rise to power of Napoleon Bonaparte, the Spanish Crown found itself too absorbed with internal security on the Iberian Peninsula to continue devoting precious resources to its colonies in the New World, which routinely ignored or evaded the laws and policies formulated by the imperial government in Madrid. Finally, in 1821, Mexico won its independence from Spain, signaling the end of Spanish imperial aspirations in Central and North America.

Empire's Wake

Efforts to determine the precise ecological impacts inflicted by thousands of years of Paleoindian and Native American subsistence activities upon the landscapes of the Great Sand Dunes and the San Luis Valley invariably suffer from a scarcity of evidence. By contrast, any analysis of the repercussions of Spanish imperial activities on those same landscapes must contend with an almost infinite variety of permutations caused by the introduction of Old World plant and animal species, religion, cultural mores, and technologies. The potential impacts of these factors, singularly or in combination, are simply too numerous to calculate or assess definitively. Further clouding the issue is the fact that the Spanish arrived in North America near the onset of what is known as the Little Ice Age. Those who traveled or dwelled in the southern reaches of the continent during the 1500s, 1600s, and 1700s encountered colder and wetter weather than those regions currently enjoy. Prior to the mid-nineteenth century, rainfall was higher, temperatures lower, and growing seasons shorter. Climatic fluctuations thus add another element to the already daunting task of determining the extent of environmental change caused by over two centuries of Spanish presence in the region.[30]

Although the San Luis Valley offered plenty of water and arable land, Spain enjoyed only marginal success in its attempt to assimilate the valley, either militarily or culturally; its remote location, rigorous winter climate, Native American opposition, and stubborn bison all combined to blunt Spanish imperial reach into the region. With the collapse of Spain's empire in the New World, the onus of settling New Mexico fell to the newly independent Mexican government. Unlike imperial Spain, Mexico initiated a campaign to settle the valley based not on plundered gold, religious proselytizing, or enslaving indigenous peoples but on a realistic, chastened, subsistence-oriented system much more interested in settling farmers and ranchers than in garnering riches or power to satiate the demands of a distant imperial government. Mexican efforts to settle the San Luis Valley encountered the same withering native resistance and brutal winter climate that had frustrated their Spanish predecessors, but persistence on the part of Hispanic farmers and ranchers eventually led to the establishment of the valley's first permanent settlement in 1851.

In 1820, a full year before Mexico achieved its independence, two Hispanic farmers, Jose Luis Baca de Sondaya and Antonio Matias Gomez, successfully petitioned the civil and military governor of New Mexico for rights to "the tract of land commonly called The Springs of The Medano and The Zapato [Zapata], and the Rito [stream, probably North Spanish Creek] which leads near the outlet of the Pedregosa Mountain [Crestone Peak] and which joins the Laguna Grande [large lake]." The petition involved approximately 230,000 acres of land surrounding the Great Sand Dunes on the eastern edge of the valley, an area of blowing sand and thick desert shrubs that, at first glance, appeared an undesirable and inhospitable place for settlement. Yet according to affidavits filed during court proceedings held in 1874 to determine the validity of the land grant, Gomez and Baca de Sondaya built a house and corral; ran horses, cattle, and burros; and planted corn, beans, and squash.[31] How did such an apparently prosperous homestead succeed in an environment that seemed singularly ill-suited for ranching and farming?

The answer lies in the region's ubiquitous water. Although plagued by arid, sandy soil in places, the area encompassed by the Gomez/Baca de Sondaya land grant contained a variety of vegetation that included blowout, Indian rice, and grama grasses, all nourished by a plethora of small streams and creeks. In addition, the area around San Luis Lake (presumably all that is left of the "Laguna Grande" of the land grant petition) supported abundant grasses, as did the areas around Medano, Zapata, and Indian Springs, where plentiful water ensured fertile soils. By judiciously applying this water through rudimentary irrigation ditches, the aspiring colonists increased available forage for their livestock herds and transformed at least a portion of the grant into productive farm and grazing land. More importantly, the Gomez/Baca de Sondaya homestead established a pattern of land use and water consumption that defined well over 150 years of Hispanic settlement in the San Luis Valley. The pattern consisted of independent, subsistence-level farms and ranches that took advantage of the region's copious water resources to irrigate land for crops and pasture for cattle and sheep, in marked contrast to the earlier attempts by imperial Spain to domesticate the valley's bison and exploit the mineral wealth of the Sangre de Cristo and San Juan mountain ranges.

Baca de Sondaya apparently sold his interest in the grant to Gomez in 1828. Native Americans left the Gomez homestead alone until 1838, when the "Utes rose against the settlers and compelled [them] to leave."[32] By this time the Mexican government had made numerous other land grants (*mercedes*) in the valley, including the Conejos, Tierra Amarilla, and Sangre de Cristo grants. These were huge tracts of land granted to various individuals who promised to settle and defend the frontier from foreign encroachment. Mexico reasoned that occupying the land would prevent an incursion by American citizens from the north and east, especially the Texans, who persisted in casting an aspiring eye on the lands of the San Luis Valley. To ensure a secure border, Mexican law required that settlement begin within two years of receiving a grant, along with the stipulation that the land be cultivated and never abandoned.³³ Such conditions proved exceedingly difficult to meet in light of the valley's relatively short growing season and constant threat of attack from roving bands of Ute and Comanche. Still, a considerable number of Hispanic pioneer settlers attempted to gain a foothold in the valley in the years after Gomez and Baca de Sondaya petitioned for their land grant. Some settlers were of purely Spanish ancestry; others were mestizos, a mixture of Spanish and Mexican Indian descent, and still others were *genizaros*, a class of mainly non-Pueblo Indians who had either abandoned or been stripped of their tribal associations and adopted Spanish lifeways.³⁴

Out of this ethnic and cultural mélange came the first wave of Hispanic settlers in the valley, searching not for gold or souls to save but for irrigable land and fertile soil. As early as 1833, colonists from Mexico tried to settle on a land grant near the Conejos River in the southwestern portion of the valley. Subsequent years saw attempts to settle other communities, but indigenous inhabitants (mostly Ute and Comanche) continued to threaten permanent settlements and successfully dislodged all who tried to remain in the region. Seasonal settlements became the norm, with colonists planting crops in the spring and summer and abandoning the valley following the autumn harvest.³⁵

Finally, with the end of the Mexican War in 1848, the U.S. government began utilizing military force to suppress Indian raids in northern New Mexico. Increasing pressure from the U.S. Army finally convinced the region's Ute Indians to agree (on paper, at least) to stop harassing local citizens. Sporadic raids continued, but by 1849 a small, crude collection of dwellings had sprouted up on Costilla Creek in the extreme southern end of the San Luis Valley. Two years later, in the spring of 1851, a small group of Hispanic colonists settled near Culebra Creek in the extreme southeastern portion of the valley. Named San Luis, this town is generally considered to be the oldest permanent settlement in what later became the state of Colorado.³⁶

Hispanic settlements in the San Luis Valley exhibited a considerable degree of Spanish colonial influence. After gaining independence in 1821, Mexico followed Spanish colonial laws and continued basing the organization of settlements on the Spanish Ordenanzas, a series of systematic regulations that dictated town and village planning. "Although the obligation of dependence on Spain is forever broken," stipulated the Ordenanzas, "these laws which regulate the obligations and rights of those who compose the new community, cannot and ought not to lose their force. . . . Thus, all . . . laws which had emanated from the kings of Spain and the sovereign authority . . . are acknowledged and respected."³⁷ The Ordenanzas dealt primarily with the actual physical layout of a community. Settlers built dwellings made of log or adobe contiguously around a central plaza for defensive purposes, with prominent lots reserved for religious or governmental buildings. In the frontier regions of the upper Rio Grande river valley, settlers modified this layout to include ranchos, consisting of small, individual households clustered adjacent to community farms and orchards, somewhat isolated but still close enough to take advantage of the protection offered by the village plaza in an emergency.[38]

For early Hispanic farms and ranches in the San Luis Valley, the digging of a community acequia, or irrigation ditch, became the top priority. To ensure a dependable water supply for homes, livestock, and crops of wheat, beans, corn, and other vegetables, the settlers of San Luis began digging the San Luis People's Ditch in 1852. The hand-dug ditch channeled water from Culebra Creek to local farms and orchards and became both the first recorded water right in the valley (as well as the state), and the oldest irrigation system in continuous use in Colorado. The ditch also supplied water for the communal public pasture, or *vega*, a nine-hundred-acre expanse on the east side of the settlement.[39] As Hispanic farms and villages began to prosper in the San Luis Valley, irrigation systems necessarily became more complex. When a series of settlements along a single stream began to experience water shortages in a dry year, overall organization of the use of water and ditch maintenance required cooperation at both the community and intercommunity levels. An elected *mayordomo de acequia* (ditch steward), who understood colonial laws governing water appropriation, oversaw construction and maintenance on a village's ditch system. Mayordomos from each community would meet on a regular basis to discuss water allocation and overall system maintenance, thus ensuring an equitable distribution of the valley's most precious resource.[40]

Within a few short years after the founding of San Luis in 1851, a network of irrigation ditches and crude roads laced most of the arable land in the southern end of the valley. Dozens of small farms dotted the landscape, and further settlement of the San Luis Valley and the northern periphery of New Mexico proceeded rapidly during the 1850s. The Conejos and Sangre de Cristo land grants attracted waves of Hispanic settlers, and a gradual expansion toward the western side of the valley led to the founding of La Loma, Piedra Pintada, Las Garritas, and other communities in the Del Norte area.[41] These small, rural villages maintained a mixed pastoral-mercantile economy that existed without extensive commerce or large-scale extractive industries such as mining, logging, or manufacturing. Instead, Hispanic communities were largely self-sufficient; villagers spun and wove their own fabric, raised their own food, and schooled their own children. Moreover, the small scale and communal structure of these early settlements, along with their limited technology, largely inhibited the overexploitation of valuable community resources.

The Spanish-speaking farmers and ranchers who settled in the San Luis Valley may have descended from the conquistadors, but they had no interest in plundered gold, or in converting or enslaving the indigenous populations. Rather, these settlers became part of a pious, pastoral Hispanic cultural milieu that emphasized cooperation and community over profit and conspicuous consumption. Hispanic settlement also set a precedent for the future large-scale irrigation of the valley, an activity that eventually had profound repercussions for the region's landscapes. Although the valley's relative isolation and lack of a permanent Native American population delayed the effects somewhat, two and a half centuries of Spanish contact radically altered both the cultural and physical landscape of the San Luis Valley. The introduction of horses, domestic livestock, firearms, metal tools, and assorted contagious diseases by the Spanish Empire significantly affected the lifeways of the Ute, Apache, Comanche, and Navajo who frequented the valley and left a legacy of Spanish influence in the region. Yet in spite of Spanish military might and aspirations for empire, the valley also managed to resist permanent settlement and imperial control until well into the nineteenth century. By contrast, the later, considerably less ambitious Hispanic settlement of the valley succeeded where the Spanish Empire had failed by introducing a system of small-scale, irrigated farming and ranching that gradually replaced countless centuries of nomadic Native American subsistence-level hunting and gathering.

Although Spanish influence left an indelible impression, it did not end the evolving transformation of the San Luis Valley. Back in 1807, in the very midst of the slow collapse of Spain's New World empire and forty-four years before Hispanic settlers founded San Luis, the valley witnessed the beginnings of yet another phase of human exploration and landscape change. To the east loomed the burgeoning American Republic fast developing its own appetites and perceptions about the landscapes that comprised its expanding realm, and once again the Great Sand Dunes and San Luis Valley silently waited on the edge of an empire.

THREE | Dunes in the Great Desert

This region, which resembles one of the ancient steppes of Asia, has not inaptly been termed The Great American Desert. It spreads forth into undulating and treeless plains and desolate sandy wastes, wearisome to the eye from their extent and monotony.
Washington Irving, 1836

The Plains are not deserts, but the opposite, and the cardinal basis of the future empire of commerce and industry now erecting itself on the North American Continent.
William Gilpin, 1860

ON JUNE 24, 1806, shortly after his return from an expedition to the headwaters of the Mississippi River, Lieutenant Zebulon Montgomery Pike received a letter from General James Wilkinson, commander of the Western Army and governor of the Louisiana Territory, detailing the objectives of an expedition to explore the southwestern reaches of the recently acquired Louisiana Purchase of 1803. Wilkinson ordered Pike to establish peaceful relations with the Kansas and Osage Native Americans and report on the region's geography, natural resources, rivers, and flora and fauna, with specific instructions to "ascertain the Direction, extent, & navigation of the Arkansaw [sic], & Red Rivers."[1] Wilkinson cautioned Pike that his mission might take him near the Spanish border in the Southwest, and he was advised to "move with great circumspection" to avoid any encounters or hostility with Spanish forces.[2]

Following Wilkinson's instructions, Zebulon Pike and his company of twenty-three men, whom he had earlier described as a "dam'd set of rascals," embarked on an epic exploration of the American West, a journey that led them across the vast expanse of the Great Plains to the rugged interior of the towering Rocky Mountains. There, in January 1807, chance and circumstance combined to produce Pike's historic rendezvous with the stunning landscape of the Great Sand Dunes, an encounter that began the American phase of discovery, exploration, and transformation of the San Luis Valley.[3]

During the century of American territorial expansion that followed Pike's expedition, successive waves of fur trappers, explorers, soldiers, surveyors, and economic "boosters" ventured into the San Luis Valley, each contributing their own descriptions and impressions to the growing body of knowledge about the region. Their journal entries, diaries, maps, and official reports added layer upon layer of information until an increasingly complex, and often vaguely contradictory, series of descriptions of the San Luis Valley emerged that deeply influenced public perception of the valley's natural resources. Unlike the Spanish that came before them, however, the Americans also left detailed records of their encounters with the Great Sand Dunes, a landscape that, judging from the dearth of historical references, the Spanish apparently found unremarkable.

Zebulon Pike's descriptions of the Great Plains, reinforced by the maps and reports of the Stephen H. Long expedition to the northern Rocky Mountains in 1819–20, helped inspire the initial public perception of the Great Sand Dunes and the San Luis Valley as part of the "Great American Desert." During the mid-1820s, shortly after the Long expedition, reports from fur trappers described the valley's bountiful water and animal resources and offered the first direct contradiction of the Great American Desert myth. After the fur trade

Figure 3.1 Charles Willson Peale, *Zebulon Montgomery Pike, from Life*, circa 1808. National Park Service, Independence National Historical Park.

DUNES IN THE GREAT DESERT | 51

began disintegrating around 1840, the government-sponsored railroad survey explorations of the 1840s and 1850s produced a more ambiguous assessment of the valley's resources. Reports from these expeditions depicted a landscape that seemed at once barren yet fertile, arid yet moist, composed of rich soil amid vast tracts of "worthless" lands. The post–Civil War economic boosters further muddled this confusing image of the San Luis Valley by employing glowing promotional rhetoric in books, pamphlets, and brochures to tout their vision of the region as an American Eden, filled with abundant water resources, excellent climate and soil, and clear potential for profit. The boosters' efforts were the final stage in a cumulative process of exploration and characterization that spanned almost the entire nineteenth century and helped inspire the intensive American settlement and exploitation of resources of the late 1800s, which forever transformed the landscapes of the San Luis Valley.

Sea in a Storm

The Pike expedition left the outskirts of St. Louis in July 1806. By the middle of October, the expedition had reached the Great Bend of the Arkansas River in present-day central Kansas, where Pike divided his company, sending a small group to descend the Arkansas while he took the remainder of his men and proceeded upriver, following the Arkansas west toward the Rocky Mountains. On November 15, Pike spotted a "small blue cloud" on the distant horizon. The "cloud" turned out to be a towering peak in the Front Range of the Rocky Mountains, later christened Pikes Peak, and Pike and his small party "gave three cheers to the Mexican Mountains."[4] Pike spent the rest of November and December 1806 exploring the region around the peak that now bears his name. By December 22, after a confusing series of journeys into the mountains to the west, Pike satisfied himself that he had located the headwaters of the Arkansas in the vicinity of present-day Twin Lakes, Colorado.[5] He then turned his attention to locating the Red River, which he assumed lay to the southwest, just beyond the crest of the rugged White Mountains (Sangre de Cristos). Whether Pike knew it or not, and historians continue to debate the matter, the true headwaters of the Red River gather in the Texas Panhandle just south of present-day Amarillo, roughly three hundred miles from where Pike and his men searched for the Red's headwaters.[6]

On January 23, 1807, Pike and his men reached the Wet Mountain Valley, southwest of present-day Pueblo, Colorado. Four days later, after enduring a bitter winter storm, the expedition "struck a brook running west, which [they] hailed with fervency as the waters of the Red River."[7] Battling deep snow, the expedition members followed this brook toward the summit of Medano Pass, which they crossed before making camp for the night. The next day, January 28, 1807, Pike and his men began the slow descent down the western side of the Sangres, where Pike noted they "discovered after some time that a road had been cut out, and on many trees were various hieroglyphicks [sic] painted," most likely a reference to markings left on aspen trees by Native Americans.[8] Continuing down the pass, the Great Sand Dunes suddenly came into view, and Zebulon Pike dutifully recorded the first, and arguably the most famous, English language description of the dunes and the San Luis Valley:

> After marching some miles, we discovered through the lengthy vista at a distance, another chain of mountains and nearer by at the foot of the White Mountains, which we were then descending, sandy hills. We marched on the outlet of the mountains, and left the sandy desert to our right; kept down between it and the mountain. When we encamped [near the confluence of Castle and Medano Creeks], I ascended one of the largest hills of sand, and with my glass could discover a large river [the Rio Grande], flowing nearly north by west, and south by east, through the plain which came out of the third chain of mountains. . . . I returned to camp with news of my discovery.
> The sand hills extended up and down at the foot of the White Mountains, about 15 miles, and appeared to be about five miles in width.
> Their appearance was exactly that of a sea in a storm, (except as to color) not the least sign of vegetation existing thereon.[9]

Pike's description, which first introduced the Great Sand Dunes into the American consciousness, is notable for including a reference to the dune's perceived lack of vegetation. Pike wrote that the dunes had "not the least sign of vegetation existing thereon," yet Lieutenant. E. G. Beckwith, a U.S. Army explorer who visited the San Luis Valley less than fifty years later, noted "artemisia" (a large sage) growing on the dunes. The current dunefield boasts a variety of plant life, including rabbitbrush, scurfpea, sunflowers, Indian ricegrass, and scattered stands of willow surrounding the various interdunal ponds, so what accounts for this vegetative discrepancy?

The vicissitudes of winter weather offer one possible explanation. Plant life on the main dunefield fluctuates with seasonal variations, becoming dormant in the winter and blooming again with the warmer temperatures of spring and summer. Consequently, vegetation on the dunes is hardly conspicuous during the winter months; the pale color of its dormant state tends to blend in with the sand dunes. Additionally, the high ridges of the dunes are too exposed to wind and weather to support much plant life, so the vegetation that does exist is concentrated in the pockets and depressions between the shifting dunes, a survival strategy that allows the plants to contend with the challenging environment of the main dune mass.[10]

Snowfall also tends to accumulate in these pockets and depressions, yet even during years of heavy snowfall, snow does not cover the dunes for long periods. Since Pike entered the San Luis Valley in January, in the midst of one of the

Figure 3.2 Depiction of Zebulon Pike looking west over the Great Sand Dunes toward the Rio Grande River, January 28, 1807. Original artwork by Heather Paul.

region's notoriously cold winters, it is conceivable that he observed no vegetation because he saw the dunes during one of those rare occasions when snowfall had completely buried them. However, Pike's reports made no mention of any snow on the dunes, even though he "ascended one of the largest hills of sand," a long, strenuous climb that surely would have revealed the presence of snow. Such a thorough reconnaissance also seems to rule out the possibility that Pike made his assessment after surveying only a small portion of the Great Sand Dunes.

The search for a definitive explanation for the discrepancy between Pike's historic observations and the reality of vegetation on the current dunefield is further complicated by Pike's role in creating the myth of the Great American Desert. His descriptions of the dunes and their lack of vegetation are remarkably consistent with his overall interpretation of other landscapes he encountered earlier in his travels. Details gleaned from his official expedition report, published in 1810, indicate that Pike did not limit his emphasis on sandy soils and scant vegetation to the Great Sand Dunes or the San Luis Valley. In fact, Pike characterized the entire southern Great Plains from Oklahoma to New Mexico as "sandy deserts" similar to the Sahara Desert of Africa.[11] "I saw in my route, in various places," he wrote, "tracts of many leagues where the wind had thrown up the sand in all the fanciful form of the ocean's rolling wave; and on which not a speck of vegetable matter existed."[12]

Pike's description of sandy soil on the Great Plains that resembled "the ocean's rolling wave" with "not a speck of vegetable matter" is strikingly similar to his comparison of the Great Sand Dunes to a "sea in a storm" with "not the least sign of vegetation." Pike may have believed that his encounter with the sprawling immensity of the Great Sand Dunes actually validated his earlier appraisal of the Great Plains. By emphasizing the aridity and sandy soil of isolated tracts of land in the southern Great Plains and the Southwest, Pike managed to characterize the entire region, including the San Luis Valley, as a vast desert. In all fairness, Pike's observations did not concentrate solely on sandy deserts and shifting dunes to the exclusion of all else. For example, after leaving the vicinity of the Great Sand Dunes, Pike described the San Luis Valley as "one of the most sublime and beautiful inland prospects ever presented to the eyes of man," where "the great and lofty mountains covered with eternal snows, seemed to surround the luxuriant vale, crowned with perennial flowers, like a terrestrial paradise." Glowing praise indeed for the seemingly barren winter landscape of the valley.[13]

Pike and his men continued their journey south through the San Luis Valley and eventually built a stockade on the north bank of the Conejos River, about four miles northeast of present-day Sanford, Colorado, where they raised a fifteen-star, fifteen-stripe American flag over territory claimed by Spain. On February 26, 1807, a detachment of one hundred Spanish cavalrymen appeared at the stockade and requested that Pike and his men accompany them to Santa Fe, where they became involuntary "guests" of the governor of New Mexico, who confiscated their belongings and interrogated them repeatedly. Pike and his men were later taken to Chihuahua, where Spanish authorities continued questioning them for several months before finally escorting Pike and the majority of his men to the U.S. border near Natchitoches, Louisiana, which they reached in late June 1807. The Spanish held five of Pike's men for an additional two years. The Spanish regional governor, meanwhile, received a reprimand from his king for releasing Pike before receiving an official apology from the U.S. government for trespassing.[14]

On his return, Pike spent the next six months editing and rewriting his journal and official report, neither of which formally referred to a "Great American Desert." That dubious distinction instead belongs to Major Stephen Harriman Long, whose map and report from his 1820 expedition to the Rocky Mountains were the first to refer specifically to a "Great Desert" stretching across the Great Plains all the way to the Rockies.[15] Characterizations of the region as essentially "waste lands" actually predate both Pike and Long, however. As early as 1805, American geographers were aware of an account written by French-Canadian trader and explorer Jean Baptiste Trudeau (also spelled *Truteau*). Lewis and Clark referred to Trudeau as "Trodow" and apparently carried portions of his journals with them on their famous 1804–1806 expedition to the Pacific. Detailing his journey to the upper Missouri River, Trudeau described the lands to the west of the river as "vast prairies which extend without interruption to the foot of the mountains of rock [Rocky Mountains]. . . . These large prairies, or great waste lands, are completely sterile; scarcely grass grows there."[16]

Other travelers and explorers who visited the Great Plains and the Southwest in the years following Pike's expedition reinforced his perception of the region as an arid landscape and, whether intentionally or not, helped establish a firm foundation for the Great American Desert myth so faithfully supported by subsequent nineteenth-century explorers and observers. Writer and future congressman Henry M. Brackenridge traveled up the Missouri in 1811 and described the lands to the west as "extensive tracts of moving sands similar to those of the African deserts."[17] In 1819 English naturalist Thomas Nuttall used the term "pathless desert" to describe the lands beyond the Missouri, while Dr. Edwin James, chronicler of the 1820 Long expedition, described the region as a "dreary plain, wholly unfit for cultivation, and of course uninhabitable by a people depending upon agriculture for their subsistence." In addition, James hoped the region might "forever remain the unmolested haunt of the native hunter, the bison, and the jackall [*sic*]."[18]

James also described the land between the Canadian River and the Rockies as "sandy wastes and thirsty, inhospitable steppes," while Long himself described the Great Plains as "a dusty plain of sand and gravel, barren as the deserts of Arabia." James's map of the Long expedition, which enjoyed wide distribution, helped further reinforce the image of the Great American Desert. Contemporary journalists and historians freely borrowed from both the map and the expedition's observations. In 1843, Thomas Farnham referred to the land from the

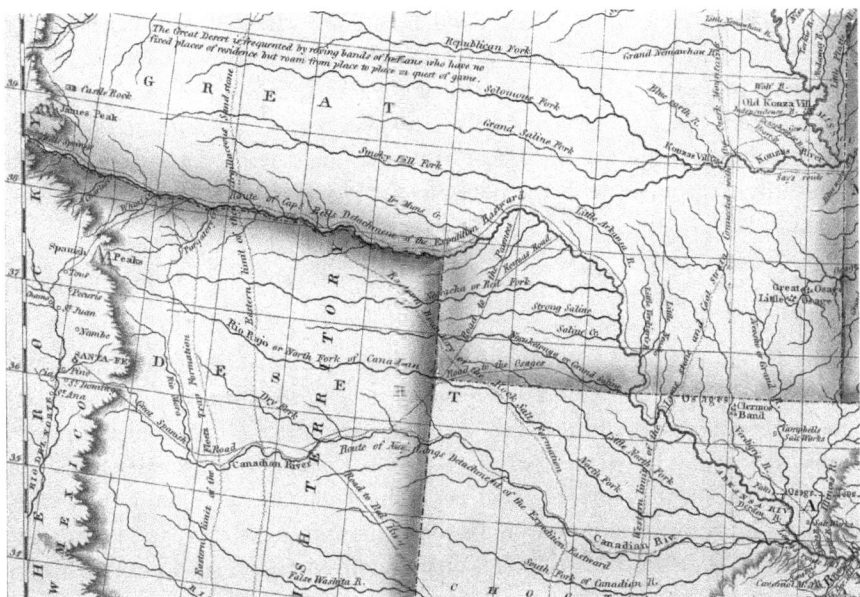

Figure 3.3 Highly influential map from the 1820 Long expedition depicting a "Great Desert" stretching eastward from the Rocky Mountains. David Rumsey Map Collection, www.davidrumsey.com/home.

100th meridian to the Rockies as "usually called the Great American Desert." By the middle of the nineteenth century, the term had firmly embedded itself in the American vernacular. It would be at least a generation before the potent image of the "great desert" disappeared completely from the popular consciousness.[19]

Contrary to the negative connotations normally associated with the term "desert," Zebulon Pike actually perceived a genuine value in the arid lands of the Great Plains and Southwest. Convinced that "from these immense prairies may arise one great advantage to the United States," Pike believed the "desert" would restrict the expansion of the American population, thereby preventing the reckless extension and possible disintegration of the Union. Furthermore, Pike felt that the desolate prairies would provide a sanctuary for the "wandering and uncivilized Aborigines of the country."[20] Ultimately, despite Pike's best intentions, the burgeoning American nation refused to be confined by the "deserts" of the Great Plains, and the "sanctuary" idea that Pike prophesied was corrupted and convoluted into the forced removal of the "aborigines" and their eventual resettlement on squalid reservations.

Zebulon Pike's vision notwithstanding, the characterization of the Great American Desert lingered in the popular imagination for the entire first half of the nineteenth century. Pike and his fellow explorers left for posterity an enduring series of descriptions, frozen in time, of the landscapes they encountered in their travels, including the Great Sand Dunes and the San Luis Valley. These descriptions, filtered through whatever particular inclination or biases these men possessed, were fundamental in establishing the perception of the region as a Great Desert. The trappers and traders who later began trickling into the American Southwest in the wake of these explorers found not a desert but, instead, a bountiful land filled with the potential for wealth far beyond the dreams of avarice.[21]

A Fortune in Fur

While Pike and others explored the mountains and borderlands of America's western frontier, a robust fur trade slowly developed along the myriad rivers and streams far to the north and east of the remote San Luis Valley. Initially confined to the lower reaches of the Missouri River, where a small number of French, Spanish, and American traders obtained pelts and robes from the Osage, Kansas, Omaha, and various other Native American tribes, the nascent fur trade soon expanded its reach. By 1807, trappers and traders began probing the upper reaches of the Missouri and its tributaries, using the river system to explore and exploit the rich fur country of the northern and central Rockies.[22] Within a few short years, demand for beaver pelts skyrocketed as top hats became an essential fashion component for well-dressed men in the urban centers of the eastern United States and Europe. Compelled by the vagaries of fashion and the potential for profit, ambitious fur trappers pushed ever deeper into boundless realms of the trans-Missouri West, generating an extensive knowledge of the shape and character of the lands they traversed.

Compared to the Missouri River country and the northern Rockies, the San Luis Valley lagged a few years behind in attracting the organized industry of commercial trapping, due primarily to its relative isolation, resistance from Native Americans, and bountiful trapping elsewhere. More importantly, the area remained under Spanish control for the first two decades of the nineteenth century, which further discouraged extensive trapping in the region. The situation changed dramatically in 1821 when the newly independent Mexican government opened its northern borders to American merchants. The San Luis Valley's relatively close proximity to the bustling trading center of Taos and to the recently opened Santa Fe Trail virtually ensured a steady influx of fur trappers and traders into the region. No longer threatened with arrest for trespassing by the Spanish, these adventurous entrepreneurs soon targeted the abundant beaver that thrived in the numerous rivers and streams draining the vast expanse of the San Luis Valley.

The considerable success of the fur trappers along the beaver-rich waterways of the San Luis Valley and upper Rio Grande basin offered the first direct contradiction of the earlier perception of the region as merely another portion of the Great American Desert. As early as 1822, trapper Hugh Glenn and his company of nineteen men harvested over 1,100 pounds of beaver pelts in the Sangre de Cristo region, which they hauled to St. Louis and sold for the tidy sum of $4,999.64.[23] The fur merchants of St. Louis likely found it difficult to reconcile the image of the Great American Desert with the reality of Hugh Glenn's abundance of pelts. Further evidence of the region's rich bounty came from the chronicler of Glenn's expedition, fur trapper Jacob Fowler. Unlike many of the trappers, who were often illiterate (or tight-lipped, at best) and not particularly interested in keeping diaries or writing personal reminiscences, Fowler kept a journal.[24] He visited the San Luis Valley in early 1822 and briefly described the region as "an oppen plain of great Exstent [sic]" and noted that "this party Has Caught Some Bever and their Is Sign of more in the River [sic]."[25] Trappers like Fowler enjoyed such great success in the San Luis Valley and the upper Rio Grande that in 1838 Mexican officials declared a six-year moratorium on trapping beaver and otter along the entire river drainage. The ban proved unenforceable, of course, but its attempted implementation illustrates both the richness and rapid depletion of the region's wildlife resources.[26]

Fur trappers of the San Luis Valley and Rio Grande basin exemplified the dominant nineteenth-century American attitude toward the natural environment: an emphasis on short-term profit over long-term sustained yield. The quest for profit blinded the trappers to the irrationality of their extractive endeavors: trapping the beaver to the brink of extinction essentially eliminated the very resource that provided the trappers with their livelihood. By 1840, Taos-based trappers had stripped most of the beaver from the Sangre de Cristos and the San Luis Valley. Such extensive resource depletion represents the "dark obverse" of the standard, romanticized version of the fur trade. Generally portrayed as a glorious episode in the historic saga of westward expansion, the fur trade also signified an early stage in the massive alteration of a variety of ecosystems across the American West.[27]

A precipitous decline in the demand for beaver pelts after 1834 marked the beginning of the end for the fur trade in North America. By the early 1840s, the fickle whims of fashion and economics, coupled with the increasing availability of Chinese silk and new technologies for felting hats, conspired to eliminate the profitability of the fur trade. Many of the trappers and traders returned east to lives of relative obscurity and anonymity, while others remained in the Rocky Mountains hunting bison for the growing trade in bison hides and meat. A few, like Kit Carson, gained fame as guides for the U.S. Army or for the coming waves of settlers. The fur trade itself passed into the realm of frontier romanticism, yet its legacy of short-term, rapacious extraction of natural resources persisted throughout the remainder of the nineteenth century. With the decline of fur trapping, the economic focus of the relentless American westward expansion shifted to the widespread exploitation of minerals, timber, land, and water.[28] In the San Luis Valley, the success of the fur trade offered an undeniable indication of the valley's prodigious water resources. The trade also established a pattern for later resource extraction and revealed the first cracks in the foundation of the Great American Desert. The coming wave of American explorers further eroded the image of aridity, but it would be almost thirty years before the stubborn myth collapsed completely.

Frémont's Disaster

In 1848, only a few short years after the fur trappers had removed their last beaver traps, the Treaty of Guadalupe Hidalgo formally ceded an enormous region of the American Southwest, including the San Luis Valley, to the United States. Compelled by visions of Manifest Destiny and eager to explore the land and resources of this latest territorial acquisition, Missouri senator Thomas Hart Benton organized an expedition under the leadership of his son-in-law, John Charles Frémont, to ascertain the feasibility of a central railroad route to the Pacific. The quest brought Frémont to the southern Rockies during the winter of 1848, and the reports and drawings generated by his expedition initiated the next phase in the evolution of public perceptions concerning the Great Sand Dunes and the San Luis Valley. Considering the difficulties Frémont and his men subsequently encountered in the San Juan Mountains, their reports did remarkably little to reinforce the prevailing notion of the region as an inhospitable province of the Great American Desert.

Following the valley of the Huérfano River westward, the expedition approached the Sangre de Cristo Mountains and ascended to the summit of Mosca Pass on December 3, 1848. From the crest of the pass they gazed into the vast "valley of the Rio del Norte," which one of Frémont's men described as "one broad, white, dreary-looking plain . . . bounded by white mountains." Descending the western side of the pass, they established their camp near the present-day Pinyon Flats campground in Great Sand Dunes National Park, where Frémont and his men endured a night of high winds, heavy snow, and temperatures that dropped to 17 degrees below zero. On December 5, they attempted in vain to march north between the mountains and the sand dunes, but subzero temperatures, deep snow, and high winds impeded their progress. Finally, expedition member Alexis Godey managed to lead the explorers over the crest of the sand hills until they reached a small creek north of the dunes, where they found shelter among the trees. Curiously, expedition reports gave scant attention to the imposing mass of sand dunes, perhaps because they were buried under several feet of snow. Frémont and a few of his men did examine the western approach to Medano Pass, and at least one man (Charles Preuss) wondered why their guide, Bill Williams, had not chosen this obviously easier route through the Sangres. On the morning of December 7, with six inches of fresh snow on the ground, the expedition struck camp and headed west-northwest toward Cochetopa Pass in the distant San Juan Mountains, where fierce winter blizzards, hypothermia, and starvation later claimed the lives of ten of Frémont's men.[29]

Figure 3.4 One of the earliest artistic representations of the San Luis Valley, drawn by Frémont expedition artist Richard Kern and dated December 6, 1848. The simple pencil sketch depicts the Sangre de Cristos behind some "snow hills," which may represent the Great Sand Dunes covered by winter snows. To the right the word "creek" is scrawled, perhaps in reference to Medano Creek. The Huntington Library, San Marino, California.

The bitterly cold weather and multiple fatalities that plagued his expedition should have convinced Frémont that the San Luis Valley and surrounding mountains could indeed be hostile places in the winter, yet astonishingly his official report claimed that "the result was entirely satisfactory. It convinced me that neither snow of winter nor the mountain ranges were obstacles in the way of the railroad."[30] Frémont's enthusiasm to promote the proposed "central route" for the Pacific railroad evidently outweighed the more troubling aspects of his trip, which other members of his party considered a complete and unmitigated disaster. The fatalities that occurred during his ill-advised foray into the rugged and treacherous San Juan Mountains in the dead of winter offered ample justification for their appraisal.[31]

In retrospect, the expedition was not a complete failure. Frémont obtained additional information about the region, and the expedition produced what is perhaps the first artistic image of the San Luis Valley. Drawn by expedition artist Richard Kern and dated December 6, 1848, the simple pencil sketch has been labeled in contemporary sources as a view "From the Sand Hills in the San Luis Valley," and "Scene apparently drawn from the sand hills near Medano Creek, San Luis Valley."[32]

Amid what appears to be riparian vegetation in the foreground, the word "creek" is scrawled. Kern's journal gives no name for the creek, mentioning only that "Godey led camp over immense sand hills with from one to six feet of

Figure 3.5 Possibly the first definitive artistic depiction of the Great Sand Dunes, variously attributed to Edward Kern or his brother Richard. Note the sand dunes piled at the base of what appears to be Mount Herard in the middle left of the image. Allegedly completed during Frémont's 1848 expedition, the drawing's apparent lack of snowfall and summerlike landscape seem to suggest that one of the Kern brothers, probably Richard, actually made the drawing during the 1853 Gunnison expedition through the San Luis Valley. The Huntington Library, San Marino, California.

snow." Kern does describe the valley as "nothing but sand, snow, sagebrush, and greasewood," one of the first recorded descriptions of sagebrush and greasewood in the eastern portion of the San Luis Valley, but Kern gave no further information regarding vegetation on or near the Great Sand Dunes.[33]

Another drawing, variously attributed to either Edward Kern or his brother Richard and titled *San Luis Valley, near Sand-Hill Pass*, is far more detailed and depicts the Sangre de Cristos towering above the dunes in the background, with what appears to be luxuriant grasses and thick piñon pines around an idyllic campsite in the foreground. The drawing's summerlike landscape and obvious lack of snow seem hard to reconcile with Frémont's 1848 winter expedition, suggesting that Richard Kern may have drawn the sketch during the 1853 Gunnison expedition through the valley.[34] Regardless, the serene setting and abundant vegetation depicted in the drawing continued the evolution of imagery and public perception of the San Luis Valley, and further weakened the concept of the Great American Desert. In a few short decades, thanks to reports from subsequent explorers and the efforts of the economic boosters, the image of an arid San Luis Valley would be miraculously transformed into the very picture of fertility, a virtual Eden of the American West.

Contradictory Images

The landscape of the San Luis Valley experienced considerable changes in the years immediately following Frémont's failed expedition. By 1851, Hispanic farmers had established the valley's first permanent settlement, San Luis, and in 1852 began digging the valley's first irrigation ditch, which drew water from Culebra Creek to nourish crops of wheat, beans, and corn.[35] That same year, in an effort to protect the settlers of San Luis from persistent attacks by Ute and Comanche warriors, the U.S. Army constructed Fort Massachusetts on the south side of the Sierra Blanca massif. Although it was abandoned only six years later when the army built Fort Garland in a more defensible location six miles to the south, the fort provided a safe haven for immigrants, explorers, and travelers seeking respite from the rigors of the trail.[36] One such visitor, Gwinn Harris Heap, visited Fort Massachusetts in 1853 as a member of an expedition commanded by Lt. Edward F. Beale. Like Frémont, Beale and his men entered the valley as part of an investigation of the region's suitability for the proposed central railroad route to the Pacific, and Beale assigned Heap the task of preparing the expedition's reports and sketches.[37]

Heap's descriptions of the San Luis and Wet Mountain Valleys seem the very antithesis of the Great American Desert image. His journal entries repeatedly refer to the abundant moisture the expedition encountered on the journey westward through the Wet Mountains and Sangre de Cristos toward the San Luis Valley:

> The rain fell at intervals all night, but the clouds dispersed before dawn, and the sun rose in a bright and clear sky; the plains, however, were concealed under a sea of snowy mist. . . .
>
> An incipient rain storm made us feel sensible that we were still in the vicinity of the Sierra Mojada (or Wet Mountains), which well merit the name, for rain fell every day that we were in or near them; on the highest peaks in the form of snow, and lower down in hazy moisture, alternating with drenching showers. This humidity gives great fertility to this region, and the country bordering on the sides of these mountains, as well as in the valleys within their recesses, are unequaled in loveliness and richness of vegetation.[38]

Heap entered the San Luis Valley in June, when seasonal runoff and abundant sunshine transform the barren winter appearance of the valley into a lush and considerably more inviting landscape. Persistent moisture and blossoming plant life apparently convinced Heap of the valley's fertility and suitability for settlement, which for "the settler . . . offer every inducement; and I have no doubt that in a few years this tract of country will vie with California or Australia in the number of immigrants it will invite to it. It is by far the most fertile portion of New Mexico."[39] Heap limited his description of the dunes to calling them "a range of sand hills of moderate height." Evidently the sight of the enormous expanse of drifting sand did little to convince Heap to amend his assessment of the region as "by far the most fertile portion of New Mexico." In a journal entry that further reinforced his perception of the region's fecundity, Heap briefly described the vegetation of the broad valley, calling it "a level plain covered with artemisia, cacti, and patches of the nutritious grama."[40]

Subsequent explorers and settlers in the American West would increasingly interpret the appearance of "artemisia" (a large species of sagebrush) as an unmistakable indicator of fertile soil.[41] Heap's observation of "artemisia" and "nutritious grama" grasses suggests that the valley's soils were capable of supporting healthy forage for both wild and domestic grazing animals. He also noted the "numerous sloughs and marshes" that covered the "bottom lands of the Del Norte," providing further evidence of the region's moisture and potential for settlement and irrigation.[42] Heap's journal entries concerning the San Luis Valley portray a landscape in marked contrast to the Great American Desert image. Reports from two subsequent explorers, however, offered slightly contradictory impressions of the region that further muddled the emerging image of the valley as a fertile garden.

German geologist and surgeon Dr. James Schiel accompanied yet another railroad survey expedition, this one commanded by Captain John Williams Gunnison. As part of the continuing investigation of the proposed central route to the Pacific, Gunnison and his men entered the valley in 1853, giving Schiel the opportunity to record his impressions of the San Luis Valley as the expedition traveled north from Fort Massachusetts toward Cochetopa Pass. Strangely, Schiel wrote nothing about the Great Sand Dunes, but he did describe moments of "wading through fields of Artemisia, the only plant growing in this sandy soil." Schiel gave no indication that he believed artemisia to be an indicator of fertile soil. In fact, he stated quite bluntly that "the San Luis Valley is mostly sandy and not fertile, bushes of Artemisia and a few sorts of cacti form the vegetation." Schiel's assessment of the valley as "not fertile" seems puzzling in light of a later passage in his journal concerning a conversation he had with one of the expedition guides: "According to his words the valley where the Rio del Norte [Rio Grande] emerges through a canyon from the Sierra San Juan is very fertile, has rich pastures, and plenty of deer and wild horses."[43]

Schiel's report contained other contradictory assessments of the San Luis Valley, including the presence of "numerous beaver dams" near the northern end of the valley. "They dam the water and force it to overflow," he wrote, "which caused the development of wide swamps to the south." Overflowing water and "wide swamps" conjure an image of a lush and fertile landscape, but Schiel then described walking "over soil which was often covered by efflorescent alkaline salts, and only at the creek did we again find pasture and brush."[44] Schiel's journal entries depict the region as "very fertile" yet "not fertile," where sandy soils, artemisia, and cacti apparently coexisted with "rich pastures," and overflowing water produced alkaline salts that covered the ground, a confusing and strangely contradictory series of descriptions that further blurred the public image of the San Luis Valley. Additional reports from the 1853 Gunnison expedition added yet another dimension to this evolving imagery.

In October 1853, after the expedition left the San Luis Valley, Paiute Indians attacked and killed John Gunnison and seven of his men, including artist Richard Kern, in present-day Utah. Lieutenant E. G. Beckwith, who survived the attack, assumed command of the expedition and took responsibility for preparing its official reports. Beckwith's characterizations of the valley, while similar in many respects, occasionally differed from those of his contemporaries. For instance, in contrast to Gwinn Harris Heap's descriptions, which mentioned the Great Sand Dunes only briefly and instead tended to emphasize the region's abundant moisture and fertility, Beckwith's report is filled with references to sand, and to the Great Sand Dunes specifically:

> Turning the southern base of the sand-hills, over the lowest of which we rode for a short distance, our horses half burying their hoofs only on the windward slopes, but sinking to their knees on the opposite, we for some distance followed the bed of the stream from the pass, now sunk in the sand, and then struck off across the sandy plain, which here extends far into the valley, and is very uneven, the clumps of Artemisia fixing in place large heaps of sand, while the intermediate spaces are swept out by the wind.
>
> The sand was so heavy that we were six hours and a half in making ten miles—the sand being succeeded, on the last two miles, by a light, friable soil, and a heavy growth of artemisia.[45]

Beckwith's emphasis on the valley's ubiquitous sand did not blind him to the region's apparent fertility. In addition to the "heavy growth of artemisia," he noted "fine fields of prairie-grass, pea vine and barley-grass," and an area where "marsh-grass grew luxuriantly." Yet Beckwith also observed that "the surface of the ground, over large spaces, is often covered with effloresced salts."[46] Beckwith's image of the valley, like Schiel's, combined elements of the barren and the fertile. Even his comments on the Great Sand Dunes exhibited elements of both, registering the appearance of vegetation on the dunes, the only discernible change since Zebulon Pike had described them in 1807:

> Rode to Williams' [Medano] Pass, the approach to which from the San Luis Valley is through a grove of pitch-pine, behind most gigantic sand-hills, rising above the plain to half the height (apparently, at least, 700 or 800 feet) of the adjacent mountain, and shaped by the winds into beautiful and fanciful forms with waving outlines, for within certain limits this sand drifts about like snow . . .

DUNES IN THE GREAT DESERT | 65

> High up on the sides are seen, at half a dozen points, single bushes of artemisia—the only vegetation seen upon them, and the only change discoverable since they were visited by Captain Pike, fifty years ago, when they were entirely destitute of vegetation, and "appeared exactly like the sea in a storm, except in color."[47]

Beckwith noted two important distinctions that differed from Pike's description of the towering sand dunes. First, the shifting dunes that Pike compared to a sea in a storm had been transformed into sand "shaped by the winds into beautiful and fanciful forms." By pointing out the pleasing aesthetic quality of the dunes, Beckwith became one of the first Americans to express in writing his admiration for the graceful lines and stark natural beauty of the Great Sand Dunes. His appreciation for the inherent beauty of the landscape helped form the basis for the twentieth-century effort to protect the sand dunes, an effort that culminated with the establishment of Great Sand Dunes National Monument in 1932. The second distinction is perhaps more significant for the environmental history of the San Luis Valley. The sand dunes that Pike described as having "not the least sign of vegetation" now boasted single bushes of artemisia, a discrepancy that surely has some logical explanation. Had the valley experienced some dramatic ecological or climatological change in the interval between Pike and Beckwith, or were the two explorers somehow mistaken in their respective observations?

Modern tree-ring analysis of piñon and ponderosa pine in the San Luis Valley reveals no evidence for severe climatic fluctuations in the interval between 1807 and 1853, which seems to eliminate a climatological explanation.[48] Artemisia seeds conceivably could have been transported to the dunes by birds or in the fur of coyotes, but the wide distribution of vegetation suggests a frequent, pervasive presence by these animals, which in reality are infrequent visitors to the inner dunes. Seeds could have been transported in the digestive systems of large herbivores, wild or domestic, but this too seems unlikely. Large herds of domestic cattle had yet to be introduced into the region, and the only wild herbivore known to base a significant portion of its diet on large sagebrush is the pronghorn, which is known to occasionally forage around the dunes.[49] But why would any grazing animal venture "high up on the sides" of the sand dunes, where plant life is sporadic at best, especially when the perimeter of the dunefield offered sufficient forage?

The most plausible explanation for the appearance of the vegetation once again relies on seasonal variations. As mentioned earlier, Pike likely saw no vegetation on the dunes because he arrived in the valley during the winter, while Beckwith, who explored the dunes in August, probably misidentified the vegetation he did see as artemisia. Artemisia is essentially a large sage, but nineteenth-century travelers in the arid regions of the intermountain American West tended to identify all shrubs as sagebrush, even though salt bush or greasewood dominated many of these arid landscapes.⁵⁰ Perhaps Beckwith, unable to make a clear distinction about the vegetation he saw on the dunes from a distance, simply assumed that it was artemisia, when in fact it may have been something else altogether, perhaps rabbitbrush.

Beckwith's final description of the San Luis Valley is rather ambivalent in its appraisal of the valley's resources:

> We here leave the immense valley of the San Luis, which is one of the finest in New Mexico, although it contains so large a proportion of worthless land—worthless because destitute of water to such an extent where irrigation alone can produce a crop, and because of the ingredients of the soil in those parts where salts effloresce upon the surface. Its lower portion is adapted to the cultivation of grain, as we have seen at the Costilla and Rio Colorado; and, if its upper part should prove too cold for cereals, its fine fields of grass on and above the Rio Grande del Norte, must make it valuable for grazing.⁵¹

The valley described by Beckwith is at once inviting and hostile, one of the "finest in New Mexico" yet full of "worthless lands," well adapted for growing grain yet plagued by alkaline soil. As a result, the evolving perception of the San Luis Valley remained decidedly ambiguous. Historic descriptions about its resources and prospects for settlement were an amorphous collection of vaguely contradictory images, each seemingly emphasizing a different feature of the landscape. The valley remained on the periphery of commercial development and pioneer settlement until the arrival of the economic boosters opened the floodgates for the waves of settlers who poured into the region in the years following the Civil War. In their effort to obliterate the myth of the Great American Desert, the boosters took all of the disparate imagery and contradictory perceptions of the valley and reformulated them into an exaggerated representation that compared the San Luis Valley to an American Eden. In the process, the boosters inspired monumental and unprecedented changes to the region's landscapes.

The Booster Juggernaut

By the spring of 1859 the small trickle of mountain men and explorers who spent the first half of the nineteenth century probing the secrets of the Rocky Mountains had turned into a raging torrent of would-be argonauts. The discovery of gold near present-day Denver in 1858 touched off a staggering case of gold fever, as tens of thousands of Americans endured the grueling trek across the Great Plains to seek their fortunes in the newly discovered goldfields of the Pikes Peak region. The innermost reaches of the lofty Rockies, long the domain of Native Americans and the occasional fur trapper, suddenly came alive with mining camps and boomtowns filled with eager prospectors and the shrewd, opportunistic merchants who supplied the materials necessary to strike it rich. Many of the "fifty-niners" who crossed the prairies carried with them guidebooks filled with articles, maps, and facts about the gold country. One such book included a speech by William Gilpin, tireless advocate for settlement and economic development in the mining country of the Rockies and future first governor of the Territory of Colorado.[52] Gilpin's speech heavily promoted the San Luis Valley and trumpeted the potential riches just waiting for discovery by eager settlers. Among the first to establish the legacy of economic boosterism that would have a profound and lasting impact on the landscape of the American West, Gilpin and his fellow boosters, whom historian Wallace Stegner called the "tribe of Gilpin," were essentially business promoters who utilized glowing rhetoric and inflated claims to lure prospective investors and settlers across the Great Plains to Colorado. By highlighting (often misleadingly) the San Luis Valley's mineral wealth and abundant water resources, and by downplaying its scant rainfall and ubiquitous sand, the boosters systematically dismantled the conception of the San Luis Valley as merely another province of the Great American Desert, whose existence had been vouched for by travelers and vaguely indicated since Zebulon Pike's report of 1810.[53]

The boosters, however, could not claim sole responsibility for settlement in the valley. The years following the establishment of San Luis in 1851 saw an ever-increasing number of Hispanic settlements. Along with the small communities came fields of crops, pasture for cattle, and a growing network of irrigation ditches. In 1856 the valley's first flour mill began operating just south of the Conejos River. Churches, schools, and mercantile stores soon followed, and the southern reaches of the San Luis Valley soon began to resemble rural northern New Mexico, with small, subsistence-level farming and ranching communities dotting most of the irrigable land in the valley, growing crops and tending livestock for family consumption or for barter with neighboring communities.[54] But throughout the 1860s, it was the boosters who seized the opportunity to publicize the San Luis Valley as a fertile wonderland bathed in sunshine, plenty of water, and a healthy climate. Their efforts to dispel the image of the Great American Desert attracted Anglo-American subsistence-level farmers and ranchers as well as the more ambitious and profitable commercial farming operations. This influx of settlers and commerce, inspired by the tribe of Gilpin, dramatically accelerated the extractive endeavors of humans in the region and forever changed the face of the San Luis Valley.

William Gilpin was no stranger to the valley. His first visit had been in 1844 while on his way back to Missouri after spending the winter at Fort Vancouver (near present-day Portland, Oregon) as a member of John C. Frémont's 1843–44 expedition to the Pacific. In the spring of 1844, Gilpin traveled to the Willamette Valley, where he assisted settlers in drafting a petition to the U.S. Congress seeking a territorial survey and creation of a territorial government. Gilpin agreed to take the petition to Washington, D.C., and set out on the Oregon Trail, heading east as far as Fort Bridger in present-day Wyoming before turning south, where he picked up the Old Spanish Trail and entered the northwest corner of present-day Colorado. He continued south until he struck the Gunnison River, which he followed south and then east until he reached the San Juan Mountains, where he crossed the Continental Divide at Cochetopa Pass and descended into the vast San Luis Valley.

At some point during his journey through the San Juans, Gilpin discovered what he thought was gold-bearing ore, which probably began his lifelong fascination with the region. He passed briefly through the San Luis Valley again in October 1846 while serving with Colonel Stephen W. Kearny's Army of the West during the Mexican-American War. In 1862, after a brief and tumultuous stint as first governor of the newly formed Colorado Territory, Gilpin purchased an option to buy the Luis Maria Cabeza de Baca Grant No. 4, just north of the Great Sand Dunes, for thirty cents an acre. Unable to complete the Baca purchase, Gilpin turned his attention instead to the massive Sangre de Cristo Land Grant in the southeastern portion of the San Luis Valley, and by 1863 he owned or had an option to purchase nearly 1 million acres of land in the valley. Notorious for being short on cash and with substantial land payments looming, Gilpin realized that he had to convince outside investors of the valley's potential for settlement and profit before his dreams of a San Luis Valley filled with settlers, ranchers, miners, and farmers could come to fruition.[55]

To facilitate the required infusion of capital and help publicize the virtues of the valley in general and his land holdings specifically, Gilpin employed the services of a number of "experts" to inspect and report on the valley's potential for settlement and exploitation. Among these experts was an able, ambitious chemistry professor named Nathaniel P. Hill, who would one day gain considerable wealth and fame for his success with gold and silver smelting in the Colorado gold country.[56] Hill agreed to tour Gilpin's land and assess its attributes, then dutifully reported that the San Luis Valley had "value independent of minerals, but if it contains gold and silver, its value can scarcely be estimated." The valley's arid climate inspired a less emphatic assessment from Hill, who noted almost in passing that "the country . . . never receives any rain."[57] Another of Gilpin's shrewd moves in his tireless promotional campaign was partnering with William Blackmore, a London promoter and financier whose enthusiasm for the valley rivaled Gilpin's. Blackmore had visited the United States in 1863–64 with a proposal to finance the Union Army's Civil War effort by selling gold bonds secured with American public lands, an idea that President Lincoln seriously considered before he vetoed the plan in the face of expected public opposition to foreign speculators. Later Blackmore helped market the stock of the Union Pacific Railroad in Europe.[58] Through his connections with the railroad, Blackmore heard of Gilpin's plan to attract investors and conceived a plan to purchase part of the Sangre de Cristo Grant and sell it to European immigrants. Before proceeding, Blackmore enlisted the aid of famed scientist and surveyor Ferdinand V. Hayden to do a resource survey of the property. Blackmore paid Hayden $500 in cash plus $10,000 in stock for his report, which he completed in December 1868.[59] Hayden responded with understandable generosity in his praise for the Sangre Grant:

> The land embraced by the Sangre de Cristo Grant . . . is by far the finest agricultural district I have seen west of the Missouri River. . . .
>
> Along the entire eastern border there is a lofty range of mountains, which seems to be charged with ores of gold, silver, copper, lead and iron. . . .
>
> I can affirm that I know of no region of the West more desirable for settlement than this just described, combining as it does all of the elements of wealth and productiveness.[60]

More glowing praise for the San Luis Valley could be found in a book entitled *Colorado: Its Resources, Parks and Prospects*, which Blackmore published in 1869 as part of the scheme to lure American and European investors and settlers to the region. The descriptions contained in the book offered the most direct contradiction yet of the valley's reputation for aridity and sandy, alkaline soils. Repeated references to an enormous lake in the middle of the valley absolutely refuted the image of the Great American Desert, including one passage written by Gilpin describing how the "San Luis Lake, extending south from the point of the foothills, occupies the centre of the parc [sic] for *sixty miles* [emphasis added], forming a bowl without any outlet to its waters. It is encircled by immense saturated savannas of luxuriant grass. Its water surface expands over the savanna during the season of the melting snows upon the Sierras and shrinks when the season of evaporation returns."[61] Another passage in Blackmore's book, written by Edward Bliss, noted that the valley's "streams which irrigate and drain this vast interior basin are numerous and abundant in volume"; Bliss also referred to the "Sahwatch [sic], or San Luis Lake, a body of water some sixty miles in extent, into which flow nineteen streams. This lake has no visible outlet, but it is supposed to discharge its surplus water through subterranean channels."[62] Blackmore himself described the valley in an 1869 article for London newspaper *The Standard* as being "watered by thirty-five streams, and has in its centre a lake sixty miles long by twenty wide."[63]

This sixty-mile lake with its numerous feeder streams also appeared on a series of booster-era maps of the region. Not coincidentally, the location of the lake corresponded almost exactly with the boundaries of what is now known as the Closed Basin, which Ferdinand Hayden had described as "one vast swamp or bog, with a few small lakes" in 1873. Because of the peculiar topography and geology of the San Luis Valley, a sizable portion of the northern end has no natural outlet for water. Instead of flowing across the surface of the valley to eventually join the Rio Grande, water coursing down the western slopes of the Sangres is confined in the Closed Basin and seeps underground, where it collects in the voluminous aquifers beneath the valley floor. Occasionally, when rainfall

Figure 3.6 1870 map of the "San Luis Parc" depicting the sixty-mile "Sahwatch Lake" of booster fantasy. The perimeter of the ephemeral lake corresponds closely with the approximate boundaries of the Closed Basin. The Denver Public Library, Western History Collection, CG4312.S35.

is considerable or the snowmelt in the surrounding mountains is particularly heavy, this abundant groundwater rises to the surface, where it can create the appearance of an enormous lake. Such was most likely the case with the sixty-mile lake described by Gilpin, Blackmore, and Bliss. In fact, Gilpin's observation that the lake "expands over the savanna during the season of the melting snows upon the Sierras and shrinks when the season of evaporation returns" verifies that the complex hydrology of the Closed Basin actually created the lake. What Gilpin and his cohorts conveniently failed to mention is that water in the Closed Basin is often alkaline, filled with soluble mineral salts that are carried upward from deep underground as water seeps to the surface. As the surface water evaporates, it leaves behind an alkaline crust now referred to as the sabkha, the "efflorescent salts" described by Schiel and Beckwith in 1853. [64]

Since the image of the massive lake figured so prominently in their vigorous promotional efforts to dispel the myth of the Great American Desert, the boosters conveniently omitted the fact that the lake actually contained a saline soup unfit for consumption or irrigation, and Hayden's assessment of the Closed Basin as "one vast swamp or bog" was understandably absent from their promotional rhetoric. Moreover, Blackmore and the tribe of Gilpin carefully avoided any mention of the Great Sand Dunes in their descriptions of the San Luis Valley. Clearly, the revelation of an enormous expanse of towering sand dunes in the midst of their garden would have done little to lure potential investors and settlers to the valley. Likewise, the Hayden and Hill reports virtually ignored the region's pervasive sandy soil and described neither the sand dunes nor the surface salts that plague a considerable portion of the valley.

Instead, the booster campaign of misinformation concentrated on exaggerated claims about the valley's abundant surface water, gushing artesian wells, and perfect climate that guaranteed huge crop harvests and the potential for untold riches. Prior depictions of the valley as an alkaline desert filled with "worthless land" slowly gave way to the booster vision of a lush and fertile garden. The arid San Luis Valley, long perceived as part of the Great American Desert, had at last become a well-watered American Eden, thanks in part to the "engaging nonsense" of William Gilpin and the promotional juggernaut of his booster minions. Ironically, Gilpin's vision for the San Luis Valley would one day prove to be remarkably prescient, as those very same water resources so vigorously touted by the "tribe of Gilpin" would eventually be harnessed to create one of the most fertile agricultural regions in the entire state of Colorado, as well as precipitate one of the most contentious battles over water in the history of the American West.[65]

Richardson's Mysterious Lake

In early October 1872, in the very midst of the booster campaign promoting the virtues of the San Luis Valley, Charles Samuel Richardson, an itinerant mining engineer and surveyor, journeyed through the region on his way to survey several mines in the San Juan Mountains of southwestern Colorado. Richardson traveled all over the Colorado Territory selling his services to mine owners who hired him to verify the legality of their claims to a share of the riches being carved out of the Rocky Mountains. Heading west toward the San Juans, Richardson passed near the heaping mounds of sand sprawled beneath the foothills of the Sangre de Cristo Mountains on the eastern edge of the valley and stopped to make a pencil sketch in one of his many leather-bound notebooks. More than merely a depiction of mountains, sand, and water, Richardson's drawing is a curious blend of reality and booster fantasy, a symbolic merging of the disparate imagery that influenced public perception of the San Luis Valley during the latter half of the nineteenth century.

Titled *The Lake and the Sand Hills*, the drawing depicts the sand dunes and the lofty Sangres as seen from the south, with brief notes describing the sand as "light brown gray" and the mountains as "a deep green up to the timberline and green gray above," and indicating "the way through the Musca [Mosca] Pass." The foreground of the drawing includes a large body of water surrounded by vegetation, with what appears to be either a person in a small canoe or a very odd-shaped log floating on the water. Richardson gave no explanation for the figure in the water, but he did describe the lake as "two miles long, at least" and noted that the water was "alkaline."[66]

Richardson's one simple drawing brings together a few key elements from an entire century of imagery devoted to describing the San Luis Valley. In the background, the crenellated battlements of the mighty Sangre de Cristo Mountains,

Figure 3.7 Richardson's Mysterious Lake. Drawn by mining engineer Charles Samuel Richardson on Wednesday, October 9, 1872, the image depicts a large lake at the southern end of the Great Sand Dunes. No such lake currently exists in this location, leading to speculation that Richardson inserted San Luis Lake into the foreground (complete with what appears to be a figure in a canoe) in order to enhance the dramatic impact of his drawing. The Denver Public Library, Western History Collection, C MSS WH361.

formidable barrier to American expansion, soar over Pike's barren dunes; in the middle distance, piñon-covered foothills rise above Beckwith's "beautiful and fanciful forms"; and in the foreground, the large, placid lake of booster fantasy, complete with what appears to be a canoeist for dramatic impact, conjures a sense of salubrity and tranquil fertility. This mysterious lake is especially intriguing. It seems at first to be verification for booster claims about the valley's abundant water resources, yet Richardson had no connection whatsoever with the boosters, and his lake sits at the base of the enormous sand dunes so carefully avoided in booster rhetoric. Little wonder that the boosters never utilized Richardson's image of the lake in their promotional literature. No such lake exists today, and none of the recent sand cores taken at the base of the dunes shows evidence for a past lake, so what exactly does Richardson's drawing represent?

Given his description of the water's alkalinity, it is highly unlikely that the drawing depicts the fresh water of Medano Creek. Although its position in the drawing corresponds exactly with where Medano Creek is actually located, just south of the main dunefield, no record is known to exist of the creek ever being mistaken for a lake, and Medano Creek's typically shallow depth makes it an unlikely location for a boating excursion. The "lake" may represent one of the interdunal ponds, but most of the ponds are located on the western fringes of the dunefield, far from the location depicted in the drawing, and none of them are even remotely close to two miles in length. The drawing might also be a depiction of San Luis Lake, a large body of water to the southwest of the Great Sand Dunes that exhibits a certain degree of alkalinity, especially during the summer and fall when seasonal runoff is at a minimum and salts concentrate in the water.[67] But San Luis Lake is almost eight miles from the Great Sand Dunes, and the lake in Richardson's drawing appears to be directly beneath the dunes. Did he employ a common technique of western artists and simply compress the distance between the lake and the dunes to enhance the dramatic impact of his drawing?

The first known photograph of the Great Sand Dunes offers no explanation. Taken in late September 1874 by renowned frontier photographer William Henry Jackson, who described the dunes as "a curious and very singular phase of nature's freak," the photograph depicts the southern end of the dunes as seen from "near Mosca Pass," an area that roughly corresponds to Richardson's drawing. Yet the photograph shows no evidence for any lake in the vicinity. Could the lake have disappeared in the two short years between Richardson's drawing and Jackson's photograph?[68]

An 1879 government survey report only deepens the mystery. That year the U.S. Surveyor General of Colorado ordered Daniel C. Oakes and Edwin H. Kellogg to sort out the confusing boundaries and legal validity of the Medano and Zapato land grant, which included an area between the Great Sand Dunes and San Luis Lake to the west. In 1820, the land had been granted to two Hispanic settlers, Antonio Gomez and Jose Luis Baca de Sondaya, who established a small ranch

Figure 3.8 The first known photograph of the Great Sand Dunes, taken near Mosca Pass by famed frontier photographer William Henry Jackson in late September 1874. The image roughly corresponds to the general location of the lake drawn by Charles Richardson in 1872, yet no such body of water is discernible in the photograph. Note the unknown figure at lower left. U.S. Geological Survey.

that included a house and corrals. In their field notes, Oakes and Kellogg state that the description of the Medano and Zapato grant "is exceedingly vague and seems to have been written in ignorance of the points of the compass," but they did confirm the existence of a "Laguna Grande" sometime in the recent past:[69]

> It is now impossible to discover what were the limits of the "large lake" sixty years ago. But from changes which we have personally noted during the past eighteen years, from the reports of trappers and guides as far back as 1857, and from the present appearance of the surrounding country, it was of much greater extent than it is now and inaccessible on account of the marshes which surrounded it. This opinion is further confirmed by all the early maps of the San Luis Valley which show a body of water many miles in extent.[70]

The Oakes/Kellogg report also suggests that a large body of water existed near the Great Sand Dunes at some point in the recent past but had since disappeared, replaced by a series of dry lake beds. "The present existence of many lakes and lake beds all more or less connected indicates that the basin now occupied by them must have been at no very distant period covered by one sheet of water," the report suggested. "In the dry lakes every stage of diminishment is easily traced. Indeed only ten years ago this basin was entirely impassable for teams and is now traveled in every direction."[71]

Figure 3.9 Rarely published William Henry Jackson photograph of San Luis Lake with the Sangre de Cristos in the background, taken in late September 1874 as Jackson headed toward the Great Sand Dunes and Mosca Pass. The lake is possibly all that remains of the "Laguna Grande" referred to in the original 1820 petition for the Medano and Zapato grant. U.S. Geological Survey.

Where had the lake gone? Aberrant climatological conditions apparently had little or nothing to do with its disappearance. The period between 1857 and 1885 was neither abnormally cool-moist nor warm-dry in the San Luis Valley.[72] Large-scale groundwater pumping remained decades away, and irrigation, although expanding, was not yet prevalent enough in the northern portion of the valley to seriously affect surface water in the Closed Basin. The intricate hydrology of the Closed Basin may have played some role in both the appearance and disappearance of Richardson's lake, but the exact extent of that role is unknown, as is the very validity of the assumption. The available evidence seems to suggest one of three possibilities: Richardson saw a lake made up of alkaline groundwater at the base of the dunes and sketched it; groundwater seeping to the surface in the Closed Basin west of the dunes created a small alkaline lake and Richardson simply included it in his drawing; or Richardson encountered San Luis Lake on his way to the mines of the San Juan Mountains and inserted an image of it into his drawing to enhance its artistic impact and appeal.

The lake does not appear in Jackson's 1874 photograph of the sand dunes, taken two years after Richardson made his drawing, so the first possibility seems unlikely. A reference in Richardson's journal to a small abandoned cabin at the head of the lake suggests that it may have been a part of the Gomez and Sondaya homestead on the Medano and Zapato land grant, but the Oakes/Kellogg report is unclear about whether the "Laguna Grande" is San Luis Lake or some other body of water.[73] Since San Luis Lake is definitely alkaline at times and certainly deep enough to accommodate a boat, it seems the most likely explanation for Richardson's lake. He may have assumed that his drawing of the Great Sand Dunes would be greatly enhanced by the presence of a large lake in the foreground, complete with an unidentified boater or curiously shaped log. In any case, the drawing illustrates the complex hydrological characteristics and peculiarities of the Closed Basin, as well as the compelling history of water in the San Luis Valley. Yet like all enduring historical mysteries, the true origins and implications of Richardson's lake may never be known.

Charles Richardson went on to a distinguished and successful career as a mining engineer in the Colorado Rockies. His prowess eventually earned him the title of chief American correspondent for the *London Mining Journal*, a position he held for seventeen years. Richardson also gained renown as one of the founders of the *Denver Mining Journal*, a short-lived publication that appeared briefly in the early 1880s. As far as can be determined, Richardson never published his *Lake and the Sand Hills* sketch, and no surviving record offers any further details about his mysterious lake.[74]

The Richardson drawing and the Oakes/Kellogg report failed to slow the booster juggernaut. Booster articles and pamphlets touting the San Luis Valley's fertile soil, salubrious waters, and healthy climate attracted unprecedented numbers of settlers to the region during the last three decades of the nineteenth century. An abundance of small farms, ranches, and towns soon sprouted all

Figure 3.10 Cover of a typical booster-era pamphlet promoting the San Luis Valley for settlement, circa 1880s. Such pamphlets helped inspire an influx of immigration into the valley during the last few decades of the nineteenth century. Note the phrase "Garden of the Rocky Mountains." The Denver Public Library, Western History Collection, C978.83 S196clvs.

across the valley floor, complete with the requisite network of irrigation ditches that carried life-giving water to the new arrivals. Miners probed every gulch and canyon of the Sangres and San Juans searching for minerals, and loggers hauled timber from the precipitous mountain slopes for fuel, building material, and the endless fence posts that defined the perimeters of progress in the region.

Monumental change had finally come to the San Luis Valley, thanks in part to the exaggerated claims of the boosters and their declaration of war against the Great American Desert, an enemy whose very existence seemed dubious at best. Not surprisingly, settlement in the San Luis Valley roughly coincided with the collapse of the Great American Desert myth, as the vaguely unfocused image of an arid, alkaline desert full of "worthless land" gave way to the booster-inspired image of an American Eden. The promotional sleight of hand employed by the tribe of Gilpin blurred the line between fantasy and reality, but the dramatic transformations that occurred in the landscapes of the San Luis Valley during the late nineteenth century proved to be considerably less illusory.

FOUR | Arid Eden

The great park of the San Luis is the principal of these wondrous plains. There is literally, according to their account, no disease in these fertile tracts, and even preexisting disease, especially of a pulmonary nature, vanishes after a brief residence. . . . Again, the rarity of the air and the absence of humidity disinfect the earth. . . . Fish, waterfowl, birds of song, and game frequent the streams and groves. Innumerable mines of the richest ores lie around.

William Blackmore, 1869

JEAN FRANÇOIS (FRANK) HERARD first glimpsed the sweeping expanse of the San Luis Valley in 1849. Flush with gold fever, Herard passed through the region on his way to the newly discovered goldfields of California. Although he enjoyed limited success with gold mining and eventually returned to Kansas, Herard apparently never forgot the inspiring vistas and bracing climate of the vast alpine valley nestled in the heart of the majestic Rocky Mountains. A little over two decades later, perhaps enticed by the glowing rhetoric of William Gilpin's boosters and their sensational descriptions of abundant water and fertile land, Frank Herard packed up his wife and two young children and left Kansas in the spring of 1872, bound for the San Luis Valley and a better way of life.[1]

Despite the death of Herard's young daughter during the journey west, the family pressed on, determined to reach Gilpin's Promised Land. When they finally arrived in the valley, "artist with an ax" Frank Herard built what "was about the best fitted log house anywhere" in an area known as Deadman Gulch, amid the foothills of the Sangre de Cristo Mountains on the eastern edge of the valley, just north of the Great Sand Dunes.[2] In 1875, after three years in the Gulch, Frank Herard learned that he had illegally located his rough-hewn log cabin on land that belonged to the massive Baca Grant. Considered a squatter by the owners of the Baca, with virtually no chance of obtaining clear title to his homestead, Herard reluctantly moved his wife Julia and teenage son Ulysses south to the greener pastures of Medano Park, a verdant oasis tucked behind the northeast corner of the Great Sand Dunes at the foot of Medano Pass.

Located a significant distance from the farms and settled communities of the southern San Luis Valley, Medano Park contained one of the most fertile habitats on the western slope of the Sangres, brimming with wildlife, clear water, ample timber, and luxuriant grasses for grazing animals. By the late 1870s, more than a hundred families of all origins, including the Herards, had settled in Medano Park and the numerous canyons north and south of the dunes.[3] Hard against the flanks of the Sangre de Cristo Mountains, the region encompassing the inviting garden of Medano Park also included the enormous expanse of shifting sands that comprise the Great Sand Dunes. In this curious, paradoxical landscape that blurred the distinction between the Great American Desert and Gilpin's American Eden, the Herards and their fellow settlers began altering the environment around them in profound and lasting ways.

Like the Herards, the homesteaders who chose to settle near the dunes were part of the massive wave of Anglo-American immigration that flooded into the San Luis Valley during the 1870s and 1880s. Some entered the valley from the north, using the toll road that Otto Mears had constructed over Poncha Pass in 1867, while others simply followed the Rio Grande northward from New Mexico. Still others used Sangre de Cristo Pass (also known as the Old Trappers and Traders Trail) to the east of Mount Blanca, where successive groups of Native Americans, Spanish explorers, and American fur trappers had blazed a well-worn path on the trading routes south to Taos. The route that saw perhaps the heaviest use was the rugged pass just east of the Great Sand Dunes that the Spanish had named Mosca (Spanish for "fly" or "mosquito"). Known to fur trappers and early explorers as Robidoux Pass (occasionally spelled "Robideau" or "Roubideaux"), Mosca Pass was an old Indian trail that had been popularized by French-Canadian fur trapper Antoine Robidoux, who crossed the pass sometime between 1835 and 1840 while pulling a two-wheeled cart filled with trade goods. Robidoux built a small trading post at the western base of the pass, which he operated for only a brief period before seeking his fortunes farther west, but the pass that bore his name soon became such an important route that local historian George Harlan later described it as "the lifeline of the San Luis Valley during the decade of 1860–70."[4]

In 1871, Frank Hastings obtained a charter from the Territory of Colorado to operate a toll road over Mosca Pass, and by 1880 Robidoux's old trading post had become a post office known as Mosco. By 1887 the post office was the center of a small community called Montville, traces of which are still visible today. Hastings later sold the charter for the increasingly popular road; by the 1890s, tolls set by the new owners were two dollars per wagon, one dollar for a horse and rider, and fifty cents for each head of cattle or sheep. Traffic was considerable, averaging between thirty and forty wagons per day, but periodic flooding and prohibitive maintenance costs continually plagued the route. One particularly heavy flood in 1911 destroyed the road so completely that it was never rebuilt, but during its heyday the road proved to be one of the most popular routes for the unprecedented influx of farmers, ranchers, miners, and plain old dreamers who poured into the once-isolated San Luis Valley during the latter half of the nineteenth century.[5]

Attracted by pamphlets and guidebooks touting the region's healthy climate and potential for settlement, these new arrivals came seeking cheap land for ranching, fertile soil for farming, and bountiful wealth in the rough-and-tumble mining camps and boomtowns of the surrounding mountains. A seemingly limitless supply of timber and wild game greeted them, as did a staggering abundance of water, an attribute that distinguished the San Luis Valley from other arid regions and made it especially attractive to prospective settlers. Unfortunately, the valley suffered many of the same problems and discord that plagued much of the late nineteenth-century American West, including violent disputes over water rights and public lands, overgrazing, environmental degradation, wildlife destruction, tumultuous economic boom and bust cycles, ethnic and racial hostility and discrimination, and shameful treatment of indigenous peoples. These issues defined the Anglo-American settlement that occurred in the San Luis Valley in the aftermath of Gilpin's booster campaign, and they established a pattern of massive resource extraction and commercial enterprise that characterized the region's development well into the twentieth century.

The valley's original Hispanic inhabitants, many of whom settled along the numerous rivers, creeks, and streams of the southern valley during the 1850s and 1860s, were searching for suitable pasturage, sufficient water, and arable land in what historian D. W. Meinig described as "a centrifugal movement from the historic settlement core along the Rio Grande . . . northward along the flanks of the Sangre de Cristo and far into the San Luis Valley."[6] For these early settlers, ranching and farming primarily provided a means of supporting a family and community, with perhaps some discretionary income for trade or barter with other villages. Hispanic resource extraction mostly focused on subsistence rather than on the generation of capital, although these settlers were not what modern observers would call enlightened environmentalists, nor were they averse to the idea of large-scale ranching and farming.[7] Some Hispanic sheep herds destined for distant markets numbered as high as 10,000 head and played a significant role in the overgrazing and erosion of prime grasslands. Many early Hispanic ranchers did, however, purposely reduced the size of their herds, both to lessen their vulnerability to Indian raids and because they could not afford adequate fencing to control large numbers of animals. As for large-scale agriculture, a lack of metal tools often restricted the size of Hispanic farms, simply because steel farm implements were either unavailable or prohibitively expensive, forcing early Hispanic settlers to utilize wooden plows, hoes, and spades to turn the soil and dig irrigation ditches. This lack of technology necessarily limited the extent of their cultivated fields and therefore the magnitude of their environmental impact on the land.[8] Finally, the pastoral culture of the early Hispanics, deeply rooted in religion and traditional concepts of community, simply did not emphasize pure profit as a motivation for farming and ranching, nor did it condone the wasteful usage of natural resources.[9]

By contrast, the new wave of Anglo-American settlers arrived in the San Luis Valley with the latest metal tools, as well as land-use attitudes conditioned largely by the realities of a market economy. Just like their Hispanic predecessors, these new arrivals recognized that subsistence farming and ranching provided the basic necessities for their families and livestock, but the impulse to generate a profit also exerted an overwhelming influence on their land-use practices. The large-scale commercial farms and ranches established by Anglo-American settlers clearly reflected their culturally based affinity for capitalism and profit. The small-scale resource extraction that characterized the previous 10,000 years of historic human activity in the San Luis Valley paled in comparison with the widespread extraction and ecological transformation initiated by the arrival of the Herards and their fellow settlers, who were operating under the assumption of inexhaustible resource abundance typical of pioneering Anglo-Americans in the frontier regions of the American West during the nineteenth century.

Livestock in the Valley

In his 1873 speech to the Colorado Stockgrowers Association, Dr. Hiram Latham, a surgeon for the Union Pacific Railroad, asserted that stock raising should not cease until "every acre of grass in Colorado is eaten annually." Latham liked to emphasize his enthusiasm for large herds of livestock by quoting an old Spanish proverb: "Whatever the foot of the sheep touches turns to gold." Latham and countless other stock promoters like him conceived of sheep and cattle as money machines, capable of transforming grass into profit all across the West's "billion acres, boundless, endless, gateless." To Anglo-American ranchers in the San Luis Valley, the region seemed like the perfect location for Latham's arithmetic of enterprise. By filling the immense valley with teeming herds of cattle and sheep, ranchers could easily transform native grasses into greenbacks, yet the environmental impact of these herds proved to be considerable. Everywhere they grazed, the land became more susceptible to invasion by other exotic species, altering the composition of local plant communities and initiating widespread erosion across the valley floor.[10]

Anglo-Americans were not the first to introduce livestock into the San Luis Valley. As early as 1820, ranchers from Taos drove "over 5,000 head of cattle . . . into the Valle de San Luis to winter," and in the years following the end of the Mexican American War in 1848, a wave of Hispanic farmers and ranchers headed north from New Mexico and began homesteading in the valley, a process that continued throughout the 1850s and 1860s.[11] Among them was Teofilo Trujillo, who arrived with his family around 1865 and became one of the first permanent residents in the Medano-Zapata area. Settling just north of Dollar Lake on the extreme western edge of what later became the Medano Ranch, Trujillo built one of the finest adobe homes in the valley and began raising cattle and sheep. He later became known for traveling around the valley in a brightly painted stagecoach. In 1879, Trujillo's thirteen-year-old son Pedro settled on an adjacent 160-acre claim less than a mile west of the Trujillo homestead, where he built an American-style two-story log home that contrasted completely from his father's classic adobe dwelling. Pedro and his father differed in their approaches to ranching as well: Teofilo continually increased the size of his sheep herd as he prospered, while Pedro preferred horses and cattle and absolutely refused to herd sheep, even warning his father that his herds would eventually cause trouble with the nearby cattle ranchers. Years later, his prediction would prove sadly prophetic when violence erupted between local sheepherders and the valley's cattlemen over grazing rights on lands in the public domain.[12]

The large sheep herds introduced by early Hispanic settlers and later Anglo-American commercial ranchers in the latter half of the nineteenth century greatly altered the region's ecosystems. From 1867 to 1887, the northern portion of the Sangres from Poncha Pass to Cotton Creek suffered from the constant trampling and nibbling of some 50,000 sheep run by Hispanic lessees.[13] "The chief evil effect of sheep grazing in the forest," claimed one 1903 report, "is the destruction of the humus and ground cover resulting in reduction of forage species and an increase in erosion."[14] One heavily traveled sheep trail in the western portion of the valley felt the effects of an estimated 100,000 sheep every spring and fall during the peak use years between the late 1880s and early 1890s. Similarly, enormous herds of cattle grew fat on the bountiful grasses that carpeted the floor of the San Luis Valley and the lush meadows of the Sangres, where several ranches ran herds as large as 20,000 head.[15] So efficient were these voracious grazers that modern researchers have found few instances of the Sangres' original vegetation unchanged by grazing.[16]

Over in the midst of the Great Sand Dunes, Ulysses Virgil (Ulus) Herard made his own contribution to the overgrazing problem. After his father, Frank, died in 1892, Ulus started a small herd of Hereford cattle. The herd multiplied so rapidly that actually counting the cattle soon posed a problem; estimates of the size of his herd ranged from 1,500 to 6,000 head.[17] Since he built no fences on his homestead, Herard's cattle were free to roam the entire length and breadth of Medano Park and often ventured onto the Great Sand Dunes to feed on the interdunal vegetation. They were not the only cattle in the area. Starting around 1865, in addition to his growing sheep operation, Teofilo Trujillo ran a herd of cattle that eventually numbered roughly 800 head in the area near Medano Springs, just west of the dunes. In the 1870s other ranchers began operating near Medano Springs and on the Zapata lands to the south of the dunes.[18] The number of cattle grazing near the dunes toward the end of the nineteenth century clearly reached into the thousands, perhaps more. If the stories of Ulus Herard's prowess as a hunter are true, these huge herds attracted a large number of predators to the area. Ulus claimed to have killed at least 100 mountain lions; an equal number of black bears, one of which he killed with a hand ax; and perhaps even a few grizzly bears.[19]

Herard also added new species to his homestead, including a population of rainbow trout from a fish hatchery that he built on the banks of Medano Creek in 1890. While it is unclear whether the hatchery produced trout for domestic or commercial purposes, its presence further illustrates the role that Herard and his fellow settlers played in the alteration of existing ecosystems adjacent to the Great Sand Dunes.[20] At some point during his tenure in Medano Park, Ulus Herard introduced a herd of thoroughbred horses to the vast sea of sand, leading some local valley residents to speculate that the legendary "web-footed mustangs" they kept seeing galloping across the Great Sand Dunes were actually the wild offspring of Herard's thoroughbreds. As with his cattle, Herard allowed his horses to wander freely, and years of roaming across the soft sand resulted in huge, web-shaped hooves, which apparently enhanced the horses' ability to escape would-be wranglers and perhaps spawned the persistent myth of the mysterious mustangs among local San Luis Valley residents.[21]

The introduction of large herds of cattle, sheep, and horses profoundly impacted the abundant wildlife populations encountered by early Spanish and American travelers in the San Luis Valley. A member of Oñate's 1598 expedition reported an "infinite number of cattle [bison]," while fur trapper Antoine Leroux stated that "thousands of buffalo and antelope were present" in the valley in 1820. Trapper Jacob Fowler described bighorn sheep, wild horses, deer, elk, and bear near present-day Del Norte in 1822.[22] With the arrival of Anglo-American ranchers, any animal that could provide food for a family or compete with domestic livestock for forage fell before the rifle and pistol. The Sangres' last native elk vanished in 1878, and by 1890 native deer and pronghorn populations had significantly dwindled, with bighorn sheep teetering on the brink of extinction. The carnage extended even to squirrels, gophers, jackrabbits, skunks, and assorted other small grazers.[23] Western devotion to making the land as accommodating as possible for herds of livestock led to a relentless campaign against any predators that might make a meal of cattle or sheep. Ranchers shot wolves on sight, but as wolves gradually learned to avoid humans, ranchers resorted to lacing animal carcasses with strychnine poison. Other predatory animals, including bears, mountain lions, and coyotes, were poisoned, trapped, and shot. In 1904, a professional hunter hired by ranchers killed the last grizzly bear in the Sangres.[24]

Large-scale ranching brought undeniable economic benefits to the San Luis Valley, but overgrazing by large herds of sheep and cattle also contributed to

Figure 4.1 Photograph of Ulysses Virgil Herard, "The Frenchman," taken circa 1930 at his homestead in Medano Park on the eastern edge of the Great Sand Dunes. Herard typified the hardy pioneers who settled in the San Luis Valley during the late 1800s and transformed the face of the land. National Park Service, Great Sand Dunes National Park and Preserve.

widespread erosion and range degradation, while another consequence of the livestock industry in the valley stemmed from the cultural clash between sheepherders, who tended to be Hispanic American, and cattle ranchers, who were primarily Anglo-American. Conflict between sheep and cattle ranchers was common in the American West during this period, and as a first-generation Hispanic American settler growing up in the San Luis Valley, Pedro Trujillo witnessed firsthand the constant tension and frequent clashes between the two groups. In January 1902, just as Pedro had warned, his father Teofilo Trujillo became the target of violent intimidation when about ninety head of his sheep were slaughtered and others driven away by employees of cattleman George Dorris. A few weeks later, the violence escalated when local cowboys burned down Teofilo's house, along with a reported $8,000 in cash. With his house gone and his livelihood, not to mention his life, continually threatened, Teofilo Trujillo finally sold his homestead in March 1902 to the owners of the sprawling Medano-Zapata Ranch for $30,000 and moved to San Luis. Pedro Trujillo sold his ranch in the same transaction and also moved away, forced off his land like his father by the same ethnic intimidation that so often characterized the cultural conflict between traditional Hispanic American agricultural and sheep-herding practices and the larger-scale, Anglo-American ranching operations of the San Luis Valley during the late nineteenth and early twentieth centuries.[25]

All That Glitters

Sheep and cattle ranchers were not the only new arrivals flooding into the San Luis Valley during the post–Civil War period. The mining boom that echoed for decades across the valley began around 1870 with the first major gold strike near Summit in the San Juan Mountains west of Alamosa. Passage of the Mining Law of 1872, which granted any U.S. citizen the right to stake a mining claim in the public domain, spurred legions of eager prospectors to seek their fortunes in the mountains that flank the San Luis Valley. Many of these miners validated their claims by digging prospect pits, which soon pockmarked the foothills of the San Juans and Sangres, including the hills adjacent to the Great Sand Dunes. They also began cutting large stands of timber for mine braces, sluices, and fuel for campfires, cookstoves, and smelters. Few had qualms about abusing the local water supplies. One livery stable in Bonanza actually straddled Copper Creek, making the task of watering the stock and flushing out the stables a foregone conclusion. While manure can occasionally be beneficial to aquatic plants and animals, such dumping caused serious disputes between miners and the ranchers who lived downstream, who felt that the miners had carelessly polluted their water supply.[26]

In addition, miners routinely dumped mill tailings and mine wastes near local creeks and streams. Anne Ellis, a longtime valley resident who lived near Bonanza as a child, recalled warnings not to drink from Kerber Creek because of

Figure 4.2 Downtown Crestone in June 1901, during the height of the Sangre de Cristo mining boom. Within a decade the boom would begin to go bust. The Denver Public Library, Western History Collection, X-7575.

concerns that toxic mine wastes had poisoned the water and caused the "tyford [*sic*] fever" that killed cattle downstream. Ambitious miners caused further damage by using dynamite to blast ore from the steep slopes or to create new roads, exacerbating erosion in the region. Successful mines attracted new rail lines and legions of eager miners, compounding such problems as timber depletion and the safe disposal of human sewage. By the late 1880s, the huge Orient Mine, about thirty miles north of the dunes, was reportedly extracting 200,000 tons of iron ore annually, spurring the expansion of railroads and causing an influx of workers and businesses into the region around Villa Grove, which further increased pressures on local resources such as water, timber, and wildlife.[27]

Over on the far eastern edge of the valley, above the sprawling tract of private land known as the Baca Grant, the Sangre de Cristos finally began giving up the gold and other precious metals that had been generating rumors and tempting miners since the first Spanish incursions into the region in the late sixteenth century. Between 1870 and 1880, various mineral strikes in the canyons and drainages on the western flanks of the Sangres attracted scores of prospectors eager to try their luck at hard-rock mining. In the summer of 1879 came news of "fabulous gold strikes" in an area known as Burnt Gulch, named for the fires set by local Indians in the early 1850s to drive game animals, as well as deter the further influx of settlers. A year later, George Adams, who had been leasing the Baca Grant from William Gilpin, officially platted the boomtown of Crestone. By 1901, Crestone boasted a bank, two newspapers, five general stores, two lumberyards, five saloons, twice-daily train service on the Rio Grande and Sangre de Cristo Railroad, "one district of ill repute," and a population nearing 2,000 people. At the height of the mining boom, from roughly 1901 to 1906, railcars running on the spur from Crestone south to the Independence mine hauled away as much as $80,000 a month in precious gold.[28]

Farther south, prospectors scoured every nook and cranny of the Sangres for precious metals, including the canyons in the foothills between Crestone and the Great Sand Dunes, where Indiana native John Duncan staked his first mining claim in 1874. Duncan came west at age twenty-three after hearing a veteran of Frémont's 1853–54 expedition to the Rockies describe the gold "lying in abundance" on the slopes of the Sangres. He entered the San Luis Valley through Medano Pass and began digging near Pole Creek, eventually hitting pay dirt on a "long finger of a mountain" known as Milwaukee Hill. By 1890, news of Duncan's promising gold strikes had spread throughout the entire region, attracting an ever-increasing parade of miners who began driving claim stakes from Cottonwood Creek near Crestone all the way south to Sand Creek on the very edge of the Great Sand Dunes. Duncan himself built a sturdy one-room log cabin near Milwaukee Hill, where by 1892 he had officially platted the mining camp of Duncan, with lots selling for between twenty-five and seventy-five dollars each.[29]

Like Crestone, the mining camp of Duncan soon became a bustling community offering all the latest amenities and conveniences for miners, including grocery stores, a livery, a newspaper, saloons, telephone service, a bathhouse offering steaming-hot baths for twenty-five cents, and a post office that featured mailboxes that could be opened from the outside so a man could collect his mail without getting off his horse. One thing the miners and businesses of Duncan did not have, however, was explicit permission to be on the private property of George Adams, who had purchased the sprawling Baca Grant from William Gilpin in 1885 (some sources claim 1886). Convinced that the miners were illegally removing minerals that rightfully belonged to him, Adams tried to shut down the mining camps and evict the trespassers. The miners, who had always operated under the assumption that the Baca Grant covered only the surface and grazing rights, considered the mineral rights to be public domain and ignored him. The dispute eventually wound up in local court, which initially ruled in favor of the miners, but Adams appealed the ruling all the way to the U.S. Supreme Court, which reversed the lower court's decision in 1897.[30]

Emboldened by the Supreme Court decision, Adams persisted in his efforts to evict the miners, employing U.S. Marshals as enforcers. Armed resistance and scattered violence ensued, forcing the marshals to dynamite several empty cabins in Duncan in order to convince the miners of their authority. In 1900 Adams sold the Baca Grant to the San Luis Valley Land and Mining Company, a wealthy eastern concern that continued eviction proceedings against the miners and their families. The new owners did pay the miners $125 each for their homes, then offered to sell them back for $10, providing the structures were moved off the Baca Grant. A number of miners took advantage of the offer and relocated their homes south toward Short Creek, which was outside the boundaries of the Baca, and established the aptly named mining camp of Liberty in the foothills just north of the Great Sand Dunes.[31] Once again, the lure of striking it rich cast a timeless spell on the region's miners, and Liberty soon became a bustling mining camp, complete with a barbershop, a general mercantile shop, a hotel, even a fledgling school district. Gold-bearing ore, however, was in short supply, and Liberty flourished for less than two decades before it suffered the same fate as so many other mining towns in the Colorado high country, falling into dusty ruin when the mines played out.[32]

While Duncan and Liberty (and, to a lesser degree, Crestone) may have failed to sustain themselves as thriving boomtowns, eager miners continued seeking riches in the foothills of the Sangres well into the twentieth century. Not even the Great Sand Dunes escaped their attention. Since at least 1873, when prospectors first investigated the sand dunes for gold near the foot of Mosca Pass, rumors had circulated about the possibility of gold hidden in the towering hills of sand, including the persistent legend about a "Spanish oxcart" laden with gold bars that had been swallowed by the blowing sand. Such legends gained some measure of validity when prospectors actually discovered evidence of "flour gold"—extremely fine particles, exceedingly difficult to collect in profitable quantities—at the dunes in the late 1920s.[33]

Figure 4.3 Hotel Liberty, circa 1907. Forced to relocate from the Baca Grant, a handful of miners established the town of Liberty just north of the Great Sand Dunes in 1900. Within two decades the mines had played out, and Liberty became a ghost town. National Park Service, Great Sand Dunes National Park and Preserve.

Even though the flour gold was present in only trace amounts, the discovery appeared in the newspapers as a "big strike." The *Pueblo Chieftain* declared, "The sand has been tested . . . and found to have in it gold running three dollars to the ton."[34] Succumbing to the barrage of publicity, hordes of miners arrived at the dunes to stake their claims to what had been described as a miner's dream: an enormous gold deposit that required no digging, drilling, or blasting and no hauling or crushing of ore. Miners could simply arrive and pan for their fortunes. A few eager miners set up sluices adjacent to Medano Creek, only to discover that the fickle stream offered little by way of a reliable water source. In fact, harvesting the minute traces of flour gold from the sand proved so arduous that most miners simply gave up and returned to the vagaries of traditional hard-rock mining in the surrounding mountains.

While a few lucky individuals did manage to strike it rich from mining in the Sangres and San Juans during the late nineteenth and early twentieth centuries, the majority of prospectors experienced less success and eventually moved on to other strikes and boomtowns when the gold and other minerals played out. In their wake, they left behind a legacy of shallow prospect pits, dangerous mine shafts, heaping mounds of poisonous mine tailings, depleted timber and wildlife resources, and dozens of severely polluted waterways, some of which took decades to clean up. Ghost towns like Duncan and Liberty now offer only silent testimony to the passage of the mining camps, where the tantalizing visions of fabulous wealth that once attracted so many miners to the region slowly gave way to the cold reality of abandoned mines and forlorn, empty buildings, filled with nothing but the shattered dreams of hopeful prospectors and their long-suffering families.

Sand and Saws

The survival of early San Luis Valley miners and settlers partly depended on securing reliable sources of timber, which the Sangres and San Juans initially furnished in abundance. Mining camps needed timber for flumes, smelters, and mine props, while local communities sought finished lumber for churches, schools, and businesses. Individual homesteads required wood for heating and cooking, as well as for constructing homes, barns, corrals, and the ubiquitous fence posts that both contained livestock and defined distant property boundaries. The arrival and expansion of William Jackson Palmer's Denver and Rio Grande Railroad in the San Luis Valley in the late 1870s further impacted Sangre de Cristo timber. Like the burgeoning settlements and mining camps, railroads and the company towns that sprouted along their routes required timber for homes, bridges, railroad ties, depots, and telegraph and (eventually) telephone poles. The railroads connected the booming livestock and farming operations of the valley to markets throughout the rest of the country. In return, the railroads brought the latest tools and durable goods, as well as legions of immigrants, into the region, a profitable exchange that encouraged further growth in the valley, thereby intensifying extractive pressures on the area's timber resources.[35]

Before the establishment of the San Isabel Forest Reserve in 1902 and the Rio Grande National Forest in 1903, the thickly forested foothills of both the Sangres and the San Juans provided huge tracts of virgin timber for loggers, who cut indiscriminately, without interference or regulation from any local or state authority, and with no clear titles or permits for the lands they so vigorously logged.[36] Between 1874 and 1900, most of the easily accessible timber on the lower western slopes of the Sangres had fallen to the ax and two-man saw, forcing crews to climb ever higher in search of desirable wood.[37] The Free Timber Act of 1878, which prohibited the cutting of live trees on public lands but allowed for the removal of dead ones, exacerbated the extraction of timber from the mountains surrounding the San Luis Valley. In order to collect sufficient timber to supply the mining camps and railroads, loggers simply circumvented the law by setting fire to vast stands of forest, which killed the trees and hastened their removal.[38] Curiously, cross-sections of piñon and juniper later collected at Great Sand Dunes National Monument show no evidence of fire-scarring, even though some specimens were over two hundred years old.[39] Perhaps the presence of so much sand blunted commercial logging efforts near the dunes. Sand embedded in trees quickly dulled even the sharpest blades, and drifting sand made transporting large quantities of this relatively inaccessible timber both exceedingly difficult and prohibitively expensive.[40]

In addition to the fires intentionally set by loggers, sheepherders routinely set huge tracts of forest ablaze in an effort to improve forage, and local cattle ranchers occasionally ignited blazes to keep their livestock from consuming the poisonous larkspur plant.[41] Such fires consumed countless acres of forest. One early valley resident, Abigail Wales Shellabarger, claimed that the middle to upper parts of every drainage on the western slope of the Sangres burned in the 1870s, except for Rito Alto Creek, where the Wales and Shellabarger families had their ranches and suppressed the spread of fire.[42] The exact ecological consequences of these fires on the health of the forests or their wildlife inhabitants is unclear, but the fact that huge tracts of the Sangre de Cristos were logged and burned clearly illustrates that increasing human settlements inflicted at least some damage on the region's ecosystems.

As commercial logging proliferated up and down the western Sangres throughout the Anglo-American settlement era, the area immediately adjacent to the Great Sand Dunes attracted a smaller-scale, more specialized type of wood gathering: the harvesting of timber by individuals purely for domestic uses such as home heating and cooking. Generally, proximity determined where settlers collected their wood from the Sangres. Folks from Alamosa and neighboring communities concentrated on the southern slopes of the Sierra Blanca massif. Residents of Hooper cut in the foothills north of the Great Sand Dunes, while those from Mosca preferred the area around Zapata Falls, Denton Springs, and

North and South Arrastra Canyons, just south of the dunes. The area immediately east of the dunes also became a popular wood-harvesting location with the completion of an all-weather road in the mid-1920s. Even in the dead of winter, this road provided access for wood haulers who drove their wagons across frozen Medano Creek to reach the Ghost Forest.[43]

Depending on the severity of the winter, settlers in the northern portions of the valley consumed anywhere from six to over twenty-five tons of wood per family annually, with an average yearly usage of roughly twelve to fifteen tons.[44] Some families burned the copious rabbitbrush and greasewood that grew near their towns and homesteads, but this fuel proved inadequate for all but the smallest homes because it burned so quickly. Piñon became the wood of choice for cooking and home heating because it produced hot, long-lasting fires. Wood haulers generally avoided Rocky Mountain juniper for firewood because it had a tendency to spark and pop when burning, which constituted a significant fire hazard in the wood-frame houses of the valley. Instead, valley residents utilized juniper for fence posts, primarily due to its resistance to rotting in the ground.[45] Wood collectors also avoided ponderosa, mostly because of the significant effort required to hack its tremendous bulk into manageable pieces. Perhaps the sheer size of ponderosa is what spared them from rapacious wood collectors and accounts for the lasting survival of the impressive collection of culturally peeled ponderosa pine trees at the Great Sand Dunes.[46]

Collecting wood for fuel and fence posts seemed simple enough, but hauling it back home presented a far more difficult task. The vast stretches of soft, wind-blown sand in the vicinity of the Great Sand Dunes dictated that residents in this area do their wood hauling in the winter, when frigid temperatures resulted in firmer roads and trails. Over the course of the winter months, crews from each community made several trips to the Sangres to collect wood, often traveling in a convoy of horse-drawn wagons in order to render aid to those unfortunate enough to get stuck in the sand. These early wood-gathering forays generally lasted a minimum of three days, with crews routinely traveling as much as fifty miles in search of suitable timber. The rugged terrain of the Sangre de Cristos effectively limited the penetration of wagons into the foothills, and most early collection efforts occurred in areas close to existing roads. In order to conserve both labor and saw blades, crews avoided live trees and concentrated instead on collecting dead, fallen timber. As the years passed, such wood became increasingly scarce. Frank Berryman, an early Mosca pioneer who came to the valley in 1888, reported that early wood-hauling expeditions found so much dead piñon that drivers often had difficulty turning their wagons around without running over dead timber.[47]

The first decades of the twentieth century introduced new techniques and machinery to the process of timber extraction; one inventive soul even used dynamite to "lift" large, standing-dead trees out of the ground. The most profound change to timber gathering came with the advent of the steam engine. In

Figure 4.4 San Luis Valley resident Herb Winner's massive, twenty-five-horsepower Cross-Compound Reeves steam tractor, "Master of Them All." The contraption, capable of hauling 40,000 pounds of wood out of the Sangres in one load, had enormous steel wheels that churned through the deepest sand drifts and effectively introduced the Machine Age to wood-hauling in the San Luis Valley, circa 1908. Kat Olance, Luther Bean Museum, Adams State College.

1906, valley resident Herb Winner used a gargantuan, twenty-five-horsepower Cross-Compound Reeves steam tractor that he named "Master of Them All" to haul 40,000 pounds of wood out of the Sangres in one load. The contraption had enormous tractor wheels that churned through even the deepest sand drifts, hauling more wood with less effort and effectively introducing the Machine Age to wood hauling in the San Luis Valley.[48] Later innovations included the introduction of truck hauling and chain saws. Wood gathering for domestic use continued to be a conspicuous part of life for valley residents well into the 1940s, when kerosene stoves gradually replaced wood-burning units and the availability of alternative fuels such as coal and propane increased.[49]

The establishment of Great Sand Dunes National Monument in 1932 seriously curtailed wood collecting in the area adjacent to the dunes, but both wood collecting and livestock grazing continued well into the 1950s, simply because much of the east side of Medano Creek remained under the control of the U.S. Forest Service and various private individuals until that time.[50] In 1995 the National Park Service initiated a comprehensive study to determine the precise effects of historic wood harvesting on the distribution of piñon-juniper woodlands within the monument. Specifically, park managers wondered if the current low reproduction rate of juniper and relatively high density of piñon could be attributed to such wood harvesting. The evidence suggested that the present distribution of tree species and stumps roughly corresponded to historical accounts of wood collectors preferring dead piñon over live piñon, while both live and dead juniper were extensively harvested for use as fence posts until the 1940s and 1950s, when steel posts became generally available. The study found no direct evidence implicating historic harvesting of juniper for fence posts as the sole cause of the species's present low density and population growth.[51] Since organic material from decaying trees is known to provide important nutrients and encourage seed germination in both piñon and juniper, the removal of such material likely played a significant role in determining current piñon/juniper distribution.[52]

Whether through mining, logging, hunting, or ranching, Anglo-American settlers in the San Luis Valley wrought substantial havoc on the region's ecosystems during the late nineteenth and early twentieth centuries. Ramshackle sawmills sprouted anywhere accessible timber grew; roads to the sawmills and mines ran through the middle of streambeds; huge herds of cattle and sheep relentlessly browsed and trampled the native vegetation; hunters and ranchers slaughtered native wildlife with impunity; and miners dumped toxic wastes that seriously polluted portions of numerous local waterways.[53] The Great Sand Dunes, however, escaped many of the harmful repercussions of the Anglo-American phase of settlement in the valley. The vast, seemingly barren sea of shifting sand offered virtually nothing in the way of exploitable resources for new settlers, even though extensive extractive activities continued to occur all around the main dunefield. Wood harvesting persisted in the Sangres adjacent to the dunes throughout the Anglo settlement period, but this had little direct impact on the dunes, and it paled in comparison to the wholesale commercial logging occurring elsewhere in the Sangres.

Cattle and sheep grazing also continued around the dunes during this period. Ulus Herard's cattle, along with other herds in the area, were well known for traversing the dunes in search of grass. In 1901 a local cowboy tracking wild mustangs on the dunes reported fighting "through tall grass, from two feet to waist high, that grew in the dunes. Tracks and signs showed that the cattle knew about this grass." In addition, in 1932 a local miner reported that "cattle were all over the dunes."[54] Still, given the sheer size of the constantly shifting dunes, the adverse effects of this grazing proved considerably less conspicuous than those created by the tens of thousands of sheep and cattle elsewhere in the greater valley. Indeed, the Great Sand Dunes successfully resisted almost every extractive endeavor initiated by Anglo-American settlers in their midst, including the dubious attempt to extract gold from the towering dunes in the late 1920s.

Gilpin's Garden Blooms

Perhaps no other human activity or extractive endeavor altered the landscapes of the San Luis Valley as much as large-scale commercial agriculture. Development of cultivated land exploded during the late 1860s and early 1870s for several reasons, not the least of which were the efforts of Gilpin's booster minions and their enthusiastic promotion of the valley's potential for settlement. Establishment of the Territory of Colorado in 1861 provided some degree of government for the valley and ensured that it would no longer be just a huge, neglected portion of northern New Mexico. The 1862 Homestead Act, which offered land in generous 160-acre chunks all across the West, including the San Luis Valley, further inspired agricultural development, as did the end of the Civil War in 1865, which encouraged army veterans, refugees, and the dispossessed to head west for new opportunities.[55] A significant number of these immigrants traveled to Colorado and the San Luis Valley, where the burgeoning mining industry and commensurate population growth presented a keen opportunity for profit. More importantly, Gilpin's fabled garden in the valley offered prospective farmers copious quantities of the arid West's most precious resource: water.

Beginning with the valley's first irrigation ditch at San Luis in 1851, access to water defined the extent of agriculture in the San Luis Valley. In 1855 the Guadalupe Main Ditch and Head's Irrigation Ditch began carrying water. By 1859, irrigation canals laced the valley around present-day Del Norte. Within a decade, more than forty ditches provided water for settlers in the greater Del Norte area. The area around Saguache attracted substantial settlement, as well as the valley's first mechanized farm equipment, and soon developed into a renowned wheat-growing center, where farmers used natural arroyos to carry water until proper irrigation ditches could be dug.[56] As the valley's irrigation network grew and more water became readily available, farmers plowed under thousands of acres of sagebrush and hardy grasses to make room for the large commercial farms that eventually made the San Luis Valley famous throughout the world. Although the farmers were hampered somewhat by a brief growing season, ranging from 90 to 142 days depending on location, portions of the valley blessed them with abundant sunshine, fertile soil, and copious water that proved ideal for the cultivation of a multitude of crops, including lettuce, spinach, cabbage, potatoes, carrots, barley, oats, cauliflower, alfalfa, and wheat. The valley's fertility became readily apparent by 1875, when the first commercial grower of potatoes in the valley produced 70,000 pounds in one season for sale to the miners of the San Juans.[57]

Extension of the Denver and Rio Grande rail network further facilitated this rapid growth of agriculture and contributed to the valley's vigorous market economy; by the 1890s, the valley boasted rail connections with Santa Fe, Salida, Denver, Durango, and Creede. Freight cars full of manufactured goods and raw materials poured into Alamosa, while outbound trains carried mineral ores, lumber, livestock, and farm products to distant markets all across the country.[58] Likewise, the railroads inspired an influx of new settlers. Companies like the Union Pacific and the Denver and Rio Grande Railroad organized excursions for newspapermen and "land lookers," sent agents to publicize railroad routes, and printed handbills and pamphlets describing the fertile soils and favorable terms offered on lands served by the iron rails. The valley even developed a fledgling tourism industry thanks in part to the railroads, and increasing passenger traffic inspired the construction of a number of small resorts, typically centered around one of the valley's soothing natural hot springs.[59]

As farmers plowed more land in the northern portion of the valley for farms and ranches and town sites, demand for water increased accordingly, inspiring a more sophisticated network of canals and ditches to provide the life-giving water. Theodore C. Henry, an ambitious land speculator and early founder of the town of Monte Vista, secured financing to build four major canals, including the massive Rio Grande Canal, which supplied water to irrigate roughly 500,000 acres. Between Alamosa and Del Norte, an enormous network of irrigation canals reached out from the Rio Grande for thirty or forty miles to the north and south. To supplement the expansion of irrigation ditches and canals, residents near the town of Moffat drilled the valley's first artesian well in 1869, a notable development that would prove to have dramatic implications for the future of water in the San Luis Valley.[60]

By the late 1880s, the valley featured a patchwork of irrigated farms, towns, and ranches that attracted people from all over the country and the world. Mormons began settling in the valley in the late 1870s. Swedish settlers chose the Monte Vista area for their new homes, while Dutch settlers claimed the Waverly area just west of Alamosa. Near the Great Sand Dunes, German families began farming corn around Hooper and Mosca but switched to raising wheat after discovering that the growing season limited corn harvests. All across the valley floor, farmers copied the German settlers and successfully adapted their crop regimes and cultivation methods to accommodate local variations in soil and access to water. The explosion in agriculture showed no signs of slowing, and prosperity seemed assured all across the valley.[61]

Yet the Great Sand Dunes continued to resist every human effort to transform them. Farmers tried repeatedly to cultivate the land around the perimeter of the main dunefield, but the waters of Medano and Sand Creeks were simply too seasonal and unpredictable for irrigation purposes. Besides, the sandy soil provided scant nourishment for growing crops, and few residents found success challenging the formidable conditions around the dunes. An 1898 newspaper article described the tribulations of one determined farmer who tried to fence his property, noting that the "limit of such reclamation seems to have been attained now and the last farmer out on the desert has reached the point where there is a warning to go no further. Three times has the farmer put up a fence to mark the end of his property. The farmer . . . has planted three nine-foot fences and not a vestige of the last one is in sight. The sand, therefore, is over thirty feet deep."[62] Drifting sand continually thwarted the farmer's efforts, highlighting the dunes' ability to resist the limitations and transformations imposed by humans. Large-scale commercial agricultural enjoyed phenomenal success in other parts of the valley, but even the smallest operations proved unsuccessful near the Great Sand Dunes, where sandy soils simply refused to succumb to the plow or the irrigation ditch. Offering few exploitable resources in its arid, windblown landscape, the Great Sand Dunes were immune to the extractive endeavors that swirled around their perimeter throughout the late nineteenth century. But the dunes' apparent resistance to exploitation mattered little; Gilpin's Garden had succeeded beyond even the boosters' wildest dreams elsewhere in the valley, where development and resource extraction continued unabated.

The success of Anglo-American settlement in the San Luis Valley was not without its price, however. Vicissitudes of commerce and fluctuating economics in far-off financial centers soon intervened to burst the bubble. The silver crash and nationwide financial Panic of 1893 dealt a severe blow to agricultural endeavors in the valley. Crop prices plummeted, and business all across the region took a serious downward turn, exacerbated by a serious drought that dried up many irrigation canals. Fields of crops withered in the relentless heat, forcing many small farms out of business. The drought itself lasted only a short while, but farmers throughout the valley and far down the Rio Grande valley suffered its effects. As early as 1890, the prodigious consumption of water in the San Luis Valley caused serious water shortages downstream in New Mexico. By the time the drought hit the valley in 1893, the Rio Grande ran dry near El Paso, Texas, and thousands of farmers along the lower Rio Grande lost their crops. In the San Luis Valley proper, the initial euphoria of the pioneer Anglo-American immigrants soon faded because of the drought. Some residents abandoned their homes and farms, leaving uncultivated, overgrown fields in their wake. A few decades of extractive land-use practices and unrestricted consumption of water, coupled with a relatively short drought, had combined to seriously deplete the valley's surface water supplies. Residents who remained were faced with the prospect of financial ruin if additional sources of water could not be found. Toward the end of the nineteenth century, the future of San Luis Valley agriculture looked grim indeed.[63]

Water Woes

The effects of the drought of 1893, combined with the diversion of water from the upper Rio Grande, destroyed countless acres of valuable crops up and down the river. The massive Rio Grande Canal near Del Norte on the western edge of the valley deprived many downstream valley farmers and ranchers of water. In 1893 the canal, consisting of roughly 210 miles of ditches and laterals, diverted almost the entire flow of the river in order to send water to its priority shareholders. The Rio Grande Canal Association, the mutual ditch company that owned and operated the canal, had every right under Colorado law to take such a large share of water. The association had filed a valid claim, and it was putting the water to beneficial use.[64]

When Colorado became a state in 1876, the new state government adopted the doctrine of "prior appropriation" for water allocation. Simply put, this doctrine, known colloquially as "first in time, first in right," gave priority water rights to the first person or persons who put the water to beneficial use. Inspired by the flurry of water claims that occurred during the California gold rush and bolstered by a frontier tradition that emphasized the superior claims of the first people to settle an area, the doctrine of prior appropriation differed significantly from "riparian" water rights, long the prevailing system in England and the eastern United States. Unlike prior appropriation, riparian water rights emphasized mere legal possession (that is, owning a piece of land adjacent to a stream or river) as the prime consideration in establishing a valid right to a water source. Thus landowners could acquire water rights simply by purchasing land adjacent to a reliable water source.[65]

Due to the peculiar circumstances of the frontier, the riparian doctrine proved unworkable in the arid American West of the nineteenth century. Common folks, settling wild frontier regions without the economic resources of wealthier citizens, needed to get water to lands not directly adjacent to rivers or streams in order to succeed as homesteaders. Realizing this, and also interested in encouraging settlement, many western states adopted the prior appropriation doctrine in some form, and nine states—Colorado, Utah, Mew Mexico, Nevada, Idaho, Arizona, Wyoming, Montana, and Alaska—eventually chose it as their sole water law.[66] Implicit in the prior appropriation system, and later made explicit by law, was the principle that the water had to be put to beneficial use. Cessation of such use meant an end to the water right.[67] Because water in the Rio Grande Canal system provided sustenance for crops in the valley, and since its claim dated from 1881, farmers downstream with later claims had no legal recourse when their irrigation canals ran dry and their crops died in the blistering heat. Coupled with the drought of 1893, the increasing settlement and cultivation of land in the San Luis Valley and elsewhere along the Rio Grande required that something be done to ensure that downstream users and later arrivals had access to sufficient water. Consequently, in 1896 an International Boundary Committee composed of representatives from Colorado, New Mexico, Texas, and the Republic of Mexico met to discuss the possibilities for sharing water from the Rio Grande. The committee signed no treaties and imposed no penalties, but the U.S. Department of the Interior did agree to restrict the development of further irrigation projects on the upper Rio Grande and stopped granting rights-of-way over public lands for the construction of new reservoirs.[68]

The enhanced restrictions hampered development of the water delivery systems and reservoirs that proved so crucial to the economic survival of the San Luis Valley. As a result, farmers were forced to search for alternative water sources. Many turned to the valley's abundant groundwater to supplement supplies from surface flows, and access to underground water took on a new significance. Artesian wells, in which water is forced upward by internal hydrostatic pressure and which thus generally require no pumps, soon renewed the promise of fertility all

Figure 4.5 Typical San Luis Valley artesian well and irrigation ditch, circa 1900. Such wells renewed the promise of fertility after the overappropriation of the valley's surface water resources during the late nineteenth and early twentieth centuries. U.S. Geological Survey.

across the valley floor. Heavily promoted in railroad and real estate pamphlets of the late nineteenth and early twentieth centuries, these wells provided cheap and plentiful water for farmers and ranchers, yet few early valley settlers actually understood the complex hydrological forces that created them.

Although it receives only seven to ten inches of rain annually, the San Luis Valley is blessed with a series of underground aquifers that are among the largest in the American West.[69] These aquifers are not, as is sometimes assumed, enormous subterranean caves or caverns containing gigantic lakes of water; rather, they consist of water-bearing layers of permeable rock, sand, soil, and gravel. More specifically, the San Luis Valley aquifer system consists of at least two readily identifiable units: an upper, shallow, or "unconfined" aquifer, and a lower, deeper, or "confined" aquifer. These aquifers are separated by volcanic rock and a layer of largely impermeable clay known to locals as the "blue layer" or the "blue clay," that limits the vertical movement of water underground. Since the separating layer of clay and rock varies in depth and thickness, determining the precise hydrologic connections between the two aquifers is difficult. Estimates of the depth of the unconfined aquifer range from 50 to 200 feet below the surface (sometimes less), while the confined aquifer may go as deep as 6,000 feet, perhaps deeper. Wells drilled into the confined aquifer are capable of flow rates between 2,000 and 4,000 gallons per minute. Currently there are some 6,000 wells drilled into the two aquifers beneath the San Luis Valley.[70]

Settlers in the vicinity of Moffat northwest of the Great Sand Dunes first tapped this enormous water supply as early as 1869, but perhaps the most dramatic historical manifestation of the valley's unique aquifer hydrology came shortly after the drought of 1893 had dried up the Rio Grande. Since the late 1880s, farmers in the Mosca-Hooper area just west of the Great Sand Dunes had relied on an irrigation technique known as subirrigation, or "bringing up the sub," to provide water for their sprawling fields of spring wheat and oats. The technique entailed inundating crop fields with immense quantities of water in order to raise the underground water table. Doing so enabled the roots of crops to absorb the water through a capillary or "wicking" action. Lacking sophisticated pumps and sprinkler systems, farmers simply diverted water from the area's artesian wells and surface streams and allowed it to flood their fields. While the technique sounded good in theory, in practice the results proved disastrous.[71]

Much to the surprise of farmers in the area, the peculiar hydrology of the Closed Basin obliterated the perceived benefits of "bringing up the sub." Containing roughly 2,940 of the San Luis Valley's approximately 8,000 square miles, the Closed Basin is separated hydrologically from the rest of the valley by a low topographic divide, which effectively prevents surface water flowing into the Closed Basin from ever reaching the Rio Grande River.[72] In fact, the actual boundaries of the lowest point in the Closed Basin correspond quite closely with the location of Gilpin's wildly exaggerated "sixty-mile lake" of 1869. Considering that Gilpin described such a lake at least a decade before significant well-drilling and cultivation occurred within the Closed Basin, it becomes painfully obvious how years of intensive irrigation in an area with no apparent natural drainage could ultimately raise the underground water table to a point that it would completely waterlog the soil and eventually seep onto the surface, which is exactly what happened. The rising groundwater flooded across the Closed Basin, evaporated, and left behind a concentration of alkaline salts, toxic to many plants. As a result, by 1896 farmers in the area began abandoning their waterlogged fields.[73]

Elsewhere in the valley, residents continued to pump groundwater for their crops and pastures. Further regulations concerning the diversion of water from the Rio Grande came in 1906, when the United States and Mexico signed a treaty that guaranteed 60,000 acre-feet of Rio Grande water annually to Mexico. Once again, farmers in the San Luis Valley had to limit their water consumption in order to ensure adequate supplies for downstream users. Despite the new restrictions, local private funding financed the construction of five reservoirs between 1910 and 1913 (the Rio Grande/Farmers Union, Santa Maria, Continental, La Jara, and Terrace Reservoirs) to meet the irrigation needs of San Luis Valley farmers. This same period witnessed the construction of eight major irrigation drains designed to reclaim some 90,000 acres of waterlogged land in the valley, yet similar efforts to reclaim land in the Closed Basin failed miserably.

Between 1911 and 1921, local residents attempted several ingenious drainage systems in the Closed Basin, resulting in some minor alleviation of problem waterlogging, but most had the unintended consequence of causing floods and alkaline deposits, forcing many farmers to move their operations away from the Great Sand Dunes and the Closed Basin to areas with better natural drainage. Their departure allowed plants such as rabbitbrush and greasewood, which were better suited to the alkaline soil, to move in and gradually fill the void. By 1915, farmers had completely abandoned immense stretches of land once hailed as possibly the most fertile wheat-growing region in Colorado. In effect, the Great American Desert had reclaimed a portion of Gilpin's Garden.[74]

Restrictions imposed by the treaty between the United States and Mexico compelled farmers in the San Luis Valley to tap the region's groundwater with increasing vigor. By 1910 at least 3,200 wells punctured the valley's prodigious underground aquifers, some of them under such extreme pressure that they sent geysers of water over fifty feet into the air when tapped. The majority of these wells supplied water to irrigate large-scale commercial farms, which produced such an abundance of quality crops that the cultivation of lettuce, potatoes, barley, and other crops eventually made the fertility of the San Luis Valley famous throughout the world.[75] As the nineteenth century passed into memory, the valley had developed into a true American Eden, verdant with lush fields of valuable crops and an abundance of water to nourish them, despite the drainage problems in the Closed Basin and the restrictions limiting the availability of surface water for irrigation. Water from the aquifers more than made up the deficit, and well drilling continued at an impressive pace. Irrigation turned previously unproductive land into valuable commercial farms and ranches all across the valley floor, with the notable exception of the area around the Great Sand Dunes and the Closed Basin. The poorly drained sump and the immense sea of sand effectively defined the eastern limits of successful agriculture, yet the remainder of the San Luis Valley had truly blossomed, at last fulfilling the promise of Gilpin's Garden.

The Great Sand Dunes remained stubbornly resistant to the San Luis Valley's profound transformations, despite facing continuous extractive pressures from assorted gold-mining and construction companies seeking profits from the immense sea of sand. Alarmed by these activities and increasingly aware of the value of wilderness and public lands, a determined group of concerned San Luis Valley citizens initiated one of the most remarkable developments in the entire history of the Great Sand Dunes. Long regarded as little more than a curious geological aberration, the dunes that had mesmerized and frustrated travelers and settlers for centuries were about to be officially recognized and protected in perpetuity, not for the benefit of a few select mining and construction companies but for the enjoyment and edification of future generations.

FIVE | Monumental Dunes

The actions of the dunes are similar to the vicissitudes of life. The peak of the sands today is the valley of tomorrow; the valley of yesterday, the mountain of today. In fact the entire aspect of the dunes is changed overnight. Unlike the "footprints in the sands of time," the footprints in these sands fade away. Scientists have aptly called these sand dunes the eighth wonder of the world.

Genevieve McLaughlin, 1928

A HEADLINE IN THE OCTOBER 15, 1922, edition of the *Denver Post* reported a truly astonishing development: "Three Lakes Discovered amid Great Sand Dunes of Southwest Colorado; Forestry Ranger Forces Way across Desert and Finds Relics That Indicate That Indians Practice Strange Rites on Banks of Shining Lakes." The brief article accompanying the headline eagerly described U.S. Forest Service (USFS) ranger Paul Gilbert's discovery of a series of interdunal ponds "gleaming like mirrors in the bleak swell and fall of the desert":

> In an area eighty miles square soft sand that clings to the feet and makes suffering the price of progress rises in clean sweeps to heights of a thousand feet, and drops abruptly in precipices ever-changing with the change of the wind. Only the determination of the explorer would impel an assault on those ramparts, and it was this determination—a strong desire to prove or disprove a persistent, incredible rumor—that sent Ranger Paul Gilbert of the nearby San Isobel [sic] National Forest plunging across the waste on a grueling trek, whose reward was the actual finding of the lakes, clear as crystal, cold as snow water, reflecting the blue and white of the sky and the sharply defined light and shadow of the dunes in pictures

worthy the brush of any painter willing to undergo biting hardship for the sake of pure color.[1]

Local residents and previous inhabitants of the valley almost certainly knew about the ponds, but news of Gilbert's discovery so impressed USFS Assistant District Engineer J. L. Brownlee that he urged for the interdunal ponds to be added "to the list of Colorado's great natural wonders." The article quoted Brownlee describing the "Indian relics" found by Gilbert (the same Paul Gilbert who had also explored La Caverna del Oro back in 1920) as "burial cairns . . . and stone bowls and rough carved pedestals," surrounded by "arrowheads enough to suggest that one day there was a hot battle there." Brownlee further noted that "plans are extremely tentative now, but some day the forest service will send an official party to the lakes prepared for the taking of photographs and possibly for excavations."[2]

Engineer Brownlee also offered his opinion concerning the potential origin of the lakes. "There is but one way that the lakes could be fed, and that is by springs from the bottom," he declared. "They cover several acres each, and the water is fresh and sweet to the taste."[3] A subsequent expedition to the dunes in 1923 by Ranger Gilbert and two San Isabel National Forest chiefs, Supervisor A. G. Hamel and Deputy Supervisor S. E. Doering, verified Gilbert's discovery of the interdunal ponds and provided additional details. "The clear water lakes are geologically quite as interesting as the Medano. It is the belief that there are twelve or fifteen of these that vary in size from a tenth of an acre to four or five acres," reported Hamel and Doering. "Their depth varies from four to fifteen feet and the water is as cold and clear as most any mountain water. These lakes are not stagnant although they have no inlet and no channel from one to the other. They are entirely surrounded by small sand dunes."[4] Furthermore, Hamel, like Brownlee before him, proposed his own theory for the presence of the ponds, explaining that "[in] my own reasoning I account for the existence of these lakes as an underground channel from Sand Creek, or as the place where the underground channels from Sand Creek and Medano River meet, thus forcing the water to the surface. The fact that the water is so clear bears out the theory that it must be moving all the time."[5]

Paul Gilbert's discovery of the interdunal ponds dramatically highlighted the extraordinary paradox of precious water existing in the very midst of what many considered to be a waterless expanse of sandy desert. The small oases of liquid were but one manifestation of the intricate hydrologic forces at work beneath the immense valley floor. Gilbert's discovery also linked him to a long historical continuum of engineers, bureaucrats, developers, ranchers, and other "experts" who spent the better part of the twentieth century trying to figure out the complex hydrology of the San Luis Valley and determine how best to allocate its water resources. No doubt Paul Gilbert would have been astonished by the fact that, more than seventy years after his discovery, the interdunal ponds would become the focus of intensive environmental research at the Great Sand Dunes, their size and number closely studied in an effort to determine the precise effects of climatological change and groundwater pumping on the continuing viability of the dunes.

Gilbert's discovery occurred during a period when San Luis Valley residents and businesses increasingly perceived the dunes not as a barren wasteland but rather as a remarkably unique landscape deserving of protection and preservation. The presence of Native American artifacts at the ponds only enhanced this evolving perception by clearly illustrating the long historical connection between humans and the Great Sand Dunes. No longer simply an incongruous, inhospitable pile of sand, the dunes now took on a new significance as a unique repository of cultural relics, a crucial thread in the broad tapestry of San Luis Valley history. As early as 1920, chambers of commerce in several towns including Alamosa, Del Norte, and Pueblo were actively calling for federal protection for the dunes, clearly recognizing what such protection could do for local businesses by attracting additional tourists to the San Luis Valley.[6]

The dunes were also becoming increasingly popular with journalists and travel writers. In 1921, a *Rocky Mountain News* article proclaimed "'Hissing Sands' of Colorado's 'Sahara' One of World's Great Scenic Wonders," while an article from the June 1922 issue of *Travel* magazine touted the possibilities for adventure at "The Desert of Hissing Sands."[7] Local boosters and other advocates for protecting the dunes likewise extolled their virtues in the lyrical language typical of the period: "[The] white heat of the mid-day sun, casting a spell of mystic charm over the shimmering sands, reminds one of the age-old fairy tale of the city beneath the sea that comes to the surface, for a day, once in a hundred years, and then again is swallowed by the waves."[8] Perhaps inspired by such flowery prose, in January 1924 the Alamosa Chamber of Commerce passed a unanimous resolution urging official protection for the dunes, noting that they were "of peculiar interest to the tourists," and advised the U.S. Forest Service to add the sand dunes to San Isabel National Forest.[9] District Forester Allen Peck acknowledged the growing local enthusiasm for protecting the dunes in an official dispatch addressed to "The Forester [Henry Graves], Washington D.C." and dated March 29, 1924. "The local communities are very anxious that this area should be reserved in some form and at one time proposed that it should be added to the San Isabel Forest," advised Peck. He then added his own opinion: "It is more properly National Monument material and can more easily be reserved under the Act of June 8, 1906 . . . than by Congressional action as part of the National Forest."[10]

Passed by Congress at the very height of the period of social activism and political reform known as the Progressive Era, the "Act" that Peck referred to was more commonly known as the Antiquities Act of 1906, which granted the president of the United States the authority to create national monuments for the explicit purpose of preserving "historic landmarks, historic and prehistoric structures, and other objects of historic or scientific interest that are situated upon the lands owned or controlled by the Government of the United States." Most crucially, the national monuments, unlike national parks, required no authorization from Congress; they could be created with a mere stroke of the president's pen.[11] Much of the impetus for the passage of the Antiquities Act came from a growing desire to protect America's prehistoric archaeological heritage during the late nineteenth century. Exciting discoveries of prehistoric ruins at various sites in the American Southwest, including the spectacular cliff dwellings at Mesa Verde in southwestern Colorado, had increasingly focused public attention on the need to protect the cultural prehistory of America from vandalism and theft. Indeed, the establishment of Mesa Verde National Park, which came less than one month after the passage of the Antiquities Act, was a direct response to the looting of Native American relics by assorted amateur archaeologists, pothunters, curio seekers, and plain old grave robbers.[12]

In light of the inherent emphasis the Antiquities Act placed on the protection of America's archaeological heritage, the presence of Paleoindian and Native American artifacts at the interdunal ponds clearly bolstered the argument for designating the Great Sand Dunes as a national monument. USFS Supervisor A. G. Hamel's official report indicated that vandalism was already occurring at prehistoric sites near the dunes, and District Forester Peck expressed similar concerns. "The examination indicates that this area has great historical value," he wrote, stressing that the dunes were "becoming very popular, with the result that much of its historical and scientific value will shortly be destroyed since many of the people visiting it are interested only in collecting specimens of the races which occupied it in the past. It should be reserved for science and thoroughly studied."[13]

Hamel and Doering's report also noted clear evidence of prehistoric inhabitants at the dunes:

> Old Indian camps, large enough to indicate from three to five thousand souls, are everywhere apparent. Every water course, spring, lake, and sinks or arroyos where short season lakes exist bear evidence. . . .
>
> In an hour's search we picked up fifteen perfect [arrowhead] specimens and one imperfect tomahawk. On both trips to the region we unearthed skeletons that had been apparently buried in the sand immediately within the campground which would indicate that they did not remove their dead far to bury them, or else that the camp was hastily abandoned after a massacre or pestilence.[14]

The presence of prehistoric artifacts and remains at the dunes subsequently attracted the attention of Smithsonian Institution Bureau of American Ethnology Chief J. Walter Fewkes, who heartily endorsed the national monument proposal, although he was careful to emphasize that he had never even seen the Great Sand Dunes. "I have never visited this region," he explained, "but archeological possibilities of the Monument seem to call for the protection and the vicinity of the Forest Reserve points to the Forest Service as its guardian."[15]

Despite the recommendation of Fewkes and other proponents of including the dunes within the boundaries of the nearby San Isabel National Forest, the USFS was "unwilling to take this action, since the area is not of forest or grazing type." Instead, in 1924 the USFS, an agency of the Department of Agriculture that administers the nation's forests and grasslands, referred the Great Sand Dunes monument proposal to the National Park Service (NPS), an agency of the Department of the Interior that manages the nation's national parks, monuments, battlefields, seashores, and other conservation and historic properties in the United States. Given that the administration of national monuments falls under the purview of the NPS, this referral made perfect sense. Curiously, though, the NPS took no immediate steps to secure national monument status for the Great Sand Dunes. Lack of funding may have been to blame, but for whatever reason, the NPS conducted no additional investigations of the dunes, nor did it initiate or solicit any follow-up reports on the USFS recommendation. The USFS even "offered to have their nearest ranger keep general supervision of the area, if desired," but this generous offer apparently failed to inspire any decisive action on the part of NPS administrators, and the matter floundered in a maze of regional and federal bureaucracy for several years thereafter.[16]

Meanwhile, in a booster effort that would have done William Gilpin proud, the Alamosa Chamber of Commerce erected road signs in 1926 to direct tourists to the dunes, and San Luis Valley citizens and newspapers in the region continued their tireless efforts to publicize the area's most renowned attraction.[17] Among the most prominent of these promoters was Dr. Frank Spencer, a professor of history and mathematics at Adams State College in Alamosa. In 1927 Spencer wrote an article for the *Alamosa Journal* in which he called the San Luis Valley a "Mecca for Tourists" and described the Great Sand Dunes as "veritable sand mountains . . . an area of barren, glittering white sand almost devoid of either vegetable or animal life, yet uncannily fascinating." Spencer's prose occasionally tended toward exaggeration: "At times the air is so charged with electricity that it leaps with livid flame from each exposed point." In another passage, Spencer even managed to reference the famous Lorelei rock on the eastern bank of the Rhine River in Germany, inspiration for a number of Germanic myths and legends: "It is a land of mirages and illusions, an Elfinland where danger lurks constantly but with all the lure of the Lorelie [*sic*]." Through such hyperbolic language Spencer enthusiastically promoted the dunes to potential tourists and predicted a bright future for the shifting sea of sand, proclaiming, "Yesterday they were almost unknown, today they are visited by hundreds, tomorrow they will be sought by thousands."[18]

This spirit of local boosterism for the dunes continued in other regional publications and newspapers of the era, although occasionally the authors of these

Figure 5.1 The Great Sand Dunes as they appeared in 1927. The effort to secure federal protection for the dunes accelerated during the late 1920s and early 1930s. National Park Service, Great Sand Dunes National Park and Preserve.

pieces chose to highlight not only the virtues and beauty of the "Weird Grey Hummocks" but also the dunes' peculiar hazards, including raging sandstorms, flocks of sheep and their herders being buried by the "restless, avenging sands," and, perhaps most ominously, the "treacherous quicksands" of Medano Creek:[19]

> Weird indeed is the Medano or so called Lost River by the Indians who inhabited this fascinating region. At the source, which is high up in the Sangre de Cristos, the water rushes over the rocks, but when the lower land is reached it sinks into the masses of sand leaving only a shallow depth of water in what would be the main course of the river otherwise. Quicksand in the river is treacherous.[20]

Exaggeration of the size of the main dunefield and the height of the tallest dunes also occurred in various articles during this period, including at least one that described how the sand hills extended "over an area of eighty square miles, and rise above twelve hundred feet above the floor of the valley. They are conceded to be second in size to the Sahara, which is the largest desert of shifting sands not found on the shores of a body of water."[21]

In reality, the Sahara Desert currently covers roughly 3.5 *million* square miles of northern Africa, while contemporary estimates of the size of the main dunefield at the Great Sand Dunes range from around thirty to forty square miles, depending on how they're measured, and the tallest dune is approximately 750 feet above the valley floor. Misleading statistics about the dunes mattered little

Figure 5.2 Medano Creek, known to local Native Americans as the Lost River, sinking into the sand on the southern edge of the Great Sand Dunes. National Park Service, Great Sand Dunes National Park and Preserve.

to the editors and authors of these articles, which were designed not to deliver definitive factual information but rather to highlight the history, mystery, and utter incongruity of the Great Sand Dunes for tourists who might be tempted to spend their hard-earned vacation dollars in the San Luis Valley. Despite the promotional efforts of Frank Spencer and other boosters of the dunes, the prospects for a national monument continued to languish at the federal level during the late 1920s. That situation might have continued indefinitely if not for the efforts of a handful of strong-willed, civic-minded local women from the PEO Sisterhood, an international women's philanthropic and educational organization founded at Iowa Wesleyan College in 1869, who took matters into their own hands and initiated a grassroots movement to secure permanent protection for their beloved dunes.

Sisterhood of Sand

Citizens of the San Luis Valley have long been proud of the role the PEO Sisterhood played in the establishment of Great Sand Dunes National Monument, a process that most historical sources agree began in June 1930. Concerned about persistent and destructive efforts to mine gold from the dunes, as well as alarming reports of local construction companies hauling away truckloads of sand for use in mixing concrete, Elizabeth Spencer, wife of the aforementioned Dr. Frank Spencer and active member of Chapter V, the PEO Sisterhood's group in Monte Vista, gave a luncheon presentation in which she told her fellow members about the dunes' unique beauty and stressed the importance of preserving the area as an important archaeological site.²² Spencer concluded her presentation by urging her PEO colleagues to support the worthy cause of creating a national monument at the dunes, a suggestion the membership enthusiastically endorsed, and before the meeting ended Chapter V president Anna Mae Darley duly appointed Elizabeth Spencer as chairman of the newly formed Sand Dunes Committee, along with Martha Jean Corlett and Myrtle Woods as Chapter V committee members.²³

Given the daunting scope of the task ahead of them, Spencer and her cohorts solicited help from the other San Luis Valley PEO chapters, including Del Norte (BH) and Alamosa (AE), as well as the Colorado State Chapter, and together they began an ambitious letter-writing campaign to recruit local, state, and national politicians to their cause, with an eventual goal of presenting their case directly to the policy makers in Washington, D.C. Among their first orders of business was contacting Monte Vista attorney George Milton Corlett, who was not only married to Sand Dunes Committee member Jean Corlett but also happened to be the lieutenant governor of Colorado serving under Governor William Herbert "Billy" Adams, for whom Adams State College in Alamosa was named.²⁴

At the behest of the PEO, George Corlett crafted a petition in support of the national monument designation, which according to several historical sources was then given to state senator A. Elmer Headlee (D–Monte Vista). The petition specifically asked Headlee "to present the attached Memorial to the next State Assembly, as a joint Memorial of the House and Senate of Colorado, to the Congress of the United States of America."²⁵ Headlee complied and introduced the petition to the Colorado legislature, which "voted to grant support for the project." Interestingly, despite its name, a legislative "memorial" does not refer solely to commemorating the dead or any other solemn occasion; rather, it indicates the "method by which the legislature addresses or petitions Congress and other governments or governmental agencies."²⁶ A number of subsequent newspaper and historical accounts of the PEO's involvement in the national monument process incorrectly referred to both the memorial and the petition as "bills" that were introduced in Congress.²⁷

In fact, confusion over the precise difference between a "petition" and a "bill" occurs repeatedly in the historical record regarding the Great Sand Dunes National Monument proposal. Technically, a petition is a request from an organization or private citizens' group sent to one or both chambers of a legislature asking them to support, or oppose, particular legislation or pleading for favorable notice of a matter that has not yet received congressional attention. Petitions are then referred to appropriate committees for further consideration. A bill, on the other hand, is a specific legislative proposal designed to become law if approved by Congress and signed by the president. To create a national monument, neither a petition nor a bill (or a memorial, for that matter) is required, only the president's signature.

Along with asking Headlee to present a memorial, the PEO also sent a petition to Senator Lawrence C. Phipps (R-Colo.) that described the dunes as "unique in character; widely known and unparalleled in scenic grandeur" and respectfully requested that Phipps prepare and introduce a "Bill, Act, or Memorial" in Congress in order to protect the dunes as a "National Monument, Park, or Playground."[28] The petition concluded on a cautionary note, warning that "if steps are not taken promptly to isolate and segregate this territory, which is now principally unoccupied, and remove the same from entry by individuals, small portions thereof will undoubtedly be entered and occupied by individuals thereby greatly lessening the general value to the community and to the Nation as a National Park and Playground."[29] In a letter to PEO Chapter V and dated December 24, 1930, George Corlett described hand-delivering the petition to Phipps, stating that he "personally handed your petition to Senator Phipps in Washington the other day and also mentioned the matter to our Congressman, Guy U. Hardy. Both of them expressed themselves as being willing to do anything they can to further the undertaking."[30]

Senator Phipps's remarkably rapid reply came on December 27, just three days after Corlett's letter to PEO Chapter V, in which the senator acknowledged receipt of the petition and indicated to Corlett that he had taken the matter up with National Park Service director Horace M. Albright "in an endeavor to enlist his aid" and dutifully reported Albright's response:

> The records in this office indicate that this area was considered in the Department for national monument purposes in 1924–25 but at that time sufficient information concerning the area was not available to properly appraise its value for this purpose. We have since, however, secured an item in our annual appropriations which will permit the examination of such areas in detail with a view to determining their merits and desirability for national park or national monument purposes.
>
> We are therefore placing this project on our list of such examinations to be made by our field investigator and shall be pleased to communicate with you further after his report thereon has been made.[31]

In fact, Horace Albright had previously dispatched a letter, dated December 26, 1930, to Roger Toll, superintendent of Yellowstone National Park and field assistant to the NPS director, instructing him to investigate the Great Sand Dunes personally and assess their potential for designation as a national monument. Largely as a result of the diligence and persistence of the PEO ladies and their supporters toiling away in the relative isolation of the remote San Luis Valley, as well as the fortuitous availability of funding from the NPS, the political wheels in Washington were already turning on the question of national monument status for the dunes.[32]

Closer to home, on January 16, 1931, the Colorado State Senate issued an official Senate Joint Memorial that described the Great Sand Dunes as an "unsurpassed" scenic attraction and requested that Congress establish the dunes as a "National monument, park and playground."[33] Within a month of the Senate Joint Memorial, additional support for the monument proposal appeared in a story in the *Alamosa Journal* describing the sand dunes as "Alamosa County's greatest asset," but also commenting on the region's lack of suitable roads. "People over the nation are waiting to come in droves to see an attraction that might be classed as the Eighth Wonder. They will not come, however, until an improved highway is built to them," the *Journal* reported. "It does not matter if there are two or three modern roads to the dunes but it is essential that there is at least one motor way that can be travelled at all seasons of the year, that is high and dry, impervious to dust and not subject to washout."[34] Coincidentally, at 5:45 A.M. on February 13, 1931, the very same day the story appeared in the *Alamosa Journal*, Horace Albright's "field investigator" stepped off the train in Alamosa to begin his tour of the Great Sand Dunes, a tour that would include a trip down those very same roads.

The Toll Report

Roger Wolcott Toll was clearly the most qualified field investigator in the National Park Service during the late 1920s and early 1930s. Born in Denver in 1883 to a pioneering Colorado family, Toll had spent his youth exploring the Colorado high country, becoming intimately familiar with the myriad geologic and geographic intricacies of the Rocky Mountains. Educated at Denver and Columbia Universities, Toll graduated in 1906 with a degree in civil engineering and worked several jobs around the country before returning to Denver, where in 1912 he became one of twenty-five original charter members of the Colorado Mountain Club. After serving as an army major during World War I, Toll joined the NPS in 1919 and rose rapidly through the ranks, eventually attaining the prestigious position of superintendent at Mount Rainier, Rocky Mountain, and Yellowstone National Parks. Perhaps Toll's greatest legacy, however, was "his superb firsthand investigations and reports on proposed areas to the park system." Areas that benefited from his expertise and investigative tours on behalf of the NPS included Death Valley and Joshua Tree in California, Big Bend in Texas, and the Everglades in Florida. So vast and extensive was his field experience that NPS director Albright credited Toll with having "explored, photographed and described in reports most of the canyons of the Colorado from the headwaters in the Rockies to the California line."[35]

On arriving at the Alamosa train station, Toll was met by none other than USFS ranger Paul Gilbert, locally renowned "discoverer" of the interdunal ponds, and together they piled into Gilbert's car and headed for Mosca, "thence easterly over a road which was opened only last year [1930] and thence over unimproved roads to Mosca Creek, reaching this creek about four miles west of the summit of Mosca Pass." Toll later wrote in his report that it "is ungently [*sic*] desired by the residents of the San Luis Valley that the improved road should be continued to the dunes and also across Mosca Pass." It is likely that Toll meant "urgently" instead of "ungently," but either way he was careful to note that both Colorado governor Billy Adams and the USFS were in favor of completing the Mosca Pass road in order to improve vehicular access to the proposed monument. Toll then incorrectly predicted that a "first class automobile highway" would be completed over the pass within five years.[36]

From Mosca Creek, Toll and Gilbert made their way to Medano Creek ("dry at this time of the year") and from there ascended "to the highest of the nearby Sand Dunes," much as Zebulon Pike had done 124 years earlier. The two men spent about two hours total in the main dunefield before returning to Alamosa, where Toll attended a Kiwanis Club weekly dinner and discussed the monument proposal with local residents. The next morning Toll and Gilbert were up early and off to the Medano Ranch (also known at that time as the Linger Ranch) on the west side of the dunes, then continued about three and a half miles to the banks of (Big) Spring Creek, where they left the car and proceeded on foot "past many Indian campgrounds to Indian Spring which is the source of Spring Creek." Toll makes no mention of any attempt to investigate these "campgrounds" further, and the "Inspection" portion of his report simply concludes with mundane details of his return trip to Alamosa, a brief stop at Fort Garland, and then on to Denver via Walsenburg. He had spent a grand total of barely thirty-six hours in the San Luis Valley, with only about four to six hours of that time actually investigating the Great Sand Dunes and Indian Spring, yet the observations and recommendations he compiled in his official report would have profound implications for the future of the Great Sand Dunes.[37]

On February 16, the day after Toll left the San Luis Valley, the *Denver Post* published a short article on his visit that contained few details beyond noting that his guide, Paul Gilbert, had "made a careful study of the Mosca Pass and sand dunes."[38] The next day, in an article headlined "PEO Sponsors Bill for Making Sand Dunes National Monument," the *Alamosa Journal*, perhaps succumbing to the common error of confusing a legislative "bill" with a nonbinding "joint memorial," optimistically reported that the "bill for making the Sand Dunes a national monument has passed both houses of the state legislature, and is now being taken up in Washington D.C. and the outlook is very encouraging, according to a letter read at PEO meeting from Elmer Headlee, State Senator of Rio Grande County."[39] Less than two weeks later, Dr. Frank Spencer again took to the pages of the *Alamosa Journal* to advocate for the creation of a national monument at the dunes:

> Finally, the greatest reason from the point of view of material advantage to this part of the state is the nation-wide publicity which the setting

aside of this region as a National Monument would immediately result. The fact alone that the National Government has been brought to recognize that we have something unique in the world, something worthwhile being preserved and studied, will bring to it an immediate interest which can be gained in no other way.

Let us concentrate on this one thing. The question of roads, improvements and care may well be left to a later time. Let us first secure the Monument, the other blessings will be added unto us in due time.[40]

Roger Toll returned to his winter office in Denver after completing the inspection tour and began compiling his official Great Sand Dunes report, which ended up including a detailed description of his time in the San Luis Valley, with sections on location, general characteristics, accessibility, land ownership and proposed boundaries, need for conservation, proposed development, history of the project, and history and geology of the area, as well as an extensive collection of maps, clippings, articles, letters, supplemental reports, and photographs, including several depicting the novel activity of skiing on the dunes. All told, the materials added up to nearly a hundred pages of information highlighting Toll's positive impressions of the dunes, with emphasis on their remarkable size, tremendous height, stunning scenic location, numerous important archaeological sites, and potential for tourism development.[41]

Noting that the "Sand Dunes themselves are fairly immune from injury," Toll recommended that areas adjacent to the dunes be preserved in public ownership to prevent further homesteading, concluding that the "natural, wild attraction of the dunes as a mysterious, desert area would be detracted from if one had to approach through the gates of various homesteads, and if the grass lands were fenced up to the very edge of the dunes."[42] Furthermore, Toll advised moving quickly to protect the dunes, noting that "a number of claims have been filed within the past few years, and this indicates that the monument should be created as soon as practicable to prevent further filings."[43] He also advised against allowing cattle and horse grazing adjacent to the dunes, citing concerns over possible pollution of the area's only reliable source of drinking water, and declared that ample grazing land elsewhere in the valley meant that "withdrawal of this area would not be an economic loss to the community."[44] In addition, Toll recommended protection for the "Indian campgrounds" and suggested establishing a "small field museum" to display and interpret the Indian "relics," as well as tell the geological story of the dunes, should adequate funding ever become available. While he questioned whether national monument status would eliminate souvenir hunting entirely, he did admit that "at least whatever protection [was] offered would be in the right direction."[45]

Toll concluded his report by recommending that the dunes be officially recognized and protected as a national monument "whether or not any funds are to be expended in their development,"[46] although his final assessment was not without a certain degree of ambiguity, perhaps even reluctance, as is apparent in his suggestion that the dunes might better be protected as a state park, stipulating that if "Colorado had an organization and funds for the establishment of a state park system, this area might satisfactorily be made a state park, and ultimately the administration might be turned over to the state, but at the present time the state cannot handle it, and it seems desirable that the land should be withdrawn from entry and established as a national monument."[47] One paragraph later, Toll emphasizes the utter uniqueness of the Great Sand Dunes, and his overall appraisal seemed decidedly more enthusiastic when he declared that among "the present national parks and monuments there is no area that resembles these sand dunes. They would be distinctive and unduplicated by other reservations. I know of no sand dunes in the United States that are superior to these dunes of the San Luis Valley."[48] This curious (if minor) juxtaposition of suggesting that the dunes "might satisfactorily be made a state park" and then describing them as literally unequaled in the United States apparently mattered little to his final assessment. The Great Sand Dunes and their remarkable setting had clearly impressed Toll, whose travels had taken him to some of the most spectacular places in America, and his opinions carried considerable weight within the National Park Service.

His task finished, Roger Toll submitted his official report to NPS director Horace Albright on April 3, 1931. Albright endorsed Toll's recommendation and in turn sent it up the chain of command to his superior, Interior Secretary Ray Lyman Wilbur, who also approved the proposal. On November 30, 1931, Wilbur requested that President Herbert Hoover authorize the withdrawal of approximately 46,000 acres of land in the San Luis Valley that included the Great Sand Dunes in order to preclude the filing of additional homestead claims in the area, which Hoover did three days later, on December 3, 1931. Barely eighteen months after Elizabeth Spencer first exhorted her PEO colleagues to support the worthy cause of protecting the dunes from further development, Executive Order 5751, "Withdrawal of Public Lands for Classification, Colorado," essentially cleared the way for the official creation of Great Sand Dunes National Monument.[49]

Back in the San Luis Valley, perhaps unaware of the political machinations occurring in Washington, D.C., the PEO Sisterhood remained persistent in pursuit of its goal, and the campaign for national monument status for the Great Sand Dunes continued unabated. Alamosa Chapter AE president Millie Velhagen contacted Representative Guy Hardy (R-Colo.) and included a photograph of the sand dunes with her correspondence. In a reply letter dated December 19, 1931, Hardy thanked her for the "beautiful card" and suggested sending one directly to NPS director Albright, along with letters from at least "half a dozen people of prominence" describing why the sand dunes would make a "notable National Monument," stating his preference that he didn't want "ordinary chamber of commerce stuff—but I mean good arguments with a little descriptive stuff in it. If anybody has thoughts along archaeological lines, this would be good."⁵⁰ On December 29, the Sand Dunes Committee contacted Sen. Edward Costigan (D-Colo.) and encouraged him to assist Hardy "in the passage of this measure in the House, and . . . use your influence in securing its passage in the Senate." Again, it is unclear whether the PEO (or anyone else in the San Luis Valley, for that matter) understood that the president had the sole authority under the 1906 Antiquities Act to designate national monuments and did not require congressional approval, but regardless, the dedication to the cause is impressive, as is the committee's willingness to resort to a little friendly coercion of Senator Costigan, evident in this coy revelation: "Knowing that your Mother was one of our 'Past State Presidents' we feel that this measure, sponsored by the PEO Sisterhood, may be of special interest to you."⁵¹

Representative Guy Hardy contacted Elizabeth Spencer on January 7, 1932, informing her that the president "has issued a Proclamation withdrawing that area from entry which means that it cannot be exploited for the present," then cautioned that the NPS had its hands full with other assessments:

> In the meantime the National Park Service is studying a large number of proposed sites for National Monuments. It has something like ninety propositions before it and only a very few will be picked for National Monuments I am told. From a half to a dozen. I do not think it would help much to have general petitions signed. I do think it might help if individuals or heads of organizations who could write intelligently on the subject to write letters directly to Hon. Horace M. Albright.⁵²

As instructed, Mrs. Spencer wrote directly to Albright on January 22, taking care to include some photographs that "will show you something of the weird beauty of the place." She then highlighted other notable attributes of the dunes, noting that the "archaeologist finds many things to prove that here was the dwelling place or camping ground of a primitive race. There are also many unusual biological specimens and electrical phenomena which need careful study by experts."⁵³ Director Albright personally responded five days later:

> As you know, our investigator Roger Toll who is also Superintendent of Yellowstone National Park has investigated the sand dunes and as a result of his investigation and report we have withdrawn from further settlement the area the sand dunes cover. By doing this we are enabled to hold the area intact and to prevent further private claims being established until such time as we have had opportunity to make a more careful survey and study of the area.
>
> Your pictures and letter will be placed in our files and will help us materially in making our final decision. The article on skiing on sand is certainly very interesting; we had never thought of the dunes in that

light. From the pictures it appears to be a fine way to travel over these enormous sand dunes.⁵⁴

Sand skiers apparently had plenty of company on the dunes in those days. For years, prospectors had been hammering claim stakes—most of which soon disappeared beneath the endlessly shifting sand—and panning or sluicing for gold at the Great Sand Dunes, but in 1929 (some sources claim 1932) the most ambitious attempt to extract gold began when the Volcanic Mining Company built a small mill, complete with amalgamation tables capable of recovering minute traces of flour gold, on the banks of Medano Creek.

The construction of the mill was an ominous indication that gold mining at the dunes had escalated far beyond solitary miners with shovels and pans and entered a new, possibly more destructive phase, especially with amalgamation tables that required mercury, or vats of cyanide to leach out the gold, or other dangerous chemical operations employed to process gold-bearing ore. In fact, the Volcanic gold mill seemed to represent the very worst fears of the PEO Sisterhood, as well as citizens across the San Luis Valley, who saw a national monument at the dunes as potentially the valley's most famous tourist attraction, with all the commensurate economic benefits for valley communities. If large-scale gold mining were allowed to continue, especially in the precise spot along Medano Creek that provided the most convenient visitor access to the main dunefield, the entire national monument proposal would be in serious jeopardy.⁵⁵ Fortunately, the original assays of gold in the Great Sand Dunes turned out to be wildly

Figure 5.3 Abandoned gold-processing mill constructed circa 1929 by the Volcanic Mining Company on the banks of Medano Creek. The effort to extract profitable quantities of gold ultimately failed, but the threat of extensive mining at the dunes helped inspire the creation of Great Sand Dunes National Monument in 1932. National Park Service, Great Sand Dunes National Park and Preserve.

Figure 5.4 The remains of a gold dredge on the banks of Medano Creek. National Park Service, Great Sand Dunes National Park and Preserve, GRSA-2152.

inaccurate; in reality, the actual content of gold in the sand was only slightly higher than normal geochemical background levels. Some minute traces of gold were recovered, but the extraction process required far too much labor to justify the meager return on investment. Moreover, Medano Creek continued to be a notoriously fickle and unreliable source of water, and before long the operators of the Volcanic Mill simply packed up their equipment and moved on, leaving behind the battered foundation of the old mill on the banks of the "Disappearing River" as mute testimony to the futility of the entire episode.[56]

If the Great Sand Dunes had in fact contained profitable amounts of gold, it is unclear what impact the presence of large-scale mining operations would have had on the potential establishment of the national monument. Perhaps mining would have been allowed to continue in a more remote area of the vast dunefield while visitor facilities were developed in the more accessible portions, but in the end it didn't matter, and once again, the Great Sand Dunes successfully resisted humanity's efforts to reap some measure of profit from them. No amount of mining or cutting or plowing could integrate the dunes into the region's market economy, and while lands around their perimeter may have produced timber, minerals, and irrigable land, the actual dunes seemed immune to the extractive endeavors occurring elsewhere in the San Luis Valley. Instead, the future held another destiny for the dunes, one in which the Great Sand Dunes would attract legions of tourists instead of miners, a destiny referenced in a letter from Representative Guy Hardy to Elizabeth Spencer, dated March 14, 1932, that hints about the prospects for success for the PEO campaign:

> This information is more or less Confidential and should not be given out to the Press of course. I just thought I would tell you that I had Hon.

Horace M. Albright, Director of the National Park Service at my home for dinner the other night and got some inside dope on this question of making the Sand Dunes a National Monument. I am going to drop the hint that I think the prospect is very good and I believe that you will hear some news on that subject before many days.[57]

Hardy's letter proved to be appropriately prescient, because three days later, on March 17, 1932, fifty-five years after geologist Ferdinand Vandeveer Hayden's *Atlas of Colorado* first formalized the term "Great Sand Dunes," President Herbert Hoover invoked his presidential powers authorized by the Antiquities Act of 1906 and signed Presidential Proclamation No. 1994, thereby creating Great Sand Dunes National Monument, the nation's thirty-sixth such designation:

> Whereas it appears that the public interest would be promoted by including the lands hereinafter described within a national monument for the preservation of the Great Sand Dunes and additional features of scenic, scientific, and educational interest . . . now therefore I, Herbert Hoover . . . do proclaim and establish the Great Sand Dunes National Monument.[58]

Figure 5.5 The Great Sand Dunes in 1932, the same year President Herbert Hoover signed the presidential proclamation that officially created Great Sand Dunes National Monument. National Park Service, Great Sand Dunes National Park and Preserve.

Nearly two years of tireless advocacy had finally paid off, thanks to the collective persistence of the PEO Sisterhood, the citizens of the San Luis Valley, the National Park Service, and local, state, and national political representatives from both political parties, not to mention the initial investigations by the U.S. Forest Service in the early 1920s. Approximately 46,034 acres of land comprising the Great Sand Dunes were now protected in perpetuity from miners, homesteaders, speculators, and developers. The official press release from the Department of the Interior described the dunes as "picturesque" and "singularly beautiful" and noted that "the boldness of their relief is remarkable," while the ecstatic PEO simply stated, "Someday our children's children will thank us for our foresight in saving this magnificent natural wonder." When news of the proclamation reached Alamosa, all the church bells in town rang together in unison at high noon.[59]

As for Roger Toll, he continued his meteoric rise through the ranks of the National Park Service, eventually investigating more than 130 sites as potential candidates for national parks or monuments, the overwhelming majority of which did not, in Toll's experienced and measured opinion, meet the criteria necessary for such designation. His contention that national monuments should possess a "genuine national interest" clearly influenced his positive assessment of the Great Sand Dunes, which he believed contained the appropriate geologic, historical, and scientific attributes to justify its inclusion among America's most treasured landscapes. Toll seemed firmly on the path of one day becoming director of the National Park Service before his untimely death in February 1936. While on a tour of potential international park and wildlife refuge sites along the U.S.-Mexico border, an oncoming car suffered a tire blowout that caused a head-on collision with Toll's vehicle near Deming, New Mexico, killing Toll instantly. He was fifty-two.[60]

The Nascent Monument

Due primarily to the economic uncertainties of the Great Depression, the new Great Sand Dunes National Monument initially received scant funding from either state or federal coffers, although Alamosa County did appropriate some funding for road improvements in the days following the official designation.[61] With no money available for permanent park staff, administrative authority over the monument came first from Southwestern National Monuments (SWNM), a special administrative unit created by the NPS in 1923 to oversee national monuments in the American Southwest, but lack of funding prevented anything remotely resembling direct oversight from SWNM, and in February 1934 the NPS reassigned administrative control over the Great Sand Dunes to Mesa Verde National Park, roughly 240 miles to the west.[62] Glen King, a member of the pioneer King family that first settled near Alamosa in 1885 and whose father was the last postmaster of Montville at the foot of Mosca Pass, became the first employee at the Monument and began compiling data on visitors by

Figure 5.6 The access road to Great Sand Dunes National Monument in 1934. National Park Service, Great Sand Dunes National Park and Preserve.

counting the number of vehicles that turned north toward the dunes from the main access road. In May 1934 the superintendent of Mesa Verde National Park sent park naturalist Paul Franke on an inspection tour of the dunes, who described how the primitive road leading to the Monument "suddenly ends four and one-half miles from the dunes and this remaining distance must be covered over two rough ruts or tracks made in the prairie. Many washes and gullies must be crossed and travel slows down to about eight miles per hour. There are no high centers, but the unevenness of the tracks tosses the car from one side to the other."[63] In addition to an improved access road, Franke's report also noted that the new monument might benefit from better signage, campgrounds, and toilets to accommodate the expected influx of visitors, which in 1932 had numbered roughly 500 but seemed to be increasing yearly. By 1937, the number of visitors had grown to nearly 3,500.[64]

Funding for park operations picked up speed in the late 1930s as the moribund economy of Depression-era America finally began showing signs of life. In

1938 plans were announced for an improved entrance road at the monument, along with a picnic area and toilet facilities. In 1939 came news of additional construction plans, including a superintendent's residence and combined headquarters / entrance station, but political squabbles in Washington threatened to derail the project, and an editorial in the January 29, 1940, edition of the *Alamosa Daily Courier* commented on the continued lack of funding for the new monument, lamenting, "Two thousand dollars is little money when it is considered alongside the vast sums spent by the government on its agencies, projects and such things as national defense. Since it is so little, it doesn't seem unreasonable for the San Luis Valley to ask Congress not to cut this sum from its appropriations at the cost of losing important development work at the Great Sand Dunes National Monument."[65] The editorial concluded with an appeal to readers to contact their senators and congressmen to "do all they can to keep the little dunes appropriation intact."[66]

Without hesitation, the PEO Sisterhood responded to the challenge with a letter dated February 6, 1940, representing fifty-three members of Chapter V in Monte Vista and addressed to Sen. Edwin C. Johnson (D-Colo.), which described the monument as "the San Luis Valley's greatest attraction, bringing thousands of visitors here to view this remarkable phenomenon." The letter continued by warning that withholding the appropriation "will mean the loss by destruction of improvements already made and will deprive the Monument of many other improvements badly needed to accommodate the increasing number of visitors. We respectfully urge you to use your influence to have the two thousand dollars restored to the appropriations bill."[67] Senator Johnson responded by noting that $850 had indeed been allocated for the monument as a new item in the Interior Bill, while Congressman Edward T. Taylor (D-Colo.), chairman of the House Appropriations Committee, who had received a copy of the same letter from the PEO, replied in somewhat more detail:

> I notice that an item of $850 for the Great Sand Dunes Monument is included, which the Park Service advises me is for salary for temporary ranger service during the heavy traveled season in that monument.
>
> In checking this matter with the National Park officials here, I find that most of the money now being used for improvements in the Great Sand Dunes Monument is allocated from the relief funds—in other words, by E.R.A. projects [Emergency Relief Allocation Act, a New Deal–era jobs project].
>
> They advise that for 1938 an allocation of $38,375 was made from the relief funds for use in your monument and that a larger sum was allocated for the present fiscal year. It is that money which is being used for improvements in the way of picnic and camp grounds, development of water, construction of residences, garages, etc.[68]

Figure 5.7 The main entrance to Great Sand Dunes National Monument in 1944, including a new superintendent's residence and park headquarters to the right. The structure was placed on the National Register of Historic Places in 1989. National Park Service, Great Sand Dunes National Park and Preserve, GRSA-2200.

The threatened reduction of funding for the monument evidently never materialized, because at some point in 1940 Howard S. Rines became the first seasonal park ranger, and in September of that year work commenced on the new superintendent's residence and headquarters / entrance station using federal funds provided by the Works Progress Administration, a structure that in 1989 was added to the National Register of Historic Places.[69]

In 1942, prior to the Allied invasion of North Africa, a small cadre of U.S. Army personnel used the windswept dunes as a proving ground for assorted equipment, including jeeps and camouflage patterns designed for use in the desert.[70] Tourist visits understandably plummeted during World War II, with barely 3,000 people visiting in 1943, but visitor numbers gradually increased after the war ended in September 1945. A series of seasonal summer rangers, including Bert Clarke, oversaw what limited operations occurred at the park from 1940 until 1946, when Ted Sowers became the monument's first full-time permanent employee, the same year future superintendent Glen Bean began working as a seasonal ranger after being discharged from the Army.[71]

Trespassing cattle continued to be a problem during the war years, despite persistent efforts to keep them from wandering inside monument boundaries. In July 1943 seasonal ranger Bert Clarke wrote a letter to local rancher George Ziegler, imploring him to keep his cattle under control. "There are about fifteen head of your cattle in trespass in the monument area," he explained, adding that he "took seven head up and through your fence on Wednesday of this week, but

Figure 5.8 U.S. Army troops gathered at the Great Sand Dunes in 1942 for training and equipment testing prior to the Allied invasion of North Africa. National Park Service, Great Sand Dunes National Park and Preserve.

this new bunch came down on me last night. They raised Cain with my water supply."[72] Another letter to Ziegler less than two weeks later clearly indicates that the problem had not been adequately addressed. "There are over thirty head of your cattle in trespass on the Monument this morning," Clarke wrote, admonishing Ziegler to "please remove this stock and make provisions to keep them out by rider or fence as they seem to have gotten in the habit of coming in the last couple of weeks." Furthermore, Clarke appealed to Ziegler's sense of law and order: "I believe that you appreciate the fact that the regulation regarding the trespass of livestock on the Monument must be enforced."[73] Whether or not Ziegler complied with Clarke's request is uncertain, but starting in the late 1940s and continuing into the early 1950s, NPS personnel began fencing portions of the monument to keep cattle from wandering into the main dunefield to graze, although the wire barriers did little to deter the herds, and cattle continued to graze within monument boundaries during the next several decades.[74]

In 1946, the nearly 9,000 visitors who came to the dunes were forced to negotiate a washboarded gravel road from Mosca that one writer described as "eighteen miles of mud, sand blow-outs, dust and gravel [that] did their best to discourage even the most dedicated visitor."[75] Once at the dunes, visitors encountered limited facilities: the adobe/stucco entrance station and residence, two water-collection systems (one in Morris Gulch for the entrance station / residence, the other at the mouth of Mosca Pass), and, adjacent to the main dunefield, a parking area with a water faucet (supplied by pipe from Mosca Pass),

Figure 5.9 Monument headquarters/entrance station in 1946. National Park Service, Great Sand Dunes National Park and Preserve.

two pit toilets, and six to eight "very heavy picnic tables made of ponderosa pine logs."[76] As limited as these primitive facilities were, the infrastructure situation could have been dramatically different, not only for the Great Sand Dunes but for the entire San Luis Valley, if a committee in far-off Los Alamos, New Mexico, had made a different decision regarding an experimental and potentially devastating new weapon.

In 1944, scientists and engineers working on the Manhattan Project designated a site-selection committee to locate a suitable testing location for their top-secret device. The committee had three main criteria for the proposed site: flat terrain, so that the weapon's effects could be measured at some distance; generally clear weather, so that the test would not be delayed by poor visibility or inclement weather conditions; and low population and relative isolation, for obvious reasons. Less critical criteria included federal land ownership and proximity to Los Alamos. The committee then came up with a list of eight potential test sites that generally met the requirements, one of which was the San Luis Valley near the Great Sand Dunes. The committee visited the first four sites on the list and stopped searching after selecting a suitable location in a desolate stretch of desert known as Jornada del Muerto (roughly translated as the "Journey of Death") in the White Sands Proving Ground (now Missile Range) between Socorro and Alamogordo, New Mexico. The committee never visited the Great Sand Dunes, even though the location met every criteria outlined in the site-selection process, and on July 16, 1945, at a place now known as the Trinity Site, scientists detonated the world's first atomic bomb, a cataclysmic explosion that very well could have happened in the San Luis Valley if the Great Sand Dunes had not been at the bottom of the committee's list.[77]

On March 12, 1946, President Harry Truman officially reduced the size of the monument from 46,034 acres to 44,810 acres, evidently due to an error in the original survey, but visitor numbers continued to climb in the postwar years, reaching over 23,000 in 1949. The increase in visitation suggests that Great Sand Dunes National Monument had firmly established itself as a popular destination for tourists who increasingly took to the nation's highways in search of recreation and adventure in the newly prosperous postwar America.[78] Continuing federal protection of the dunes also ensured that large-scale gold mining efforts no longer posed a threat, although some placer mining continued in the Sangre de Cristo foothills adjacent to the dunes, including one claim in the vicinity of the old Herard homestead near Medano Creek, and newspapers continued to report on the potential riches of sand in the dunes throughout the 1940s. In 1942 the *Steamboat Pilot* had declared that the dunes held "$5,990,440,000 worth of gold," and a locally published promotional book, using more optimistic calculations, proclaimed in 1949 that the dunes held "eight billion dollars in gold," but by then the luster of gold at the dunes had faded, so efforts to extract mineral wealth from the vast sea of sand gradually became a distant memory.[79]

"Terrible roads" continued to plague visitors to the monument during this period, prompting Wilbur Foshay, secretary of the Alamosa Chamber of Commerce, to ask Senator Edwin C. Johnson for federal assistance. "What I am writing you about, and I hate to bother you when I know how busy you are

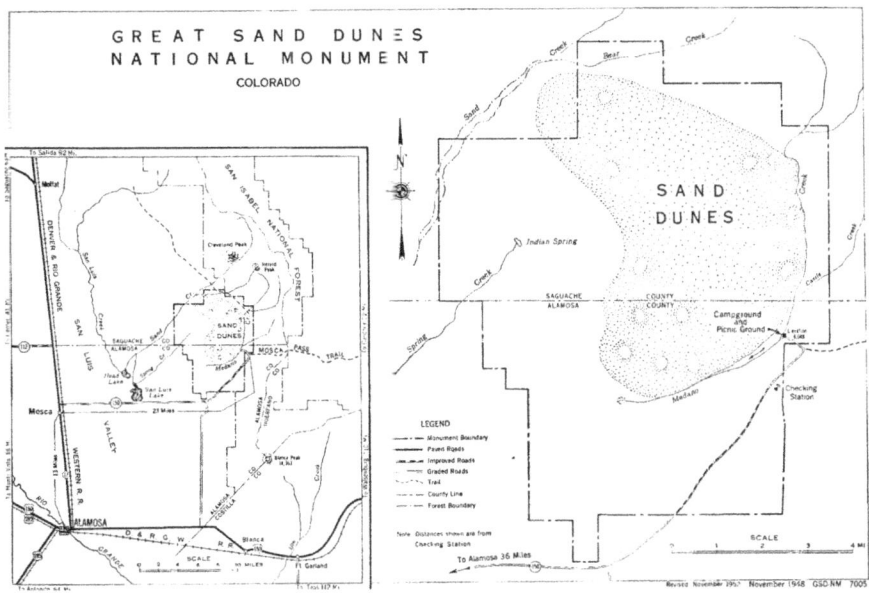

Figure 5.10 1948 map of Great Sand Dunes National Monument depicting boundaries, roads, principal features, and geographic location in the San Luis Valley. National Park Service, Great Sand Dunes National Park and Preserve.

on other things," implored Foshay in his December 1947 letter, "is to get your suggestions as to what we can do to try to get some help from the federal government on the building of that portion of the road that is their responsibility, which should come from the Park Service, and what we can do to get federal aid on the road into the Dunes."[80] The plea for federal aid apparently fell on deaf ears, because in May 1950, the same month that former seasonal ranger Glen Bean took over as the monument's superintendent, Foshay wrote Senator Johnson another letter to complain not only about the lack of decent roads, but about the lack of visitor amenities in general at the Great Sand Dunes, using Mesa Verde National Park for convenient comparison:

> Mesa Verde has 19 miles of fine improved oil surfaced road from U.S. 160 to the main headquarters, to say nothing of the roads to the ruins. They have a fine lodge where meals can be secured, sleeping accommodations from camping grounds to super deluxe cabins, and the Park Service maintains an all year around staff there with fine homes and all conveniences.
>
> Compare this with the Sand Dunes. They have a house where the Superintendent lives. He is the only permanent man. There are no accommodations for the visitors, and they have to travel twenty-three miles from State Highway 17 on a dirt road that is never any good, and yet with none of the accommodations or attractions provided at the Dunes as are provided at Mesa Verde, nearly 30 percent as many people visited the Dunes.
>
> If overnight accommodations, a dining hall and other concessions, with a good oiled road, were available, the Sand Dunes would out distance Mesa Verde two to one, and yet we can't even get folders that are lying on the shelves of the Park Service, to send out to people.[81]

Despite such pleas for aid, the federal government continued to neglect the roads and other facilities at the Great Sand Dunes, which compelled Representative John Edgar Chenoweth (R-Colo.) to write to NPS director Conrad L. Wirth in August 1952, informing him "that the present condition of the roads in the monument is very bad. The popularity of this monument is increasing. It is unfortunate that the visitors find the roads in such poor condition. The situation is the subject of considerable comment."[82]

Trespassing cattle also continued to be a problem during the 1950s, as indicated in a 1954 letter from then-superintendent Harold Schaafsma to former superintendent Glen Bean in which he explained, "As I look out the office window I can count over one hundred head of trespass stock grazing placidly, so you can see I too have made no progress in this respect." Schaafsma also queried Bean about his own experiences, wondering, "Was the situation this bad when you took over . . . or has it become progressively worse . . . ? On your next trip try to come over if you can keep from running over the cows."[83] By 1956, his persistent cattle problem notwithstanding, Schaafsma did have some success improving the road situation at the monument, largely as a result of additional funding made available under Mission 66, an ambitious ten-year federal program designed to upgrade National Park Service visitor facilities and services nationwide in time for the fiftieth anniversary of the NPS in 1966. Schaafsma referenced Mission 66 in a memo to the NPS assistant regional director dated November 12, 1956:

> Our needs for improvement and development are dire and most critical. Present employees must reside in quarters only suitable for summer occupancy. With the long overdue small addition to our staff expected in 1957 we have no provision for housing them. In view of the fact that the only other major improvement at Great Sand Dunes since its establishment in 1932 was the oiling of the road, we would like to strongly recommend that our program under MISSION 66 be completed as a "package deal." Our requests for improvements and development and very conservative and all of them are already urgently needed.[84]

Schaafsma's request for additional funding apparently got through to the right people, because the pace of much-needed infrastructure improvements began to accelerate at the dunes during the late 1950s and early 1960s. In 1956 the dunes endured another boundary change, this one authorized by President Dwight D. Eisenhower, which reduced the monument to roughly 37,000 acres, but the change had little impact on the continuing upgrade to visitor and employee facilities. A 1956 NPS summary of Mission 66 construction

Figure 5.11 Monument entry road and sign as they appeared in 1957. National Park Service, Great Sand Dunes National Park and Preserve.

Figure 5.12 The new Great Sand Dunes National Monument Visitor Center, circa 1962. National Park Service, Great Sand Dunes National Park and Preserve, GRSA-2198.

Figure 5.13 Pinyon Flats Campground, shortly before it opened in 1964. National Park Service, Great Sand Dunes National Park and Preserve.

projects slated for funding at Great Sand Dunes National Monument included an estimated $920,000 in federal aid, and by 1957 the REA (Rural Electrification Administration, a New Deal–era agency) had finally provided full electrical power to the monument, while additional housing units and a new 50,000-gallon water tank were in place by 1959, all courtesy of Mission 66 funding.

Another significant improvement came in 1960 when the state of Colorado provided over $90,000 to pave the twenty-two miles of road from Mosca to the dunes. Mission 66 funding also spurred construction of a new visitor center beginning in 1961, followed by the Pinyon Flats campground in 1964, with a capacity of 1,000 campers a night, as well as a new amphitheater for outdoor performances and interpretive programs that opened in 1966. Visitor numbers reflected the continuing upgrade in monument facilities during this period, jumping from 52,700 in 1960 to over 157,000 in 1966.[85]

With the completion of the Mission 66 program and the celebration of the fiftieth anniversary of the National Park Service in 1966, the Great Sand Dunes had undergone their most profound period of transformation since Herbert Hoover first authorized the monument in 1932. Stray cattle still regularly trespassed on the dunes, as they had for decades, but large-scale mining, wood gathering, and additional homestead claims had all been halted by the federal government. Another layer of protection came on October 20, 1976, when President Gerald Ford, under authority granted by the Wilderness Act of 1964, signed Public Law 94–567, designating 33,450 acres of the monument as wilderness, which, among other restrictions, prohibited mechanized and motorized vehicles on the dunes; it also laid the foundation for additional acreage in the Sangres to receive the same protection in the future. Nearly 250,000 people visited the monument in 1976, meaning that in less than fifty years, the Great Sand Dunes had gone from being the subject of a grassroots preservation project initiated by the local PEO Sisterhood to one of the most popular attractions in Colorado. The remote sea of sand once visited by only scant numbers of local San Luis Valley residents out for a picnic lunch or hunting for arrowheads was now attracting hundreds of thousands of visitors every year from all over the world.[86]

Yet, as had been the case for so many centuries, the Great Sand Dunes still seemed immune to the dramatic changes swirling in their midst. Even as increasing numbers of tourists and the facilities necessary to accommodate them introduced changes both evident and unseen to the landscape, the dunes themselves remained more or less unaltered. The ceaseless winds of the San Luis Valley continued to sculpt and shape the heaping mounds of sand into forms as enchanting to modern tourists as they had been to Zebulon Pike, while the mesmerizing surge flow of Medano Creek continued to gurgle along with mystical regularity, indicative of the bizarre interface of sand and water that had been enchanting human visitors to the dunes for millennia. Toward the end of the twentieth century, however, an ominous new threat emerged in the San Luis Valley—a threat that came not from overgrazing or wood gathering or gold mining or motorized dune buggies, but from a series of ambitious plans for pumping and exporting the valley's groundwater resources, an integral component of the very hydrological system that had attracted humans to the region in the first place. After enduring centuries of extractive activity with relatively few ill effects, the Great Sand Dunes would at last be confronted with a menace that could potentially jeopardize their very existence.

SIX | Sand and Water

If there is magic on this planet, it is contained in water.

Loren Eiseley, 1957

FOR WELL OVER 10,000 YEARS, the San Luis Valley provided ample water for the historic parade of humanity that passed through its borders. The Rio Grande and its numerous tributaries attracted Paleolithic hunters seeking sustenance in the game animals that congregated in the marshy wetlands of the prehistoric valley. Centuries later, Native Americans hunted amid the verdant grasslands that punctuated the valley floor, integrating the unique mixture of sand, water, and soaring mountain summits into their origin myths and cosmologies. Spanish conquistadors followed the Rio Grande north in their relentless quest for gold and empire, and when they reached the San Luis Valley, they gave names to its lofty peaks and network of rivers and streams. American explorers penetrated its rugged mountain passes and immense interior, searching for railroad routes, resources to extract, and hegemony in the far western realms of a rapidly expanding republic. In the middle of the nineteenth century, tenacious Hispanic settlers established the valley's first permanent communities and began digging rudimentary irrigation ditches, while later Anglo-American homesteaders plumbed the valley's rivers and streams and pumped its plentiful groundwater to irrigate their expansive farms and ranches.

Water, so precious in the arid interior of western North America, is readily available both on and below the surface in certain sections of the San Luis Valley, providing the catalyst for centuries of successive waves of human exploration and settlement in a land that averages only seven to ten inches of rain annually. Yet by the beginning of the twentieth century, less than fifty years after the establishment of the valley's first permanent settlement, water users had overappropriated the Rio Grande to the point of exhaustion, rendering it unable to meet the insatiable demands placed on it by the rapid expansion of commercial agriculture and ranching introduced by Anglo-American settlers in the late nineteenth century. Their sprawling farms and ranches required tremendous amounts of water, distributed by an ever-expanding network of irrigation canals and ditches that consumed more water than even the mighty Rio Grande could supply.

The response to the problem of water demand exceeding supply had profound implications for the future of water resource allocation in the San Luis Valley. Hampered by decreasing supplies from the Rio Grande and unable to compete with those who held more senior water rights, farmers and ranchers increasingly turned to groundwater supplies in an effort to secure the water they needed, resulting in a dramatic increase in groundwater pumping that continued to accelerate throughout the century. At first the extensive pumping seemed harmless. The valley's bountiful aquifers appeared to have an almost inexhaustible supply of water and had the added benefit of being easy to tap. Toward the end of twentieth century, however, the negative repercussions of groundwater pumping became increasingly apparent. Within the span of only a few generations, the pumping that had become so pervasive across the valley floor inspired a series of proposals to export the valley's liquid wealth for profit—proposals that, if implemented, potentially threatened not only the agricultural economy of the San Luis Valley, but also the long-term stability of the Great Sand Dunes' sand transport and recycling system, with unknown consequences for this unique and irreplaceable landscape.

San Luis Valley residents of the late nineteenth and early twentieth centuries evidently saw no tangible link whatsoever between groundwater pumping and the imposing sea of sand on the eastern fringes of the valley. Widespread irrigation had created lush and fertile farmland nearly everywhere across the valley floor except the barren, windblown expanse of the Great Sand Dunes, creating a perception of two separate and distinctly different landscapes: Gilpin's irrigated garden juxtaposed with the Great American Desert, coexisting but unconnected. By the last decade of the twentieth century, the validity of that dichotomy would at last begin to disintegrate, replaced by an increasing awareness that the landscapes of the Great Sand Dunes and the greater San Luis Valley were in fact intimately connected, in deep and lasting ways. More importantly, this realization raised compelling questions about the significance of water in an arid environment, its role in the San Luis Valley's economic viability, and the future of water resource allocation in an increasingly populated American West.

Circles of Rain

In 1929 Colorado, New Mexico, and Texas signed a temporary compact stipulating that upstream users, including the large commercial farms in the San Luis Valley, maintain the status quo on water delivery until a permanent compact could be ratified. After nine years of hashing out the details, the three states finally agreed on an equitable distribution of water and signed the Rio Grande Compact on March 18, 1938. In essence, the compact imposed water delivery requirements on Colorado and New Mexico by establishing a sliding scale that regulated the amount of water that legally had to be available to downstream users according to the amount of water entering the entire river system in a given year. In other words, during dry years, Colorado's water delivery obligation to downstream users decreased, and during times of higher water flows, it increased correspondingly.[1] In order to determine the exact amount of water to be delivered downstream, compact negotiators designed a complex system of measurements and calculations to address such issues as water quality and annual water credits and debts between the compact states. The compact also contained a clause that provided for the elimination of credits and debts in the event of an overflow at the Elephant Butte Reservoir in New Mexico, completed in 1916 as part of a 1906 treaty agreement between the United States and Mexico. Still, dramatic yearly fluctuations in both water consumption and water availability virtually assured continuing acrimony among water consumers throughout the Rio Grande watershed.[2]

In 1966, Texas and New Mexico filed suit against Colorado for failing to meet its obligations under the compact. Costly court battles ensued and ended up in the U.S. Supreme Court, which ruled for a continuance of the case so long as Colorado met its annual commitments to downstream users. With constant litigation over water allocation looming in their future, concerned San Luis Valley residents realized the need for a local organization to protect the valley's water rights and ensure that the historically established framework of agriculture could continue. At the urging of valley citizens, the Colorado General Assembly established the Rio Grande Water Conservation District in 1967, with a mandate to promote water resource development within the San Luis Valley, determine water policy, coordinate legal and engineering issues, and develop water projects with local, state, and federal agencies.[3]

By the end of the 1960s the San Luis Valley boasted a mélange of agencies that managed the region's water, including local water conservancy districts, irrigation and drainage districts, water and sanitation districts, nonprofit and voluntary water associations, and assorted ditch and canal companies. Further complicating matters were the assorted federal agencies that had a presence in the valley, including the Bureau of Reclamation, the Bureau of Land Management, the U.S. Forest Service, the U.S. Fish and Wildlife Service, and the National Park Service, each with a different agenda and all with some stake in the administration of the valley's increasingly appropriated water resources.[4] The numerous state and federal restrictions placed on the Rio Grande and other San Luis Valley surface water resources were not only decidedly complex; they were also fast becoming onerous for San Luis Valley farmers, many of whom simply found it easier to sink new wells than attempt to wade through the confusing bureaucratic gauntlet in order to secure water rights to increasingly depleted surface flows. For farmers reluctant to jump on the groundwater bandwagon, the introduction of a revolutionary new sprinkler system offered a compelling reason to reevaluate the benefits of groundwater pumping. The innovative device would prove to have profound significance, not only for farms in the San Luis Valley, but also for agriculture across the face of the planet.

In 1947, farmer Frank Zybach of Strasburg, Colorado, attended an irrigation field day on a farm near Prospect Valley, north of Strasburg. He watched as workers repeatedly hauled pipes that had been fitted with sprinkler heads on

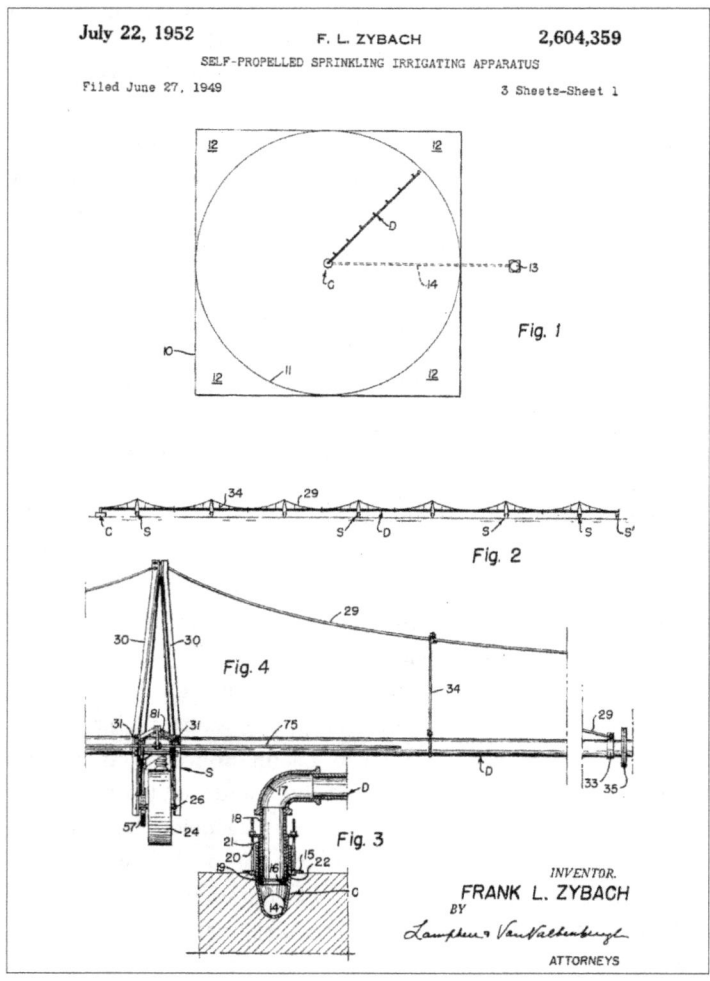

Figure 6.1 Inventor Frank Zybach's 1949 patent application for his "Self-Propelled Sprinkling Irrigating Apparatus," more commonly known as a center-pivot sprinkler. The U.S. Patent Office granted his patent in 1952, and the device subsequently revolutionized irrigation across the world. U.S. Patent Office.

posts from one section of a field to another and admired how efficiently the sprinklers could deliver water to hilly terrain, but the tedious process of slogging through the mud to relocate the pipes looked to Zybach like too much work. An inveterate tinkerer, Zybach decided right then and there that he could design something better. Within a year he had built a working prototype of the first "center-pivot" irrigation system, and in 1949 he applied for a patent on his "Self-Propelled Sprinkling Irrigating Apparatus." On July 22, 1952, the U.S. Patent Office granted Frank Zybach Patent Number 2,604,359 for his invention. In the decades that followed, the ingenious system literally changed the

Figure 6.2 Satellite image of the San Luis Valley depicting a large number of the valley's approximately 2,600 center-pivot sprinklers. Each "circle of rain" represents roughly 120 acres of crops. The Closed Basin appears as a light-colored band to the east (*right*) of the crop circles, with the San Luis Lakes, Great Sand Dunes, and Sangre de Cristos farther east. The Rio Grande River is visible as a dark streak running from center left to center bottom. Image © 2012 Google and © 2012 TerraMetrics.

landscape of the Great Plains, the American West, and indeed the entire world. Essentially a row of powerful sprinklers mounted on an elevated, wheeled metal frame, the center-pivot rotated in an enormous circle around a central well, with a portion of the water pressure diverted to drive the wheels. Capable of traveling over sandy hillocks or rugged terrain with relative ease, the device required no land leveling or ditch digging. Instead, the raised phalanx of sprinklers threw water in great sweeping arcs that nourished crops like gentle circles of rain falling from the sky.[5]

Because of the historic success of flood irrigation, center-pivots were not readily adopted in the San Luis Valley; as late as 1964, the valley had only two center-pivot systems. Gradually, however, as the new technology caught on, center-pivot sprinkler systems transformed fields that had once been giant, square-shaped checkerboards into vast, verdant circles of green that, from the air, appeared like huge symmetrical lily pads floating on an enormous, tranquil pond. The center-pivots used water far more efficiently than earlier irrigation methods, and their success resulted in an increase in the number of wells drilled and outfitted with center-pivots. By 1973 over 200 center-pivot systems had been installed; by 1996, the number had risen to 2,132, utilized on at least half of the valley's roughly 600,000 acres of irrigated land, and by 2010, approximately 2,600 of the devices tapped both the confined and unconfined aquifers in the San Luis Valley.[6] This huge number of wells understandably increased the acreage of land under cultivation, which in turn inspired local water engineers to attempt further augmentation of the valley's water supplies. In a case of historical irony that no doubt would have intrigued William Gilpin, the Closed Basin, site of the booster's mythical "sixty-mile lake," became the focus of an ambitious effort to increase the flow of the Rio Grande by pumping even more of the valley's groundwater.

Reclaiming the Basin

Manipulation of the San Luis Valley's water resources reached new heights with the ingenious and controversial Closed Basin Project. As early as 1915, the U.S. Department of Agriculture and the U.S. Bureau of Reclamation had concluded that water could be successfully "salvaged" from the Closed Basin. By 1938, drafters of the Rio Grande Compact recognized the potential for utilizing "unused and wasted" water in the Closed Basin to supplement the delivery flows stipulated by the compact. Local water engineers studied the possibility and even made plans to install drains in the basin, but the newly christened "San Luis Valley Project" diminished in importance during World War II and the Korean War and subsequently languished in bureaucratic darkness for years afterward. The Bureau of Reclamation resumed its studies in the early 1960s and determined that 60,000 to 190,000 acre-feet of water a year could be salvaged from the basin. In 1972 Congress authorized funding for the reclamation proposal, but again the project languished until 1975, when Congress appropriated $135,000 for "further studies." After yet another delay, Congress approved funding for the project in 1980, and in 1981 construction on the project finally began.[7]

Compared to other massive federal reclamation projects undertaken in the arid American West during the twentieth century, the Closed Basin Project seemed to be a relatively modest proposal. Studies indicated that the Closed Basin continually lost a tremendous amount of water through a process known as *evapotranspiration*, a combination of evaporation of water that had seeped to the surface and transpiration of underground water through the leaves of plants. Due primarily to the region's scant rainfall, the majority of plants in the Closed Basin are *phreatophytes*, deep-rooted plants that obtain most of their water from underground water tables. Indeed, greasewood (known locally as *chico*) has been found in the valley tapping groundwater up to fifty-seven feet below the surface. Along with greasewood, other phreatophytes, including rabbitbrush and saltgrass, annually transpire an estimated 1 million acre-feet of water in the San Luis Valley, much of it from the Closed Basin.[8] Closed Basin Project engineers proposed to salvage this water by lowering the water table in the basin, thereby reducing the effects of evapotranspiration. Water recovered from the Closed Basin would be transported through a series of lateral canals to a main canal, which would then carry the water to the Rio Grande in order to help meet Colorado's downstream water delivery requirements. Completed in 1993, the project had the capacity to recover about 104,000 acre-feet of water annually, with 170 "salvage" wells, 82 monitoring wells, and a forty-two mile canal that carried water to a point south of Alamosa in the Alamosa National Wildlife Refuge, where it emptied into the Rio Grande.[9]

Figure 6.3 The Closed Basin Project Main Conveyance Channel, looking east toward the Sangre de Cristos and Great Sand Dunes National Park. Photo by the author.

San Luis Valley residents initially expressed skepticism over the perceived benefits of the Closed Basin Project. Some feared that the project would lead to an extension of the Great Sand Dunes by turning the Closed Basin into an immense "dust bowl." Others feared diminished pressure in their artesian wells or were convinced that the federal government would seize private property and water rights in order to complete the project. To counter such doubts, local officials pointed out that the legislation authorizing the project several contained safeguards to prevent such undesirable effects. By law, pumping for the Closed Basin Project could not lower well levels more than two feet, or measurably reduce artesian well pressure, anywhere outside the project boundaries, hence the battery of solar-powered monitoring wells placed around the perimeter of the Basin. Project officials also allayed fears that lowering the water table would turn the Closed Basin into a dust bowl by insisting that, although plant life might diminish somewhat, especially in areas immediately adjacent to well locations, most phreatophytes in the basin would simply adjust to the lower water table conditions, and any other adverse effects on plants would be negligible.[10] An official publication from the Bureau of Reclamation confronted the issue head-on, stressing how "some individuals have suggested that operation of the project will result in a vast extension of the Great Sand Dunes. In fact, the project will, to an extent, counteract the environmental damage of 100 years of overgrazing and diversion of salt-laden irrigation runoff into the Closed Basin. Lowering the water table and decreasing the amount of salt in the soil will eventually create an environment closer to what was there 100 years ago."[11]

Bureau officials gave no indication as to precisely how they determined the dynamics and composition of the Closed Basin environment of a century earlier, but they took great pains to emphasize the positive benefits of the Closed Basin Project. Beyond its role in helping to satisfy Colorado's downstream commitments, the project would also deliver water to the Alamosa National Wildlife Refuge and Blanca Wildlife Habitat Area, as well as to the San Luis Lakes area, in order to stabilize water levels, enhance wildlife habitat, and increase recreational opportunities. In addition, project supporters noted a number of "spin-off" benefits generated by the strict application of federal laws requiring extensive studies of the Closed Basin's ecology, archaeology, history, geology, and hydrology, which helped mitigate damage to cultural sites providing detailed information about the overall environmental impact of the project.[12]

Less emphasized by project supporters were the potential adverse effects of the Closed Basin Project on the landscape of the Great Sand Dunes. National Park Service officials expressed concern that the project's main water conveyance channel, the forty-two-mile canal running north–south through the middle of the basin, might seriously restrict the movement of sand across the valley. Such a restriction conceivably could have reduced the amount of sand heading for the dunes and perhaps completely disrupted the dune-building process, with unknown consequences for the dunes' existence. In an effort to determine the

Figure 6.4 One of the solar-powered monitoring wells that surround the perimeter of the Closed Basin Project to measure its effects. Photo by the author.

role the canal might play in restricting sand movement in the basin, NPS staff erected a number of "sand traps" on the western fringes of the monument to measure the frequency, direction, and quantity of sand migration toward the dunes, but to date no measurably adverse impacts have been detected.[13]

Ultimately, any critical evaluation of the impact of the Closed Basin Project on the Great Sand Dunes must be tempered by one crucial fact: although designed to salvage roughly 104,000 acre-feet of water annually, the project has yet to be operated at full capacity, due primarily to a number of relatively high-water years that caused the Elephant Butte Reservoir in New Mexico to overflow, which eliminated Colorado's water debt to Texas and New Mexico as stipulated by the Rio Grande Compact. In 1994, the year after the project was completed, the main conveyance canal carried only 37,000 acre-feet of Closed Basin water to the Rio Grande, a little over one-third of its designed capacity. If (or when) the project ever does salvage the maximum amount of water it was originally designed to, assessments of its impacts, both positive and negative, may require a more thorough evaluation.[14]

Profits, Protests, and Precious Water

American Water Development, Incorporated (AWDI), opened a new chapter in the tangled saga of San Luis Valley water politics in 1986 when it filed an "Application for Underground Water Rights or, in the Alternative, for the Determination of Rights to Nontributary Groundwater outside of Designated Groundwater Basins" in the Colorado's District Court Water Division No. 3. In plain language, AWDI sought to establish the right to withdraw 200,000 acre-feet (approximately 65 billion gallons, or roughly 40 percent of the average annual flow of the Rio Grande at Del Norte) of groundwater annually from the San Luis Valley and sell it to slake the thirst of Denver and the rapidly growing Front Range region of Colorado. The source of AWDI's liquid largesse was to be the Baca Grant, a sprawling tract of land hard against the base of the Sangre de Cristos, just north of the Great Sand Dunes.[15]

The Baca Grant itself has a long and often convoluted history stretching back to February 18, 1820, when rancher Don Luis Maria Cabeza de Baca (who claimed to be a direct descendant of the famed Spanish explorer Álvar Núñez Cabeza de Vaca) petitioned Spanish colonial authorities on behalf of himself and eight partners for a large tract of land known as Las Vegas Grandes containing roughly 500,000 acres in the vicinity of the Gallinas River near present-day Las Vegas, New Mexico, about 240 miles south of the Great Sand Dunes. In January 1821, after his original partners had dropped out, Baca re-petitioned for the same tract of land, this time on behalf of himself and his "seventeen sons," a number which may have arisen from a mistranslation of the Spanish word *hijos*, meaning "children," since from the historical record it is unclear whether Baca actually had that many sons; several were his sons-in-law. Local authorities apparently approved Baca's petition in 1823, but the precise legal status of the grant remained in limbo for the next several years. By 1826 Baca had obtained full legal title to the land and constructed a small *rancho* on the banks of the Gallinas, where he and his family began tending large herds of sheep, cattle, and horses that were frequently targeted by Native American raiding parties.[16]

In 1827 (at least one historical source claims 1833), Don Luis Maria Cabeza de Baca died from a gunshot wound suffered in a dispute with a Mexican soldier over some contraband fur pelts found in Baca's possession, leaving his son Juan Antonio Baca to continue ranching operations until 1835, when Navajo raiders killed him and stole all of the grant's livestock, forcing the rest of the Baca clan to abandon Las Vegas Grandes. That same year, a group of settlers, many of whom had been occupying different portions of the Baca land for several years, successfully petitioned for a large tract of land that covered essentially the same area as the original Baca grant. This development sparked an additional rush of settlement on the property and years of litigation that finally ended a quarter century later: in 1860 the U.S. government, which had assumed jurisdiction over the area in 1848 when the Treaty of Guadalupe Hidalgo ended the Mexican-American War, agreed to allow the heirs of Luis Maria Cabeza de Baca to select five separate tracts of vacant land called "floating grants" totaling some 500,000 acres elsewhere in the Territory of New Mexico in compensation for abandoning their claims to the original grant. Among their choices was a large tract of approximately 100,000 acres of land in the San Luis Valley just north of the Great Sand Dunes that included a portion of the vast sand sheet as well as sand dunes, lush meadows, forested foothills, and 14,165-foot Kit Carson Peak. This tract came to be known as the Luis Maria Baca Grant No. 4. The Baca heirs used it to pay off their lawyer, John S. Watts, who became the first in a long series of owners that later included Territory of Colorado governor Alexander Cameron Hunt and the ubiquitous William Gilpin, who tried and failed to acquire the Baca Grant in 1862.[17]

Gilpin was finally able to purchase the Baca Grant outright in 1877 and subsequently leased the land to George Adams, whose improvements to the property included erecting miles of fencing, constructing new ranch buildings, and digging an extensive network of irrigation ditches. Adams sold his interest in the Baca in 1900 to the San Luis Valley Land and Mining Company, which promptly changed its name to the San Luis Valley Land and Cattle Company (SLVL&CC) and began promoting the grant for land development with a pamphlet that highlighted the property's scenic setting amid towering mountains and abundant water for farming and ranching.[18] Not much came from the promotional effort, but ranching operations continued to improve until 1930, when Alfred Collins, a major stockholder in SLVL&CC, arrived and took over management of the property. Although he had limited experience with ranching, Collins initiated further improvements to the sprawling grant and gained some small measure of renown (and no small measure of profit) by raising purebred

Hereford cattle. In 1950, Newhall Land and Farming, a California development company, bought the Baca property and held it until 1962, when it sold the property to the Arizona-Colorado Land & Cattle Company (later known as AZL Resources Inc., a large agribusiness conglomerate). In 1971, AZL created the Baca Grande subdivision to sell individual home sites on a portion of the property.[19]

Finally, the Baca Grant wound up in the possession of Canadian oil developer and former United Nations diplomat Maurice Strong, who discovered the property among the assets of AZL Resources after he acquired a controlling interest in the company in 1978.[20] Strong assembled a group of wealthy investors and formed AWDI in 1986 with a board of directors that included, among others, former U.S. Environmental Protection Agency chief William Ruckelshaus and former Colorado governor Richard Lamm. Recognizing the potential for exploiting the Baca Grant's plentiful groundwater supplies, as well as one of the most senior water rights in the state of Colorado, AWDI proposed drilling a battery of over a hundred wells into both the unconfined and confined aquifers at depths between 200 feet and 2,500 feet below the surface, some with a capacity to pump 5,000 gallons per minute. This water would then be transported via pipeline over Poncha Pass to the thirsty urban centers of the booming Front Range, resulting in a considerable profit for the fiscally ambitious AWDI.[21]

Provoked by the audacity of what they perceived as a classic western water grab, a broad coalition of San Luis Valley ranchers, farmers, environmentalists, and other concerned citizens rallied in vehement opposition to AWDI's proposed plan. Newspapers columnists decried "a rape in the making," while local residents, in a commendable display of civic virtue, voted overwhelmingly (8,700 to 136) in favor of a proposal to tax themselves in order to raise funds for a pending legal battle against AWDI, to be spearheaded by the Rio Grande Water Conservation District.[22] Residents of the San Luis Valley believed that AWDI's proposal would result in another Owens Valley, California, where the infamous Los Angeles water grab began in 1905, and feared that their own agricultural economy would be obliterated if plans to deplete the valley's aquifers were allowed to move forward. AWDI officials countered that their proposal would benefit the valley by creating jobs and encouraging new industries.[23]

Former Colorado governor Richard Lamm joined the fray by pointing out the potential benefits of AWDI's proposal, which he asserted was "as good as it's going to get." Lamm also made sure to emphasize that he stood to make no profit from the proposal, even though he had joined AWDI's board of directors for a reported $10,000 a year in compensation, by clarifying that he was "not a shareholder or owner of AWDI. I joined the board because I believe this project is the most environmentally benign way to provide additional water to the Denver area in a cost-effective manner. I also believe it will benefit the people of the San Luis Valley."[24]

The crux of AWDI's claim to the valley's groundwater rested on two major contentions. First, AWDI argued that it could do whatever it pleased with the waters of the Baca Grant, because the water rights associated with the Baca Grant originated in an unrestricted grant from the Spanish Crown in 1821. Since the U.S. government had officially recognized the grant with the 1848 Treaty of Guadalupe Hidalgo ending the Mexican-American War, AWDI argued that neither state nor federal water statutes applied.[25] The second contention proved somewhat more complex and had tremendous implications for the long-term viability of the Great Sand Dunes and the San Luis Valley.

The state of Colorado recognizes two types of groundwater: tributary and nontributary. State water courts define tributary groundwater as hydrologically connected to "flowing streams," meaning that pumping such groundwater would have a direct, measurable effect on surface stream flows. By contrast, nontributary groundwater has no apparent hydrological connection to any surface flow, therefore the pumping of such groundwater would have virtually no effect on nearby rivers or streams.[26] According to the statute that defines nontributary claims, any attempt to secure a nontributary right to groundwater has to prove that groundwater withdrawal would affect "natural surface streams" less than 0.1 percent in one hundred years. Although AWDI originally filed for claims based on both tributary and nontributary rights, it eventually dropped the tributary claim after none of its potential investors would contract to put the water to beneficial use until the actual rights to the water had been legally secured. However, based on its ownership of the Baca Grant and its claim that the grant's groundwater was in fact nontributary, AWDI felt confident that it could withstand any legal opposition.[27] The National Park Service, the Colorado Division of Wildlife, the Rio Grande Water Conservation District, and several other local grassroots organizations including the Friends of the Dunes and Citizens for San Luis Valley Water believed otherwise, joining the overwhelming majority of local residents in fierce opposition to the plan.

Several crucial factors were working in their favor. In 1989 the NPS had secured a Federal Reserved Water Right for the Great Sand Dunes, a designation that protected the natural flow of surface streams within the boundaries of the national monument. Then, in 1991, hydrologists discovered an unequivocal connection between the waters of Sand Creek and the underground aquifers, a development that seriously jeopardized implementation of AWDI's plan. Since the hydrologic connection between Sand Creek and the underground aquifers had been proven, it didn't matter if AWDI filed its claim based on tributary or nontributary water rights; any lowering of the aquifers as a result of pumping by AWDI would compromise the natural flow of Sand Creek, thereby causing injury to a federally protected water right. Furthermore, opponents of AWDI pointed out that Sand Creek is a vital component in the sand transport and recycling system that is crucial to the long-term stability of the Great Sand Dunes, and any threat to the creek's ability to carry sand could arguably pose a threat to the dunes' very existence.[28]

A clear scientific benefit emerged in the midst of AWDI's proposal: the controversial plan to export the valley's groundwater compelled geologists and hydrologists on both sides of the issue to launch unprecedented studies of the precise relationship between the underground aquifers and the Great Sand Dunes in an attempt to determine the potential adverse effects of groundwater pumping. Not surprisingly, the results of their investigations provoked even more controversy. Experts hired by AWDI produced a complex (and highly questionable) computer model, along with supplemental exhibits, expert testimony, and scientific data that indicated few if any adverse impacts from the proposed groundwater pumping. In response, opposition attorneys produced their own data and expert witnesses, including at least one, Adams State College geologist Dion Stewart, who came to the staggering conclusion that the dunes "would dry out like sugar and be carried away by water and wind if groundwater—the 'glue' that holds the dunes together—is removed." Stewart believed that the very groundwater that AWDI wanted to pump to Denver and elsewhere was anchoring the sand from below through a process called "capillary rise." According to Stewart, if AWDI's plan had been approved, "the entire erosion process could take only three generations, the wink of an eye in geological time. It would occur so quickly that people could witness massive destruction of the dunefield in their own lifetimes."[29] Although intrigued by Stewart's theory, which became very popular with the local press, the NPS remained wary about the lack of concrete data proving a definitive connection between groundwater and dune stabilization and decided instead to focus on the proven connection between Sand Creek and the underground aquifer, as well as the potential injury to the Federal Reserved Water Right if AWDI were allowed to pump the valley's liquid wealth. As Superintendent Bill Wellman said at the time, "It's tough to go to court with just a bunch of theories." The proven hydrologic connection between Sand Creek and the underground aquifer thus became the basis for NPS legal arguments in the case against AWDI.[30]

After a five-week trial in the fall of 1991 that had, in the words of presiding District 3 water judge Robert Ogburn, "all the drama of drying paint," Ogburn ruled against AWDI, asserting that the Spanish land grant argument had no validity and that AWDI's other evidence "was not credible" and "distorted fact in its efforts to establish that the water it sought was non-tributary." On the contrary, Ogburn ruled, the water is "tributary beyond a reasonable doubt." Attorneys for AWDI appealed Ogburn's decision, citing among other issues his apparent lack of objectivity, but in early May 1994 the Colorado State Supreme Court nonetheless affirmed the ruling and upheld a motion filed by AWDI opponents to recover nearly $3 million in legal and scientific costs. Stung by the defeat, AWDI continued appealing all the way to the U.S. Supreme Court, which declined to hear the case, effectively rendering the AWDI plan dead in the water.[31]

As for Maurice Strong, he distanced himself from AWDI as soon as opposition to the water-export plan began to organize in earnest, reportedly because he felt that AWDI was no longer interested in using the water in a "socially and environmentally responsible way," as well as a "deep division of interest" between himself and his fellow investors over Strong's unwavering support of his wife Hanne's ambitious attempts to transform portions of the Baca Grant into an interfaith spiritual community for assorted religious organizations, metaphysical groups, and New Age practitioners. Strong ended up selling most of his shares in the company for a reported $2.25 million, which he donated to a charitable foundation, but his name will forever be associated, in the San Luis Valley at least, with the much-despised AWDI. And in the heady days of celebration that followed in the wake of Judge Ogburn's decision, the broad coalition of citizens toiling in the long struggle to protect the valley's water resources rejoiced in the fact that the AWDI trial had finally proven that the disparate landscapes of the Great Sand Dunes and the irrigated San Luis Valley, the Great American Desert and Gilpin's Garden, were, in fact, intimately and inextricably connected.[32]

Expanding the Sandbox

Another positive and enduring development to emerge in the aftermath of Judge Ogburn's initial ruling against AWDI came in September 1993 when Great Sand Dunes National Monument superintendent Bill Wellman and several members of his staff convened a two-day planning session with NPS representatives from a number of regional science, natural, and cultural resource staffs, along with six scientists and land managers from outside the NPS, with the goal of generating a comprehensive resource management plan for the national monument. Scientific inquiry had been ongoing at the Great Sand Dunes for years, with much of the research focused on developing a better understanding of the complex dynamics of the main dunefield, but AWDI's plan to export the valley's water inspired a renewed sense of urgency to the effort. Admittedly, the AWDI proposal had caught the NPS "flat-footed" when it came time to answer specific questions about how pumping the valley's groundwater might adversely affect the dunes, largely due to a lack of accurate scientific data concerning the precise geologic and hydrologic processes at work in the creation and continual replenishment of the Great Sand Dunes. In reality, the monument simply lacked the appropriate scientific and technical staff and equipment to gather such data when AWDI's proposal first came to light in 1986.[33]

By the time the NPS planning session began at the dunes in 1993, the monument boasted a nascent resource management staff that included biologist Fred Bunch and geologist Andrew Valdez, who together with supporting NPS personnel planned a series of investigations specifically designed to gather and monitor useful scientific data about the complex geological and hydrological forces that had shaped and reshaped the Great Sand Dunes over time, information that was critical to the evolving process of formulating a proactive (as opposed to reactive) management plan for this irreplaceable landscape. The resulting Resource Management Strategy document, compiled in 1994 and referred to by Bunch as the "Bill Wellman Blueprint," essentially outlined a five-step process for resource management at the Great Sand Dunes: (1) define the system, (2) understand the system, (3) monitor the system, (4) manage the system, and (5) evaluate actions taken. This pioneering approach, which helped focus scientific inquiry at the dunes and which the NPS hoped to fully implement within five years, profoundly impacted the future of resource management at the monument.[34]

Ever since the official establishment of Great Sand Dunes National Monument in 1932, the NPS had understandably focused its limited resources solely on managing the main dunefield and its immediate vicinity. Given the monument's relatively modest visitation numbers in comparison with more popular NPS units, this concentration of resources was both prudent and necessary; it allowed monument personnel to maximize visitor access and enjoyment, as well as the effectiveness of their meager funding and staffing levels, by improving access roads, constructing the campground and ancillary tourist facilities, and offering interpretive services that described the complex processes responsible for creating the towering sand dunes that are the dominant feature of the park. Considerably less attention was paid to the piñon-juniper-covered foothills and lofty peaks of the Sangre de Cristos that comprise the vital watershed for the dunes, primarily because a large portion of the watershed remained under the jurisdiction of the U.S. Forest Service, while the sprawling sand sheet and sabkha (salt flats) to the west of the main dunefield, which were still largely private property, remained outside the purview of park management and virtually unknown to park visitors. The NPS had essentially been managing the Great Sand Dunes as an island, separate and disconnected from the vast San Luis Valley that surrounded them, for nearly sixty years.

With the implementation of the scientific data collection protocols outlined in the 1994 Resource Management Strategy, the prevailing standards for how to manage the Great Sand Dunes beyond the main dunefield began to evolve as resource management staff attempted to "define the system." A dizzying array of instruments and equipment—including stream gauges, sand traps, monitoring wells, weather stations, and the latest Geographic Information System (GIS) technology—and the data they collected confirmed what U.S. Geological Survey (USGS) scientists had first suspected years earlier when a young USGS geologist named Sarah Andrews, working in conjunction with legendary geologist Ed McKee, conducted extensive mapping surveys of sand deposits at the Great Sand Dunes in the late 1970s. Andrews and her colleagues concluded that the main dunefield was only one of three distinct sand deposits that she labeled "sand provinces"; the other two were the sand sheet and the sabkha. Fifteen years later, additional data not only confirmed the validity of Andrews's classifications but also added the Sangre de Cristo watershed to the mix, which contributes the majority of water to the hydrologic system and also influences the wind patterns responsible for the dunes' tremendous height. As a result, the main dunefield, which had for so long been managed as an isolated, independent geologic anomaly, was instead confirmed as merely one feature (albeit a massive one) of a much larger "aeolian system" composed of four distinct components: the Sangre de Cristo watershed above the dunes, the sand sheet that surrounds the dunes, the mineralized hardpan of the sabkha to the southwest of the dunes, and the main dunefield itself. Perhaps most crucially, the data clearly indicated that this complex, interconnected system of landscapes, spanning nine distinct ecosystems from the alpine tundra of the Sangres to the desertlike floor of the valley, was linked together by a single precious, irreplaceable resource. It seemed at first like a staggering contradiction, but the facts were incontrovertible: the sprawling Great Sand Dunes, home to the tallest dunes in North America, quintessential embodiment of the Great American Desert, clearly, unequivocally depended on water for their very existence.[35]

In an elegant hydrologic cycle that has been continuing for thousands of years, water that falls on the Sangres as rain and snow collects in numerous alpine lakes or trickles its way across the alpine tundra, nourishing the few hardy plants and flowers able to survive in that inhospitable environment, before cascading down the rocky western slopes, through majestic stands of subalpine spruce and fir, past beaver ponds and quaking aspen, through mountain meadows and ponderosa forest teeming with wildlife. Continuing its inexorable journey downward, the water flows among the scattered piñon-juniper groves that dominate the lower foothills of the Sangres, through verdant riparian zones edged with towering cottonwoods, eventually exiting the mountains to mingle with the waters of Medano and Sand Creeks, where the high water table prevents the water from sinking too quickly into the sand, allowing the creeks to continue eroding and transporting grains of sand to the west for the wind to gather and eventually redeposit back on the Great Sand Dunes, just as it has for untold centuries, before finally sinking into the subterranean aquifers, with some of the water eventually emerging from the natural springs on the western edges of the dunefield to nourish the wetlands and vast grasslands of the valley floor. Each component of the system is linked to the others, each component dependent on water for its sustenance and long-term viability. A much larger and more complex Great Sand Dunes ecosystem, one that stretched far beyond the main dunefield, had finally been scientifically proven and officially recognized, along with the realization that disruptions to any single part of the system, such as fluctuations in the water table that might occur from groundwater pumping, could have potentially devastating impacts on the other parts of the system.[36]

Aside from confirming water's critical role in this emerging "big picture" perception of the Great Sand Dunes the enormous accumulation of data initiated by the 1994 Resource Management Strategy also provided answers to

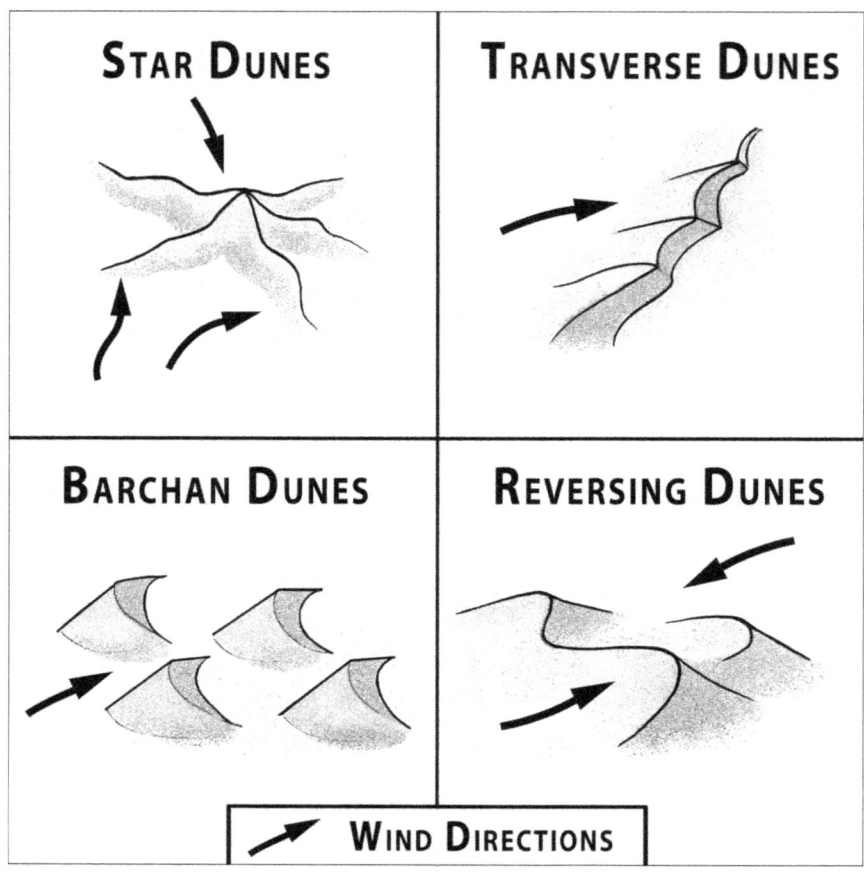

Figure 6.5 Examples of the various dune types that comprise the main dunefield at Great Sand Dunes National Park and Preserve. National Park Service, Great Sand Dunes National Park and Preserve.

several smaller but equally intriguing mysteries. For instance, analysis of samples taken from monitoring wells located near the natural springs on the west side of the main dunefield revealed a complete lack of tritium, a radioactive isotope of hydrogen produced as a result of nuclear testing during the 1950s. Throughout much of the American West, trace amounts of tritium appear in virtually all surface waters originating as precipitation after that time. Since the water flowing from the springs contained no tritium, and since studies have shown that the surface water flowing in Medano and Sand Creeks (most of which comes from rainfall and snowmelt) eventually emerges from the springs, resource management staff concluded that the water had taken at least fifty years—perhaps much longer—to work its way from the creeks to the springs through the dunes' complex hydrological system.[37]

Other instruments and processes, such as drill cores and *optically stimulated luminescence* (OSL) measurements, provided park staff with clues about the approximate age and origin of the sand in the dunes, while field observations, wind measurements, and aerial surveys provided a deeper understanding of the various types of dunes that make up the nearly thirty-square-mile main dunefield. *Reversing dunes* are the most common, caused by the valley's prevailing southwesterly winds, combined with occasional strong "reversing" winds from the northeast. The northeast corner of the dunefield, constantly subjected to swirling winds from every direction, contains a large complex of sand dunes with three or more ridges known as *star dunes*. Along the southern and eastern edges of the dunefield, where southerly winds are common, *barchan* and *transverse* dunes occur, while *parabolic* dunes tend to dominate the sand sheet, where intermittent clumps of surface vegetation and the prevailing wind direction combine to produce dunes that migrate toward the main dunefield.[38]

Park staff also compiled data from stream gauges and monitoring wells to predict the potential negative impacts of groundwater pumping, not only on the main dunefield but also on the sand sheet and sabkha, two of the four primary components of the much larger Great Sand Dunes ecosystem. The sabkha, located in a broad depression to the southwest of the dunes where groundwater is at or just below the surface, is an expansive deposit of sand hardened into a thick crust by alkaline minerals left behind by evaporating water. Early settlers in the San Luis Valley gathered this mineralized crust for use in baking and as a laundry detergent, and at some point during the nineteenth century a small community called Soda City even appeared, complete with residents who collected and pressed these minerals into blocks that were then shipped by rail to distant markets. Soda City itself has long since vanished, but the sabkha endures as evidence of the complex hydrological forces at work both above and beneath the valley floor. While the sabkha contributes very little (if any) sand to the main dunes, it does remain a potential source and continues to influence the movement of sand within the larger Great Sand Dunes ecosystem; it also supports a variety of plant and animal life in the wetlands that appear and disappear with fluctuating groundwater levels.[39]

Situated between the sabkha and the main dunefield is the sprawling sand sheet, the largest component of the Great Sand Dunes ecosystem and the primary source of sand in the main dunefield. Like the sabkha, the sand sheet depends on groundwater for its very existence, but instead of barren stretches of alkaline salt flats, the sand sheet is anchored by scattered grasses and intermittent shrubs and punctuated by parabolic dunes that surround the main dunefield on three sides, as well as wetlands that provide critical habitat for elk, pronghorn, bison, burrowing owls, and short-horned lizards. Although the sand sheet covers an enormous area, the vegetative cover is actually rather sparse and therefore easily impacted by grazing, fire, and other natural and human-caused disturbance. The vegetation on the sand sheet is also very sensitive to changes in precipitation and groundwater levels, as are the wetlands, rendering them especially vulnerable to the dramatic fluctuations that most experts predicted would have occurred if AWDI had been allowed to implement its plan for pumping vast quantities of groundwater from beneath the valley floor.[40]

Taken together, the unprecedented flood of scientific data generated as a result of the 1994 Resource Management Strategy and its implementation of the "Bill Wellman Blueprint" provided more information about the dunes and their immediate environs than had been accumulated in the nearly two centuries since Zebulon Pike first described them as "a sea in a storm" in 1807. Analysis of this data by NPS personnel revealed a landscape of surprising complexity, composed primarily of four distinct regions (watershed, active dunefield, sand sheet, and sabkha) that were interconnected and interactive with one another, each dependent on a delicately balanced set of geologic and hydrologic forces that had combined in a much larger aeolian system to create and sustain the Great Sand Dunes over thousands, perhaps tens or even hundreds of thousands of years. Crucial to the entire process is the water that continually courses through the system like blood flowing through a living body, both in surface streams and in the subterranean aquifers beneath the valley floor. For the National Park Service staff charged with protecting this unique and strangely beguiling landscape in its entirety, now equipped with the knowledge that the larger Great Sand Dunes ecosystem essentially extended from the crest of the Sangres to the farthest reaches of the sabkha on the valley floor, any disruption to the water or to the natural processes that occurred in the sand deposits within the system that could potentially jeopardize the continued existence of the dunes was simply unacceptable, a position that put them on a direct collision course with the next seemingly inevitable scheme to tap the valley's groundwater for profit.

AWDI Redux

In late 1992, in one of his last acts before leaving office, Colorado senator Tim Wirth managed to insert a few sentences into an omnibus water bill, later signed by President George H. W. Bush, that effectively prohibited any development in the San Luis Valley that could adversely impact any surface or groundwater quality or quantity, any wildlife or wildlife refuges, Great Sand Dunes National Monument, or the Closed Basin Project, or increase the costs or impede the delivery of water by the state of Colorado pursuant to the Rio Grande Compact.[41] Despite these sweeping restrictions on removing water from the San Luis Valley, as well as the incontrovertible evidence establishing a definitive connection between the valley's groundwater and the Great Sand Dunes, attempts to exploit the subterranean aquifers beneath the Baca Ranch continued in the aftermath of the AWDI trial. In 1994, after the Colorado Supreme Court upheld Judge Ogburn's ruling against AWDI, company shareholders decided to put some 97,000 acres of the Baca on the market. Among the interested buyers was The Nature Conservancy, one of the world's leading conservation groups, which had identified the Great Sand Dunes, along with the Baca Ranch and the adjacent Medano-Zapata Ranch, as biological "hotspots" worthy of protection in the early 1990s.[42]

Another potential buyer, local rancher Gary Boyce, had a more personal connection to the property. Born in Del Norte and raised in Monte Vista, Boyce had deep roots in the San Luis Valley and claimed that by the time he was eight years old he had ridden his horse "over every trail on the [Baca] ranch." He eventually left the valley to seek his fortune, which he apparently found in abundance with thoroughbred horses and various oil and real estate deals, returning to the valley in 1982 as a wealthy man with the means to begin buying up portions of a large ranch adjacent to the Baca, where his grandfather ran sheep when Boyce was a young boy.[43] In 1990, ostensibly opposed to AWDI's water-export plan, Boyce began publishing a short-lived newspaper called *The Needles* that mercilessly skewered AWDI and its backers, but the paper lasted only seven issues before Boyce inexplicably pulled the plug. Then, in the spring of 1992, Boyce formed an enterprise called Stockman's Water Company and publicly announced plans for developing the valley's water resources. Proof of Boyce's seriousness came when he hired high-profile water expert Jeris Danielson, who had recently been fired as Colorado's state engineer for suggesting that Colorado sell some of its water to California, and together they began formulating a strategy to unify valley ranchers toward the common goal of realizing a profit from their water rights.[44]

For years, Boyce had apparently been resenting farmers (more specifically, potato farmers) in the southern reaches of the San Luis Valley for taking what he thought was more than their fair share of water, both from the Rio Grande and from the underground aquifers. The Closed Basin Project, which he felt unfairly favored the farmers in the south at the expense of the ranchers in the north, was also subject to his ire. In fact, Boyce believed that virtually all of the "plumbing" in the valley—the canals, the ditches, the laterals, even the reservoirs—was manifestly unjust because engineers had installed "most of the spigots in Rio Grande County, and most of the drains in Saguache County." To redress this perceived inequity, Boyce proposed offering ranchers an option to market some of their water rights through Stockman's Water Company, which would then find buyers for the water for what he deemed a "token fee."[45]

Precisely how many local ranchers took Boyce up on his proposal is unclear, but in 1994, with AWDI apparently anxious to unload the property, Boyce approached senior executives at Farallon Capital Management seeking help in purchasing the Baca Ranch. Farallon, a large, San Francisco–based hedge fund that manages capital for high-net-worth individuals and institutions (primarily college endowments and foundations), expressed interest in the potential for profit from the Baca. To finance the purchase, Farallon subsequently formed Vaca Partners as a fifty-fifty limited partnership with Yale University, which had a reported $500 million of its multi-billion-dollar endowment fund invested with Farallon. Yale's involvement in the venture remained out of the public eye, at least for the time being. Vaca Partners in turn entered into an agreement with Boyce's Stockman's Water Company and another investor, Peter Hornick of New York, to finance the Cabeza de Vaca Land and Cattle Company, LLC, which successfully outbid The Nature Conservancy and purchased the Baca Ranch from AWDI in 1995 for a reported price somewhere in the neighborhood of $15 million. Vaca Partners maintained a 50 percent controlling interest in the property through its investment in Cabeza de Vaca Land and Cattle.[46]

In 1996, Boyce announced his own speculative proposal to tap the groundwater beneath the Baca Ranch, which San Luis Valley residents immediately dubbed "Son of AWDI" and "AWDI Two." Boyce himself preferred to call it the "No Dam Water Project." Claiming that his project would "usher in an era where people's water needs can be met without social and environmental trade-offs," Boyce emphasized a few important distinctions between his Stockman's Water Company plan and the recently defeated AWDI proposal.[47] Unlike AWDI, Stockman's would go to water court with contracts from Front Range water districts in hand, which Boyce said would fulfill the court's requirement of a proven beneficial use for the water. Furthermore, he emphasized that the No Dam Water Project would only withdraw 100,000 to 150,000 acre-feet of groundwater annually, as opposed to AWDI's 200,000 acre-feet. Boyce guaranteed the project would not deplete the aquifers, claiming that the proposal would "cause absolutely no harm to the San Luis Valley and [would] make all the difference in the world to the state of Colorado."[48] In addition, Boyce learned from AWDI's mistakes and promised to restore fifty miles of riparian habitat in the Closed Basin, establish a 50,000-acre wildlife reservation and a 10,000-acre mountain wilderness area that included two 14,000-foot peaks (Crestone and Kit Carson), as well as protect 2,000 acres in the northern portion of the Great Sand Dunes. The result would be a permanent buffer for wilderness in the Sangre de Cristos and Great Sand Dunes National Monument. Boyce also pledged to use the Baca Grant's surface water rights as liquid collateral to establish a $3 million trust fund as a guarantee that his wells would not adversely affect his neighbors' water supply. Finally, Boyce promised to stop farming a large portion of the grant in order to allow an additional 25,000 acre-feet of water to help recharge the underground water supply.[49]

The audacious plan, expertly packaged and presented, nevertheless raised intense suspicion among residents and business owners in the San Luis Valley. Many resented Boyce's implication that if he were denied his right to export water from the Baca, he would have no choice but to sell the land to less benevolent developers, who might show little or no concern for the environmental health of the ranch or its neighbors. Others questioned Jeris Danielson's motives. As state engineer, Danielson had claimed that AWDI's proposal would hurt the Closed Basin Project. Later, working for Boyce in Stockman's Water Company,

Danielson asserted that water from the Baca's surface flows would be "more than enough to augment any drawdown, any injury."[50] Because Boyce shrewdly avoided filing an official application for the groundwater under the Baca, opponents had no way to verify such claims, and the precise details of his proposal, as well as their potential impact on the valley's water resources, remained elusive. For Ralph Curtis, director of the Rio Grande Water Conservation District at the time, these "unknowns" were what made Boyce's plan so worrisome. "At the moment, everything is working pretty well in the valley, at least in water terms. And it could be that Boyce won't hurt anything. We just don't know," Curtis explained. "It would be fair to say that a lot of the opposition to Boyce comes from plain old fear of the unknown, and I share those fears. In ways, the AWDI fight was easier, because we knew what they proposed to do. Boyce hasn't filed anything official, so we don't really know what we might be up against."[51]

Boyce and his backers further muddied the waters when they spent nearly $500,000 for paid staffers and circulators to canvass the state gathering enough valid signatures for petitions seeking to place two controversial initiatives on the 1998 Colorado statewide ballot, plus an additional $400,000 for political advertising in support of the campaign.[52] The first initiative, most likely inspired by Boyce's long-held belief that potato farmers and others in the southern end of the valley were taking more than their fair share of water, would have required that certain agricultural, industrial, and municipal water users install flow meters at their own expense to measure precisely how much water they were using.[53] According to Christine Canaly, the local organizer who led the nonprofit Citizens for San Luis Valley Water in the fight against AWDI, the metering requirement was "written so that if a water meter clogs—which is frequent, considering the amount of sand that pumps can bring up here—then the farmer has to shut it down immediately until he gets a new meter."[54] Such delays could be potentially devastating to farmers already struggling with the valley's notoriously short growing season. Canaly concluded that the proposed meter initiative was nothing more than a thinly veiled attack on the local farmers who opposed Boyce's water export plan, noting that "if he [Boyce] was really interested in measuring water usage, they could measure electrical power consumption, the way they do in the lower Arkansas Valley. There's a very close correlation between electrical consumption and the amount of water pumped. Requiring water meters on the wells is just a way to make life harder for farmers here."[55]

The second initiative would have required the Rio Grande Water Conservation District to pay—retroactively and with interest—for any water pumped from beneath state trust lands in the valley for use in the Closed Basin Project, a proposal that "would pretty much bankrupt the district," according to Canaly.[56] State trust lands cover nearly 3 million acres of surface area in Colorado, including over 146,500 acres in the San Luis Valley, all of it overseen by the State Land Board (SLB).[57] The SLB is required by the state constitution to hold assets from the trust lands in a "perpetual, inter-generational public trust for the support of public schools," which are the beneficiary of roughly 93 percent of the revenue generated from trust lands, the vast majority of which comes from mineral and grazing leases, while the remaining 7 percent is slated for prisons, state parks, and public buildings.[58] Staffers gathering signatures for Boyce's petitions took full advantage of the SLB's fiduciary duties by telling prospective signers that funding for Colorado public schools would increase dramatically if the initiatives passed, while opponents of Boyce's water export plan concluded that the initiatives were an obvious attempt to break the collective financial backs of the valley's farmers and powerful water establishment, including the Rio Grande Water Conservation District, which would have virtually no chance of successfully opposing Boyce's plan in District Water Court if its coffers were empty. Boyce denied that he was trying to bankrupt the district but admitted, "If that was a result of this law, I wouldn't shed a tear."[59]

Despite widespread and vocal resistance, Boyce and his partners successfully placed both initiatives, officially labeled Amendments 15 (water metering) and 16 (state trust lands), on the 1998 ballot, with a statewide vote scheduled for November 3. For San Luis Valley native Lewis Entz, a longtime Boyce foe who at the time was a state representative (R-Hooper), the very fact that these measures were on the ballot in the first place was "the worst abuse of the ballot-initiatives process" that he'd ever seen, the first time "an individual company has had enough money to put issues on the ballot to destroy the economy of the San Luis Valley." Virtually every newspaper in the state came out in opposition, as did a lengthy list of environmental groups and agricultural organizations and a bipartisan alliance of Colorado politicians.[60] Closer to home, Citizens for San Luis Valley Water once again rallied the local troops, raising over $1 million through donations from banks, businesses, private citizens, and the aforementioned environmental and agricultural groups, while a tight-knit network of valley residents began reaching out across the state, contacting distant friends and relatives and urging them to vote against the amendments.[61] Citizens for San Luis Valley Water program director Christine Canaly remembered it as "an amazing coalition of people" who collectively had decided that "this was way bigger than the rest of us individually, so it was really important to come together as a community. What Boyce thought he could do was to manipulate the state of Colorado to vote against the San Luis Valley."[62] In a crushing blow for Boyce and his financial partners, Colorado voters overwhelmingly rejected both amendments by a whopping three-to-one margin, with the six counties of the San Luis Valley voting 96 percent against Amendment 15 and 97 percent against Amendment 16, a resounding defeat that Boyce blamed on uninformed voters.[63] His opponents, among them Karla Shriver, coordinator of Citizens for Colorado's Water, disagreed, declaring that the "voters of the whole state of Colorado have sent a very strong message to Stockman's Water Company that they saw through Stockman's plan to try and dismantle the intricate water management system in the San Luis Valley."[64] Representative Lewis Entz offered a more provincial explanation: "This just shows you that when people come down here and try to start trouble, we work them over."[65]

Suitably worked over, his dream of tapping the untold liquid wealth beneath his beloved Baca Ranch all but dried up, Boyce nonetheless insisted that the water wars of the San Luis Valley were far from over, claiming, "I think it's just the beginning, and I don't see how I'm going to put the genie back in the bottle." But the latest battle over water had also clearly taken a toll on Boyce and his Stockman's Water Company: Boyce admitted, "Frankly, I'm tired of this political stuff, and I'm glad the election is over. It's time to get back to business."[66] As was the case with AWDI, Boyce's unsuccessful attempt to disrupt the intricate and long-established water management and allocation systems of the upper Rio Grande watershed had galvanized a spirit of community and common purpose among the citizens and businesses of the San Luis Valley, who united in solid opposition to the plan. Yet the triumph over Stockman's, like that over AWDI, had come at considerable cost: millions of dollars donated and spent on expensive attorneys and costly advertising; countless hours consumed by meetings, phone calls, e-mails, and faxes to every corner of the state urging voters to reject Boyce's plan; acrimony between opponents of the plan and those few brave souls in the valley who supported it; and years of anxiety and worry over the fate of the valley's precious water resources, which was perhaps the unkindest cut of all. Despite the hard-fought victories, San Luis Valley residents had ultimately failed to win any guarantees that the next battle over the region's water wouldn't simply reignite next year, or the year after that. As long as the sprawling Baca Ranch remained in private hands, the threat remained.

The National Park Service, which had been intimately involved in both the AWDI and Stockman's water controversies, likewise had no guarantees against future threats from water developers and remained keenly aware that the historic ranch on its northwestern boundary would continue to be a problem for resource management at the dunes, especially if the next in a long line of private owners simply continued previous owners' efforts to turn a profit from the Baca's water rights. The contentious battles over the water resources of the San Luis Valley did have at least one positive benefit for the National Park Service, however. Forced to defend its water rights in court against AWDI, the NPS had launched unprecedented scientific investigations into the biology, geology, history, and hydrology of the tallest sand dunes in North America, as well as the landscapes and ecosystems that surround them. In the twelve years between AWDI's ambitious proposal in 1986 and the defeat of Stockman's Water Company at the ballot box in 1998, the NPS accumulated more useful information about the Great Sand Dunes than had been gathered during the previous two centuries. For the scientists and resource management staff tasked with analyzing and applying these data to the NPS mandate of protecting and preserving the Great Sand Dunes in perpetuity, several definable goals emerged. First, since science had proven unequivocally that the very existence of the Great Sand

Dunes depended on water, critical safeguards already in place for the surface water supplies within the national monument had to be maintained. Second, because the data indicated that the main dunefield was only one component of a much larger aeolian system, the other three components of that system—the watershed (which already enjoyed protected status as wilderness under USFS management), the sand sheet, and the sabkha—would require some additional degree of federal, state, or local protection to ensure the long-term viability of the Great Sand Dunes. And third, to guarantee that the dunes were never again jeopardized by water miners targeting the aquifers beneath the valley floor, the vast Baca Ranch simply had to be eliminated as a future water development and exportation threat. How these goals were accomplished is but the latest chapter in the compelling history of the Great Sand Dunes, a remarkable tale involving political maneuvering at the highest levels, cooperation and compromise among the strangest of bedfellows, and an unlikely series of negotiations and land deals that culminated with tens of thousands of acres of what had once been private property being transferred into the hands of the federal government. In the process, the protective boundaries surrounding the sprawling sea of sand that has been beguiling humans for well over 10,000 years were finally expanded to include the mountains above, the waters below, and the sand deposits to the west and southwest that now comprise the natural and historic landscapes of America's fifty-eighth national park.

SEVEN | Racing for a National Park

National parks are the best idea we ever had. Absolutely American, absolutely democratic, they reflect us at our best rather than our worst.

Wallace Stegner, 1983

IN EARLY JULY 1994, less than two months after the Colorado Supreme Court issued its final ruling against AWDI, Great Sand Dunes National Monument superintendent Bill Wellman and U.S. Representative Scott McInnis (R-Colo.) stood together on the precipice of a tall sand dune overlooking the confluence of Medano and Castle Creeks, where historians believe Zebulon Pike and his men spent a cold night in late January 1807. At a mutually agreed-upon moment ("Ready, set, go!"), the two men hurled themselves off the knife-edge lip of the dune and began bounding down the leeward side. Stride for stride they descended, triggering small avalanches of sand as they rapidly picked up speed, until Wellman gradually opened a slight lead, his long, graceful gait a clear indication of his extensive experience in dashing down the Great Sand Dunes. The good-natured race was a close one, but by the time they crossed an imaginary finish line, laughing and exhilarated from the exertion, Superintendent Wellman had triumphed by "2½ lengths, a sand-slide victory." For Scott McInnis, who was visiting the dunes with his family that day at the behest of his good friend Bob Zimmerman, an Alamosa County commissioner, the race with Wellman symbolized the beginning of his personal involvement in a unique legislative and political odyssey that culminated with the official creation of Great Sand Dunes National Park and Preserve.[1]

Elected to Congress in 1992 as the representative for Colorado's enormous Third Congressional District, which includes the entire San Luis Valley, McInnis was well aware of the valley's economic struggles as one of the poorest rural

Figure 7.1 Colorado congressman Scott McInnis (*left*) and Great Sand Dunes National Monument superintendent Bill Wellman race down a sand dune above Castle Creek in 1994. Fred Bunch, Great Sand Dunes National Park and Preserve.

areas of the state. He had spent a decade in the Colorado General Assembly, often relying on Commissioner Zimmerman to act as his "eyes and ears" for information about the latest developments and general mood of valley residents. For his part, Zimmerman remembered first discussing the idea of upgrading the national monument to a national park with McInnis, who at that time was a state representative, as early as 1989, primarily as a means of stimulating the local economy. As a county commissioner understandably concerned with the health and welfare of his constituents, Zimmerman reasoned that the enhanced status and greater recognition associated with a national park might translate into an increase in tourist visitation, and therefore tourist dollars, to the San Luis Valley. Roughly five years later, during their outing on the dunes near Castle Creek, Zimmerman reiterated his conviction to McInnis that the Great Sand Dunes "ought to be a national park."[2]

Although McInnis and Zimmerman later received considerable credit in the local and national press for being among the very first people to advocate for upgrading Great Sand Dunes National Monument to national park status, both men consistently downplayed their seminal role in the process, preferring instead to spread the accolades to a wide variety of worthy individuals, organizations, and communities.[3] In fact, precisely who first dreamed up the idea is a matter of conjecture, although the National Park Service has a long history of overseeing the transformation of national monuments into national parks. Indeed, some of the most renowned and scenic national parks in America—including Acadia,

Bryce Canyon, Joshua Tree, Olympic, and Zion—originally began as national monuments. Even the awe-inspiring Grand Canyon, among the most famous national parks on the entire planet, went through several different designations and levels of protection, including forest reserve, game preserve, and national monument, before achieving its current status as a national park in 1919. Such changes to National Park Service units do not happen easily or randomly. Boundary expansions and upgrades in the official designation of federal lands typically require an act of Congress and often years, even decades, of political wrangling before consensus can be reached and boundaries actually altered on the ground.[4]

National monuments and national parks differ primarily in the manner in which they are established, as well as the reasons for their initial creation. National monuments are most often created by presidential proclamation granted under authority of the Antiquities Act of 1906, although Congress occasionally creates national monuments on its own. Congaree Swamp in South Carolina and Petroglyph in New Mexico are two such examples. Monuments typically preserve a single, unique, and nationally significant feature, whether historic, archaeological, scientific, or natural, and therefore tend to be smaller in size (although this is not always the case) and often lack the diversity of resources that usually characterize national parks. Management of national monuments also occasionally falls under the jurisdiction of federal agencies other than the NPS, including the U.S. Forest Service, Bureau of Reclamation, and U.S. Fish and Wildlife Service (USFWS). National parks, on the other hand, can only be created by an act of Congress, are managed solely by the National Park Service, and are intended to protect outstanding scenic features, historic sites, or natural phenomena for their inspirational, educational, and recreational values. Hence national parks tend to be larger than monuments (although, again, this is not always the case) and usually contain a wider diversity of resources. National parks also loom much larger in the public imagination and therefore tend to attract more visitors (and thus typically more funding) than national monuments.[5]

The U.S. Congress has historically upgraded the status of national monuments to national parks for a variety of reasons, including protecting vital wildlife habitat, prohibiting mineral or other resource extraction, preserving desirable scenic views and cultural resources such as archaeological sites, or protecting such resources from the potentially harmful effects of commercial or industrial development outside existing park boundaries. The economic health of gateway and other communities surrounding national parks and monuments, however, is not among the criteria for changing or upgrading designations of NPS units, despite the fact that the presence of national parks and monuments definitely impacts the local economies of these communities. A study published in 2001 by the Colorado State University Department of Agriculture and Resource Economics examined the economic impact of redesignating eight national monuments as national parks, including Channel Islands, Great Basin, Joshua Tree,

Death Valley, and Great Sand Dunes, and concluded that such changes in designation "may imply (real or perceived) differences in the availability of services, in promotional expenditures by the National Park Service, in allowable land uses, or in the uniquely attractive features of the site. Therefore, these changes in public lands designations may result in greater visitation to a given site."[6]

For Great Sand Dunes National Monument specifically, the study estimated an annual increase of roughly 24,000 visitors after redesignation as a national park, with a commensurate increase of approximately $2.3 million in additional economic activity in the San Luis Valley.[7] For the National Park Service, whose mandate for the nation's public lands is "to conserve the scenery and the natural and historic objects and the wild life therein," such potential economic impacts are essentially beyond its realm of influence and therefore have little or no bearing on any decision to redesignate a particular monument to a national park. Yet for local communities that largely depend on tourism for their economic well-being, such redesignation can be extremely significant, as was the case in the San Luis Valley, where supporters of the national park proposal used the potential for additional revenue from the expected increase in tourism to generate widespread community support.[8]

The NPS has also occasionally opposed transforming monuments into parks, as was the case in 2009 with Pinnacles National Monument, which the NPS felt lacked "the full range of resources usually found in national parks."[9] In 1994, despite the best intentions of Bob Zimmerman, Scott McInnis, and other interested parties, a similar charge could have been leveled at Great Sand Dunes National Monument. Beyond the main dunefield and adjacent areas of archaeological interest, which together totaled less than 40,000 acres, the monument simply lacked a sufficiently diverse array of resources necessary to qualify for national park status, even though an abundance of scientific data had proven that a much larger aeolian system was responsible for the dunes' existence—a system that encompassed a broad range of ecosystems stretching from the mineralized hardpan of the sabkha to the crest of the Sangre de Cristos. Compounding the problem was the fact that of the four main components of that aeolian system (main dunefield, sand sheet, sabkha, and watershed), large portions of two—the sabkha and the sand sheet—belonged to the privately owned Baca and Medano-Zapata ranches, while management of much of a third component—the watershed—remained under the jurisdiction of the U.S. Forest Service. Even the recent threat posed by AWDI to the groundwater under the dunes was not enough, by itself, to justify upgrading the monument. When the Farallon-financed Cabeza de Vaca Land and Cattle Company outbid The Nature Conservancy and purchased the Baca Ranch from AWDI in 1995, then announced its own water development scheme, the brief window of opportunity to create a new national park had seemingly closed, perhaps permanently. It would take several years and an extraordinary series of events and determined individuals to open it again.

Unlike the original campaign launched in 1930 by the PEO Sisterhood to establish Great Sand Dunes National Monument, the effort to upgrade the monument to national park status required much more than convincing a few key individuals of the merits of such a proposal. Granted, the PEO had to rally a number of prominent local and national politicians to support its cause, and Roger Toll's 1931 report needed to persuade NPS director Horace Albright that the Great Sand Dunes were worthy of federal protection. Albright then had to convince his superior, Interior Secretary Ray Lyman Wilbur, that the dunes deserved monument status. Finally, President Herbert Hoover had to agree with Wilbur's recommendation. Hoover indicated his assent in March 1932 by signing the presidential proclamation creating Great Sand Dunes National Monument. The entire process had taken barely eighteen months, and most of the key decisions were made by individuals who had never seen the remote, sparsely populated San Luis Valley, much less the Great Sand Dunes.

By contrast, expanding the boundaries of the existing national monument and creating Great Sand Dunes National Park would require the approval of both houses of Congress, as well as the signature of the president, not to mention the full support of the communities and businesses of the San Luis Valley, which probably could have killed the proposal at any time if local citizens had suspected ulterior motives on behalf of any of the parties involved, especially when it came to threatening the valley's water resources. The well-publicized AWDI battle had clearly demonstrated the enormous potential for profit that existed in the water of the San Luis Valley, water that was worth hundreds of millions if not billions of dollars on the open market. Convincing valley residents that the federal government was the best steward for such a valuable commodity could prove difficult, if not impossible, in a region of the arid American West where the old adage "Whiskey is for drinking, and water is for fighting" had many strong adherents. In addition, acquisition of key parcels of private land adjacent to the monument, specifically the Medano-Zapata and Baca Ranches, was absolutely essential for the protection of the entire aeolian system, yet neither parcel was for sale. Like an enormous jigsaw puzzle, all of the key pieces—bipartisan political cooperation both locally and in Washington, D.C., widespread community support in the San Luis Valley, and the successful transfer of private land into public hands—had to be in place before the larger goal of a national park could be realized. Fortunately for the future of the Great Sand Dunes, another piece of that puzzle appeared in 1998 with the arrival of a new superintendent.

Steve Chaney began his career with the National Park Service in 1975, serving with various park units, including Buffalo National River in Arkansas and Mammoth Cave National Park, before being promoted in 1990 to chief of natural resources for the Rocky Mountain region, headquartered in Denver. In that capacity, Chaney had worked with Great Sand Dunes superintendent Bill Wellman during the height of the AWDI battle and was well versed in both the intricate hydrology of the dunes and the internal bureaucracy of the NPS, earning a reputation for knowing how to get things done.[10] When Wellman accepted the position of superintendent at Organ Pipe Cactus National Monument in 1997, Chaney took over the leadership position at the dunes, effective January 18, 1998, and expressed his admiration for Wellman's tenure, noting that "Bill Wellman did a heck of a job protecting the park's resources. If I can maintain the park as well as my predecessor, I will consider my work a success. I knew Bill Wellman and the kind of program he ran here, and I think this is an operation that has been well-run and is running smoothly. I violated one of my tenets about following a superintendent who's done an outstanding job."[11]

With his departure, Superintendent Wellman left behind the groundbreaking 1994 Resource Management Strategy, as well as an experienced staff that had accumulated unprecedented amounts of scientific data while successfully navigating its way through the AWDI controversy. On arriving at his new post, Chaney scheduled a series of briefings with senior staff on the latest developments at the monument. One such meeting was with NPS geologist Andrew Valdez, who had spent years gathering and analyzing information about the Great Sand Dunes and recalled that whenever he gave talks to college groups, "I'd say that our current park boundary goes around the main dunefield, but if we were designing this park today, we would extend it to include these other areas. At about that time we got a new superintendent named Steve Chaney, and so when he first showed up, I gave him that same talk."[12]

Valdez emphasized to Chaney that the main dunefield was "just the tip of a great sandy iceberg," merely one part of a much larger aeolian system that also included the sand sheet, the sabkha, and the Sangre de Cristo watershed, and suggested that the best way to protect the entire system was to expand the boundaries of the monument. Intrigued, Chaney requested additional information from Valdez in order to refine the proposal, which he then took to NPS regional director John Cook, who expressed approval. Valdez later noted that this was certainly not the first time that an expansion of the monument boundaries had been proposed, but it may have been the first time that such an idea actually moved beyond a "conceptual" phase:

> We had this proposal to expand our boundary, but there was some real substance behind it. Zimmerman had talked to McInnis before, but it was all just a concept, and lots of people have had that idea, like park rangers who wanted the park to go from Mount Blanca to Crestone, but these were just concepts, like, "wouldn't it be nice if this happened?" Based on the physical system, we came up with what I thought was a real legitimate reason to expand, more than just a concept, and therefore it was very defendable, and there was definitely a lot of political interest in it because of the water battle and the water issues we were facing. . . . The whole basis of this was to protect the aeolian and hydrological system.[13]

With a new superintendent at Great Sand Dunes and a regional NPS director fully briefed and supportive of the idea, as well as solid scientific evidence for justification, the intriguing proposal to expand the boundaries of Great Sand Dunes National Monument began to seem less like a dream and more like a viable option, especially after the Stockman's Water Company initiatives lost badly in the Colorado general election of November 1998. Several sizable hurdles still needed to be cleared, however, before the idea could be fully realized, including the formidable prospect of preserving an enormous parcel of historic, biologically rich, yet still very much privately owned land located on the very doorstep of the monument.

Medano-Zapata

Long considered among the finest ranchland in the San Luis Valley, if not the entire American West, the windblown expanse of the Medano-Zapata Ranch consists of roughly 100,000 acres of land spread across several different parcels located immediately south and west of the Great Sand Dunes. Although often referred to as a single unit, the Medano-Zapata historically consisted of at least two separate and distinct ranches—the Medano to the north of present-day County Lane 6N, and the Zapato (later changed in the twentieth century to Zapata) to the south—as well as a number of smaller parcels that a succession of owners consolidated over the years to create one of the oldest continuously operated ranches in the valley. Blessed with an abundance of wildlife and epic scenery, the Medano-Zapata has been called a "Rocky Mountain Serengeti," harboring over five hundred species of flora and fauna including bison, elk, pronghorn, mule deer, coyote, and a plethora of migratory waterfowl coexisting in a variety of habitats ranging from verdant meadows to stark shrublands, drifting sand dunes, and lush wetlands, all set against the staggering backdrop of the Great Sand Dunes and the majestic Sangre de Cristos. The ranch also contains large portions of the sand sheet and sabkha identified by the National Park Service as crucial components of the larger Great Sand Dunes aeolian system. The incredible ecological diversity of the property inspired The Nature Conservancy to identify the Medano-Zapata, along with the Baca Ranch and the Great Sand Dunes, as a globally significant biological "hotspot" in the early 1990s.[14]

The Medano-Zapata also lies at the very epicenter of human history in the San Luis Valley. Several of the most significant Clovis and Folsom archaeological sites in North America, most notably Stewart's Cattle Guard, are located on or near ranch property, while a variety of Native American cultures, including the Pueblo, Navajo, Ute, and Apache, were intimately familiar with the area, where they hunted game, collected feathers or sand for ceremonial purposes, or mined the nearby mountains for gold and turquoise. Spanish explorers made repeated references to the region's abundant wildlife and wetlands, and in January 1807, Zebulon Pike and his "dam'd set of rascals" departed the Great Sand Dunes and made camp in a large copse of cottonwood trees near Zapata Creek, somewhere in the vicinity of the present-day Zapata Ranch headquarters.[15]

The modern history of the Medano-Zapata dates to the arrival of Anglo-American cattle ranchers in the region during the post–Civil War period, many of them searching for conditions similar to those described by government surveyor Ferdinand V. Hayden during his 1868 trek through the San Luis Valley:

> As a stock-growing region, it is evident that this district could not be surpassed. The purity and dryness of the atmosphere, and the absence of deep snows permit the rich grasses to dry up gradually in August, and retain all their nutritious matter; and cattle, horses, and sheep thrive all through the winter without special care. . . . I am confident that the time is not far distant when some of the choicest stock on this continent will be raised in this valley.[16]

Such glowing praise proved irresistible to many prospective ranchers, including two brothers from Ohio, William W. and Valentine B. Dickey, who came to the San Luis Valley in the early 1870s and began consolidating smaller landholdings in the area, including the future site of the Medano Ranch headquarters, which had originally been homesteaded by Edward Hull in 1875 and which the Dickey Brothers purchased in 1876 for $2,000. In April 1877, Valentine Dickey filed his own homestead claim for 160 acres just west of the old Hull homestead, and in June of that year the "Medano Springs" post office was established with William Dickey as postmaster, the first documented indication that the Dickey Brothers were using the site as a headquarters for their ranch holdings. Their operation reportedly included driving large herds of cattle to the valley from Texas, fattening them on the region's plentiful grasses, and then selling the beef to markets such as the booming mining town of Leadville. The brothers became veritable "cattle kings," running as many as 20,000 head of cattle on 9,000 acres of deeded land, some of which was acquired by purchasing "the land titles of poverty-stricken Mexican settlers in the Zapata district," as well as roughly 90,000 acres of leased state land.[17]

After enjoying considerable success in the cattle business, the Dickey brothers sold their landholdings in 1882 to New Yorkers William Wells Durkee, son of Eugene Durkee, who had founded the Durkee spice empire in Buffalo in 1850, and Niel [sic] G. Adee, who by 1878 was reportedly operating "the handsomest ranch in the entire valley" on Zapato Creek near the base of the Sierra Blanca, a location that indicates his spread was most likely what is now known as the Zapata Ranch. With the purchase of the Dickey properties, Adee and Durkee consolidated the Medano and Zapata ranching operations and further increased their holdings to approximately "130,000 acres under fence, 90,000 acres being in one field," with one stretch of their fencing running an astonishing nine miles in one string.[18] Adee and Durkee, who formed the Medano Springs Land and

Cattle Company in 1886 to operate the ranch, apparently had few qualms about using fraud and threats of violence to achieve their aims: at least one historic source claimed they used "sham homestead claims and intimidation to drive Hispanic settlers from the vicinity in order to further expand the holdings of the ranch." Their partnership continued to prosper until March 1887, when Adee committed suicide for reasons that remain unclear.[19]

Undaunted, Durkee continued ranching operations after his partner's death, purchasing additional land, improving infrastructure, and continuing to invest in cattle, horses, and mules, but by the early 1890s the Medano Springs Land and Cattle Company had fallen on hard times. In 1894 George H. Adams, the pioneer cattleman who had amassed a considerable fortune as proprietor of the vast Baca Grant to the north of the Medano-Zapata, became trustee for the ranch company, apparently for the purpose of "winding up the firm's affairs." Adams purchased the Medano-Zapata property outright in 1898 but held it only briefly before selling it in 1901 to cattlemen Loren B. Sylvester and Richard W. Hosford, whose own brief tenure as owners is notable only for the dubious distinction of acquiring the Teofilo Trujillo properties after Trujillo and his son Pedro were forced off their land in a dispute with local cattle ranchers in 1902. Declining cattle prices forced Sylvester and Hosford to sell their interest in the Medano-Zapata in 1907 to attorney Henry C. Flower, who in turn sold the ranch to George W. Linger in 1912. In contrast to the open-range operations of previous owners, whose cattle was fattened on the natural grasses of the ranch, Linger raised Herefords as a fed-cattle business, supplementing his herds with corn feed and cottonseed cake. By 1918, the property had increased to more than 25,000 acres of deeded land and 100,000 acres of leased land. George Linger died in a plane crash in 1921, but the Linger family continued ranching operations until 1925, when his children—sons Howard, Albert, and Lyman, along with daughter Margaret—reorganized as Linger Bros. & Co. and bought out the interest of their brother Earl.[20]

Under the guiding hands of the Lingers, the Medano-Zapata became one of the most successful and respected ranching operations in the entire valley from 1912 until 1947, when the family sold the property to Texas oilman Malcolm Stewart, Sr., for a reported $900,000. Stewart continued operating the ranch much as the Lingers had, living with his wife at the Zapata headquarters while his son Malcolm, Jr., and his wife resided at the Medano. The elder Stewart died in 1959, but the Stewart family maintained operations until the 1970s, when several portions of the ranch were sold off. In 1988 Otaka International, Inc., a Japanese corporation whose principal investors were also involved with golf and hotel properties in Hawaii, purchased the Medano-Zapata from Malcolm Stewart, Jr., with plans to construct a golf course and resort. A year later, in August 1989, Otaka International sold the ranch to Rocky Mountain Bison, Inc.; Otaka principal partner and renowned architect Hisayoshi (Hisa) Ota eventually emerged as sole owner of the property. Ota, who had been instrumental in developing the Great Sand Dunes Country Club and Inn at the Zapata Ranch, introduced a large herd of bison that eventually numbered in excess of 2,000 head to the property.[21]

The Zapata's combined bison and cattle operations thrived initially, as did the golf resort, at least until the mid-1990s, when the course's remote location and short golf season slowly combined to undermine its profitability. Never much of a golfer himself, Hisa Ota tried leasing the golf operation, but success remained elusive, and eventually he decided the time had come to sell his beloved ranch, although he was wary of putting it on the open market, where unscrupulous buyers might have a chance to purchase and develop the property in potentially damaging fashion.[22] Ota had reportedly been approached several times over the years by international nonprofit conservation organization The Nature Conservancy (TNC), which was interested in obtaining conservation easements on the property, so when Ota finally decided to sell, his choice of preferred buyer was clear:

> It is such a unique, sensitive, and beautiful property. I really love this place, and want to make sure it's taken care of in perpetuity. Selling on the open market was not acceptable—I had no guarantee it would be

Figure 7.2 Entry gate to the Medano Ranch headquarters. The Nature Conservancy purchased the Medano-Zapata Ranch in 1999 in an effort to preserve the historic property. Photo by the author.

Figure 7.3 Outbuildings at the Medano Ranch headquarters, looking northeast toward the snowcapped Sangres. Photo by the author.

protected. I've been a member of The Nature Conservancy for several years, and have always liked what the organization has done. My wife [Kris] and I dreamed that this place would be used for conservation or educational purposes, and I wanted to make sure there would some public access. That's a good thing for the local community.[23]

On January 21, 1999, The Nature Conservancy issued a press release announcing that it had signed a purchase agreement with Rocky Mountain Bison, Inc., to acquire some 30,000 deeded acres, along with leases for an additional 70,000 acres of state and federal land that comprised the historic Medano-Zapata Ranch, for an estimated $6.4 million, well below the property's $10.5 million appraised value. "With these acquisitions, The Nature Conservancy will be protecting one of the most extensive natural landscapes left in Colorado," said the Conservancy's Colorado director, Mark Burget. To finance the purchase, the organization took out an "internal" loan in order to complete the deal with Hisa Ota quickly, then secured three different $1 million grants from the Gates Family Foundation of Denver, El Pomar Foundation of Colorado Springs, and the Great Outdoors Colorado (GOCO) Trust Fund. The remainder of the purchase price and additional funding for operations and educational programs at the ranch came from The Nature Conservancy members, partners, and private donations.[24]

The deal ultimately closed on June 30, 1999. The Nature Conservancy agreed to continue the existing ranching operation and to preserve the vast sand sheet and sabkha portions of the ranch bordering Great Sand Dunes National Monument. The Zapata Ranch headquarters, which had been placed on the

National Register of Historic Places in 1993, became the epicenter for TNC's ranching operation and educational programs, as well as lodging for overnight guests who came to experience firsthand the rigors and thrills of authentic cattle and bison ranching amid the scenic splendor of the Great Sand Dunes and the stunning backdrop of Sangre de Cristos. The disposition of the Medano Ranch headquarters remained unresolved at the time, but TNC immediately began developing a management plan designed to protect the ranch's significant natural, cultural, and historical assets. As for the golf course, it closed in March 2000, much to the chagrin of local duffers and despite its distant views of the world's largest sand bunker at the Great Sand Dunes. For the National Park Service, TNC's purchase of the Medano-Zapata Ranch removed forever the threat of potentially damaging development along the main approach corridors to the Great Sand Dunes, as well as protected vital elements of the larger aeolian system responsible for the continued ecological health of the dunes. Perhaps most encouraging for the future of the dunes, the purchase also happened to coincide with the synergistic convergence of several key individuals and events that were most critical for the eventual creation of Great Sand Dunes National Park and Preserve.[25]

Parks and Politicians

On May 6, 1999, several weeks prior to The Nature Conservancy's formal announcement of its purchase of the Medano-Zapata Ranch, newspaper editor Gary Taylor's regular opinion column "A Higher Plain" appeared on the front page of the *Valley Courier* with the subtitle "The Sand Dunes as a National Park." According to former superintendent Steve Chaney, Taylor's column marked the first time the nascent proposal to redesignate the monument as a national park officially appeared in the local press with an unqualified statement of support. Taylor, who earlier that week had met with Chaney to discuss the idea, provided a brief summary of the proposal and described his hopes for the future, noting, "Chaney is talking to stakeholders about the Sand Dunes becoming a national park. I heard his pitch early because I asked. He and I both hope that if a congressional legislative effort is mounted to make it a national park, that the history here of coordination and conciliation between business and environmentalists will continue."[26]

Thus began the public phase of what Chaney later described as three unique but overlapping interests gaining momentum and converging during 1999, a process that resulted in the eventual creation of Great Sand Dunes National Park and Preserve. Spurred by the costly and contentious groundwater battles that had recently rocked the valley, the first interest—eliminating the Baca Ranch as a commercial water development and exportation threat—was driven by a variety of San Luis Valley citizens, water districts, businesses, environmental organizations, and assorted state and federal agencies, including the National

Park Service. Attaching the groundwater's value to the protection of a national park was seen as the most effective, longest-lasting, and least expensive way to reduce the seemingly never-ending threat to the valley's precious groundwater.[27]

The second interest—elevating the monument to national park status as a means of stimulating the economy of the San Luis Valley—was driven mainly by the local tourism industry, community leaders and politicians, and various chambers of commerce seeking to improve the standard of living in one of the poorest regions of Colorado. These groups also felt that legislation to establish a new national park would be more politically attractive to Congress than a bill that merely expanded the boundaries of the existing national monument. Moreover, a great many valley citizens simply felt that the dunes were of such splendor, grandeur, and national significance that they undoubtedly deserved the added notoriety and prestige of the "national park" title, although the NPS was reluctant to support such a designation without the sufficient size and diversity of resources typically associated with national parks.[28]

The third interest—achieving national park status as a means of ensuring permanent protection for the entire Great Sand Dunes aeolian system—was driven primarily by the National Park Service, as well as various local conservation and environmental groups. As scientific knowledge and understanding of the dunes advanced during the 1980s and 1990s, it became increasingly apparent that permanent protection for this system was the key to the long-term ecological health of the Great Sand Dunes. Since the watershed portion already enjoyed protected status under the auspices of the U.S. Forest Service, and since The Nature Conservancy had pledged protection for large portions of the sand sheet and sabkha within the boundaries of the nearby Medano-Zapata Ranch, the immediate concern of the NPS focused on the Baca Ranch, which had been ground zero for the recent groundwater battles. While protection of the Baca's resources could have been achieved in a variety of ways, NPS officials believed that federal acquisition and management of the enormous ranch was the most realistic, feasible, and effective way to ensure long-term protection of the Great Sand Dunes aeolian system, which also dovetailed nicely with the desire to eliminate the Baca Ranch as a continuing water development threat.[29]

These three overlapping interests, each driven by slightly different motivations, nonetheless shared a single common objective: the creation of Great Sand Dunes National Park, along with a specially designated national preserve that would continue to allow elk and bighorn sheep hunting in large portions of the mountains looming above the dunes. Among the more remarkable aspects of this push to establish a new national park at the Great Sand Dunes was how quickly these disparate interests began coalescing, starting with a town meeting held by Sen. Ben Nighthorse Campbell (R-Colo.) in Alamosa on June 1, 1999. As part of a regular series of "meet and greet" public appearances held statewide in order to stay connected with voters, Senator Campbell was in town to discuss local, national, and international affairs with his constituents when local reporter Erin

Smith, who covered the San Luis Valley for the *Pueblo Chieftain*, posed what she later described as "the most important question in my entire journalistic career." Smith asked Campbell, whose own bruising legislative battle to secure national park status for Colorado's Black Canyon of the Gunnison National Monument had been grinding along for well over a decade, if he supported a similar redesignation for the Great Sand Dunes. Campbell replied that he "would not be opposed" to creating a new national park at the dunes if the people of the San Luis Valley wanted it, although he cautioned that the process could take years, as had happened with Black Canyon. Regardless, Campbell's statement of support for the idea was a pivotal milestone that was now part of the public record, so the complex political process of creating Great Sand Dunes National Park and Preserve had officially left the starting gate.[30]

On August 12, the *Pueblo Chieftain* reported that Mike Hesse, chief of staff for Rep. Scott McInnis, told a meeting of area residents that if the San Luis Valley supported elevating Great Sand Dunes National Monument to national park status, McInnis "likely will give the move his blessing," which at the time must have seemed a rather tepid declaration of enthusiasm from the politician who would eventually end up introducing the actual Great Sand Dunes National Park bill in the U.S. House of Representatives. In reality, McInnis and his staff were merely gauging the level of local support for the proposal before committing precious time and resources.[31] On August 20, Sen. Wayne Allard (R-Colo.), who had first heard about the national park idea from Adams State College professor Dion Stewart in 1998 during a public meeting in Alamosa, announced his own support for the redesignation before taking a guided tour of the monument with Steve Chaney. Chaney later told reporter Erin Smith that as superintendent he had been exploring the need to upgrade the monument to a national park and "was amazed when the *Pueblo Chieftain* ran an article" back in June with comments from Senator Campbell about the possibility.[32] After an initial burst of publicity and political support during the summer of 1999, the proposal simmered for a few months until October 21, when Congress finally passed legislation sponsored by Sen. Ben Nighthorse Campbell (who had been trying for nearly thirteen years) and Rep. Scott McInnis that officially transformed Black Canyon of the Gunnison National Monument into a national park. Less than three weeks later, the process of similarly elevating Great Sand Dunes National Monument shifted into high gear.[33]

Sensing the momentum and keenly aware of the favorable political climate, McInnis announced on November 11 that he planned to seek national park status for the Great Sand Dunes, which had attracted nearly 300,000 visitors in 1999. McInnis said he was "on a roll" after the success with Black Canyon and confident that the process could be repeated at the dunes because the "time is right, the political mood is right, and the Great Sand Dunes is definitely right for a national park designation."[34] Broad regional approval appeared on November 26: a *Denver Post* editorial declared, "The idea deserves the public's whole-hearted support and the full backing of Colorado's congressional delegation." Another crucial endorsement came on November 29 when Representative McInnis announced that Interior Secretary Bruce Babbitt had called to inform him that his proposal had the full backing of the Clinton administration, which had enthusiastically supported similar proposals throughout its entire tenure in the White House. Such support, which McInnis called "highly significant," would undoubtedly help clear passage of the proposed legislation in Congress. Exemplifying the bipartisan nature of the political cooperation coalescing around the proposal, the Republican lawmaker admitted that he had "gone head-to-head with Babbitt on some issues, but he has always been very professional on issues like this."[35] Momentum continued building in early December when Senator Allard announced plans to push the national park idea in the Senate's next legislative session. McInnis had originally approached Senator Campbell to spearhead the Senate effort, but Campbell demurred on account of his long struggle with the Black Canyon proposal and suggested that McInnis ask Allard instead. Close on the heels of Allard's announcement came news that Interior Secretary Babbitt would personally visit the dunes on December 18 after an invitation from Allard and McInnis.[36]

The first cracks in the seemingly solid foundation of bipartisan political support also appeared early that December. After reading an article in the *Valley Courier* that described a possible purchase of the Baca Ranch by the federal government as part of the national park proposal, the Saguache Board of County Commissioners drafted an official Resolution of Opposition. Disgruntled that they had not been previously consulted about the proposal, the county commissioners objected not only to the potential loss of tax revenues for Saguache County, which contains well over half of the Great Sand Dunes along with the entire Baca Ranch, but also to the potential "detrimental" effects of federal ownership of additional lands in the San Luis Valley, as well as the plan's lack of specifics about the future of the valley's water resources. Soon thereafter, a number of valley residents also began expressing concerns over pollution, increased traffic, lack of public input, and harmful effects on natural areas caused by improved access, more campgrounds, and additional tourist infrastructure.[37]

Despite these and assorted other objections, a majority of valley citizens and businesses continued to express support for the proposal. The proponents included native son and fifth-generation rancher Ken Salazar, whose family roots in the region go all the way back to the 1850s. Born in Alamosa and raised in the San Luis Valley, Salazar had been elected Colorado's state attorney general in 1998 and would later serve as a U.S. senator, as well as secretary of the interior in the Obama administration. Salazar was intimately familiar with the valley's economic woes and continuing battles over natural resources. On December 16, Salazar sent a letter to Babbitt, Allard, Campbell, and McInnis expressing his unqualified support for the national park proposal, adding details about "the oldest water rights in Colorado," his concerns over "Colorado's obligations under the Rio Grande Compact," the long-term "protection of vested water rights," and the nature of "surface and groundwater relationships." Salazar clearly grasped the crucial role of water in the valley, as well as the essential necessity of including the Baca Ranch in the proposed boundary expansion. He emphasized that acquiring the Baca "would preserve the water source and the complex system that keeps the dunes where they are" and thus "bring an end to the costly water battles that would be fought over any new proposed water project on the Baca Ranch."[38]

With Ken Salazar in the mix, five powerful politicians, along with The Nature Conservancy's Colorado director, Mark Burget, were now scheduled to meet at the Great Sand Dunes to discuss the feasibility of turning the national monument into a national park. The relatively remote dunes had never before hosted such an impressive assemblage of local, state, and national dignitaries, and as monument staff busily prepared for the gathering, geologist Andrew Valdez was sitting at his desk composing a short speech for the occasion when his phone rang. The call came from Great Sand Dunes law enforcement personnel seeking assistance with a horse that had recently died. Without hesitation, Valdez stopped working on what was perhaps the most important speech of his entire career, a speech he was scheduled to deliver to a roomful of distinguished visitors during what was sure to be among the most significant and historic meetings in the entire history of the Great Sand Dunes, and hustled off to help his fellow rangers pull a dead horse from a stall. If the dunes had waited this long to become a national park, he figured, writing his speech could wait as well.[39]

Summit in the Sand

The morning of December 18, 1999, dawned typically cold and crystal clear in the San Luis Valley. With a radiant winter sun slowly rising over the Sangre de Cristos, the historic meeting, dubbed the "Sand Dunes Summit" by Senator Campbell and variously described in the local press as the "Summit at High Dune," "Summit at the Dunes," and the "Summit in the Sand," commenced in the cozy confines of the Stewart House at the Zapata Ranch. Interior Secretary Babbitt joined assorted NPS and USGS personnel for coffee and donuts while viewing maps and discussing sand deposits, dunes geology, and the role of water in the dunes system. When the time came for his presentation, geologist Andrew Valdez described how the main dunefield was just the "tip of the iceberg" of a much larger sand system, and how incredible it would be to protect the entire sand system from where it began in the dry lakes area of the sabkha all the way to where it ended at the crest of the towering Sangres. After his presentation, as the group slowly dispersed for the short drive to the dunes, Valdez found himself chatting with Interior Secretary Bruce Babbitt, who expressed keen interest in the presentation Valdez had just delivered. Babbitt's curiosity was no coincidence. Before entering politics, he had earned a bachelor of science degree in geology

Figure 7.4 The Summit in the Sand. Historic meeting at the Great Sand Dunes on December 18, 1999, with (*from left*) Rep. Scott McInnis, Interior Secretary Bruce Babbitt, Sen. Wayne Allard, Sen. Ben Nighthorse Campbell, and Colorado attorney general Ken Salazar discussing the possibility of creating Great Sand Dunes National Park. Fred Bunch, Great Sand Dunes National Park and Preserve.

from the University of Notre Dame, as well as a master's degree in geophysics from the University of Newcastle, England. On a purely scientific level, Babbitt understood and appreciated, perhaps more than any of the other power brokers assembled that day, that protecting the Great Sand Dunes ultimately required protecting the entire aeolian system, and the best way to accomplish that goal would be to create a new national park.[40]

After meeting at the Zapata Ranch, the group reconvened at the day-use parking area next to the dunes, along with additional dignitaries, guests, and members of the local press. Secretary Babbitt, Senators Allard and Campbell, Representative McInnis, and Attorney General Salazar eventually separated themselves from the rest of the gathering and hiked together out to the dunes in the bright morning sunlight. No official record exists of precisely what the men discussed, but later reports and interviews indicated that the Great Sand Dunes National Park proposal dominated the conversation, along with a brief discussion on the controversial and long-delayed Animas–La Plata Water Project in southwestern Colorado. Photographers had a field day as the five powerful politicians stood shoulder-to-shoulder against a magnificent backdrop of windswept dunes, with the snow-covered peaks of the mighty Sangres looming in the sharp distance. It had been 192 years since Zebulon Pike climbed these same dunes, 67 years since Herbert Hoover protected them with a stroke of his pen, and 5 years since Bob Zimmerman told Scott McInnis that the dunes should be preserved in perpetuity as a national park. On this historic day, that possibility seemed closer than ever.[41]

Following the photo op and private discussion about the future of the Great Sand Dunes, the distinguished visitors and guests reassembled at the visitor center for what one local reporter described as a "jovial conference." The five dignitaries, along with TNC Colorado director Burget and Superintendent Chaney, addressed the gathering of about one hundred people and announced their unanimous support for the plan. Secretary Babbitt pledged to put between $30 million and $40 million toward the goal of turning the monument into a national park, noting that he was overwhelmed that he could listen to the comments of three Republican lawmakers and sum up his views with only one word: "Amen." The officials also discussed the continuing threat of water exportation, and Senator Campbell's statement "We're not going to let that happen to this valley" inspired a burst of applause from the audience. The Nature Conservancy's ongoing, sensitive, and very secretive negotiations to purchase the Baca Ranch from Cabeza de Vaca Land and Cattle Co. also came to light, although details were decidedly scarce.[42] Asked to comment on the negotiations, Babbitt advised patience and reminded everyone that The Nature Conservancy "has been in the lead in those discussions. The one thing I have learned from past efforts of this type is that you get the best results when you let the lead horse go out there and pull the wagon. I would urge all of us to remain quiet. Let The Nature Conservancy do their work."[43]

Conservancy director Mark Burget likewise refused to comment in detail on negotiations for the Baca Ranch and instead heaped praise on the lawmakers for their "remarkable, wonderful, historical bipartisan cooperation" in attempting to secure national park status for the dunes, a sentiment echoed by Rep. Scott McInnis, who noted, "When we agree, we can move mountains—protect mountains is a better way of putting it." Senator Allard cautioned that some difficult issues still needed to be addressed, including the "uncertainty of the Baca purchase," as well as concerns over water rights, revenues, boundaries, and private property rights. He also announced a series of town meetings to be held in the San Luis Valley the following January, where citizens could express their views on the proposal. "Take the time, give us your input," Allard said, assuring the assembled masses that the process would not move forward without their voices being heard.[44]

It seems not everyone in attendance that day shared the spirit of bipartisan cooperation. Standing outside the crush of media, Saguache County commissioner Bill McClure reiterated his concerns about the possible loss of tax revenue and stated that his county still felt slighted about not being notified beforehand. He noted, "You have all indicated that this has been an on-going issue for many months. Unfortunately, Saguache County has not been notified other than through the press, and we are somewhat upset. We fully support the water issue. . . . But we think we need to address involving Saguache County, such as the loss of our taxes, and then the input by citizens of Saguache as [to] whether they want a Park in the county."[45]

Attempting to assuage McClure's concerns, McInnis replied, "You guys haven't been left out. It has taken us five months to get everything on track; nobody's going to get the short stick," and emphasized that the tax issue, which he called a small part of a "huge, huge project," would be worked out.[46] Despite McClure and his fellow commissioners' reluctance to support the park proposal, most of those in attendance agreed that the historic meeting had been an unqualified success, and in the quiet aftermath after everyone had left, Superintendent Steve Chaney sat in his office with Ron Everhart, deputy director of the NPS Intermountain Region, discussing their next move. As Chaney later remembered it, Everhart essentially told him, "Steve, what you need to do is talk to every Rotary Club, Kiwanis Club, Lion's Club and garden club in southern Colorado and northern New Mexico that you can talk to, and focus entirely on that for the foreseeable future." Chaney took Everhart's advice to heart, and shortly thereafter embarked on what he described as an "extended road show" across the entire region; by the end he had given at least fifty presentations on the national park proposal to anyone who would listen. Little did Steve Chaney—or anyone else, for that matter—realize at the time that the strongest opposition to the plan would come not from the small towns and communities of the San Luis Valley, but from one of Colorado's very own representatives in Congress.[47]

EIGHT | **Unimpaired for Generations**

The service thus established shall promote and regulate the use of the Federal areas known as national parks, monuments, and reservations . . . which purpose is to conserve the scenery and the natural and historic objects and the wild life therein and to provide for the enjoyment of the same in such manner and by such means as will leave them unimpaired for the enjoyment of future generations.

National Park Service Organic Act, 1916

While the Great Sand Dunes National Monument is a wonderful attraction for Coloradans and visitors alike, I do not believe the monument rises to the level of a national park. The national park designation is a special significance that should not be assigned to every natural resource.

Colorado representative Joel Hefley, 1999

ON DECEMBER 20, TWO DAYS AFTER the historic Sand Dunes Summit, the office of Rep. Joel Hefley (R-Colo.) issued a press release announcing his official opposition to the proposal to elevate Great Sand Dunes National Monument to national park status. Aside from the fact that Hefley simply did not believe the monument deserved to be redesignated as a national park, he was concerned that the proposal might circumvent the involvement of the National Park Service in the process. In 1992, Hefley had successfully included legislation in the Omnibus Budget bill requiring the NPS to "evaluate and study whether a natural resource meets the requirements for a national park designation." Hefley introduced his legislation in response to the political maneuvering of Rep. Joseph M. McDade (R-Pa.), who in 1986 had convinced his colleagues in the House of Representatives to approve $8 million in funding to turn Steamtown USA, which had been described as a "second-rate" railroad museum, into a full-fledged National Historic Site, an endeavor that ended up costing taxpayers an estimated $66 million.[1] Hefley regarded McDade's influence peddling as pork-barrel politics at its very worst, and vowed to end such nonsense by insisting that the NPS follow a precise process for evaluating any future attempts to acquire new lands or reclassify existing park units, then present its findings and recommendations to Congress, a process that had not yet occurred with the Great Sand Dunes proposal. "It is my hope that if the proposal for reclassifying the Great Sand Dunes National Monument proceeds, it will only be through the proper legislative process," explained Hefley. "Congress should not be passing legislation that authorizes a new national park without following this process and receiving the input of the U.S. Park Service."[2]

Rep. Scott McInnis responded immediately, challenging Hefley's initial assessment of the dunes as lacking in suitability for national park status by declaring, "[If] there ever was a place deserving of national park standing, in my estimation it is the Great Sand Dunes. Much more than simply rising to the level of a national park, the Great Sand Dunes' delicate and diverse ecosystems are so unique, so delicate and so majestic that anything short of a national park designation fails to do it justice."[3] Hefley, however, remained unconvinced by McInnis's rhetoric. He honestly felt that the dunes were little more than a day-use area for picnickers, describing the typical visit as follows: "You drive in, take a look at it, maybe have a picnic, and you drive out." He compared the effort to upgrade the monument to the so-called "pork parks" that were created only to satisfy powerful lawmakers.[4] Hefley did agree with McInnis that the federal government should purchase the Baca Ranch to gain control over its water rights, but added that the land could be added to the monument without granting national park status. Still, he offered to work with his colleagues on the issue. "Who knows, maybe they can convince me, but given my experience and personal knowledge, I am opposed to it," he said, sounding somewhat conciliatory. McInnis acknowledged that "Joel could probably round up a couple votes and kill it in committee if he really wants to kill it," adding that such a move would be "really sad for Colorado, especially if a Colorado congressman killed the bill."[5] Hefley's position on the powerful House Natural Resources Committee, which had oversight over any park legislation, as well as his status as a ranking member of the Subcommittee on National Parks, Forests and Public Lands, gave him more than enough political influence to do just that.[6]

Despite Hefley's persistent and very public opposition, work proceeded apace on drafting a legislative bill for the proposed Great Sand Dunes National Park. Renowned Denver attorney David Robbins, widely recognized as one of the shrewdest legal minds in Colorado, composed what he termed a "very preliminary" bill, which he shared with Allard, McInnis, and their staffers. Robbins, who had served as general counsel for the Rio Grande Water Conservation District since 1981 and was known for his vigorous defense of the valley's water resources, possessed an encyclopedic knowledge of Colorado water law in general and the intricacies of San Luis Valley hydrology in particular, having played a key role in defeating both the AWDI and Stockman's water exportation plans. Among the multitude of issues he addressed in the draft bill, however, were at least two that did not involve water. The first issue, raised by a small but devoted and very vocal alliance of San Luis Valley hunters and by various outdoor advocacy groups, such as the Rocky Mountain Bighorn Society, the Colorado Bowhunters Association, and the Rocky Mountain Elk Foundation, focused on concerns that the new national park would prohibit the hunting of elk and bighorn sheep in the Rio Grande National Forest and Sangre de Cristo Mountains above the dunes, an activity that historically had been allowed under the jurisdiction of the U.S. Forest Service. The Colorado Division of Wildlife, responsible for managing the state's game herds, also expressed apprehension about the proposal, citing concerns over its ability to manage the region's bighorn population, the largest in the state, if hunting were prohibited. While attempting to address this issue in the draft legislation, Robbins and his colleagues recalled a similar situation occurring in 1994 when Death Valley National Monument was being considered for elevation to national park status, a proposal that had also met opposition from local hunters. The innovative solution, elegant in its simplicity, was the creation of a "preserve" where hunting and other recreational activities typically not allowed in national parks would be permitted, even though the area would continue to be administered by the National Park Service. Robbins proposed a similar compromise at the Great Sand Dunes by including language in the legislation authorizing the creation of a "Great Sand Dunes National Preserve" adjacent to the new national park. If approved, the legislation would transfer approximately 42,000 acres of national forest land from the USFS to the NPS, which would continue to allow hunting, fishing, and trapping in the area in accordance with existing state and federal wildlife laws.[7]

The second issue was the nagging problem of the estimated $68,000 in annual property tax revenues that Saguache County had been collecting from the current owners of the Baca Ranch, tax revenues it would lose if the federal government purchased the ranch as anticipated. Federal agencies do not pay property taxes to counties, but most agencies offset the tax losses by making Payment in Lieu of Taxes, or PILT. However, Saguache County had already exceeded its PILT ceiling, which was based on the amount of federal land in the county, creating a potential financial conundrum for county commissioners and federal officials alike.[8] A possible solution appeared in early January 2000 when Mike Blenden, manager of the U.S. Fish and Wildlife Service's Alamosa and Monte Vista National Wildlife Refuges, proposed designating roughly 36,000 acres of the Baca Ranch as a new wildlife refuge, then compensating Saguache

County for its lost tax revenue under the Refuge Revenue Sharing Act, a law dating back to 1935 that allowed for a portion of the revenues generated from such sources as timber sales, grazing fees, and various permit fees on USFWS refuges to be paid to counties in lieu of lost property taxes. Once again, as with the national preserve proposal, an elegant solution to a seemingly intractable problem had presented itself—a simple, well-intentioned strategy for ensuring that Saguache County received compensation for its lost tax revenues—but the Saguache County commissioners remained deeply suspicious of the federal government and steadfast in their opposition to the park proposal.[9]

In early February 2000, the *Pueblo Chieftain* reported that a broad coalition of environmental groups, including the Wilderness Society, National Wildlife Federation, National Parks and Conservation Association, Colorado Environmental Coalition, National Audubon Society, Colorado Mountain Club, and Sierra Club, had announced their official endorsement of the proposed legislation, which pleased Scott McInnis no end and inspired his press secretary, Josh Penry, to announce that the endorsement "speaks volumes about the quality of the legislation."[10] That same month *Colorado Central Magazine*, "The Monthly

Figure 8.1 Cover of the February 2000 edition of the *Colorado Central Magazine* depicting the worst fears of some San Luis Valley residents if the proposal to create Great Sand Dunes National Park came to pass. Mike Rosso, *Colorado Central Magazine*.

Magazine of Post-Millennial Rapture," published an illustration on its cover that seemed to represent the worst-case fears of San Luis Valley residents. Titled "A New National Park?," the drawing depicted a throng of cars and tourists, along with gas stations, hotels, and fast-food joints, cramming the entrance to the new national park with the dunes and the Sangres in the background, along with a flying saucer hovering over the entire scene as a symbol of the valley's dubious reputation as a hotbed of alleged extraterrestrial activity. Comically exaggerated, the image nonetheless embodied the type of development and commercialization that some local folks worried would happen if the national park became a reality. Perhaps most concerned were the citizens of Crestone, who feared that rampant development associated with the expanded national park would destroy the quiet peace and sacred energy of their idyllic mountain sanctuary, especially if rumors of plans to construct a new northern access road and entrance to the park through Crestone came to pass.[11]

In an attempt to quell such rumors and calm the fears of valley residents, McInnis, Allard, Salazar, and their respective staffs continued holding meetings during the first few months of 2000, but by the end of March it became clear, at least to the politicians, that the time had come to move forward. After all, it was an election year, and if the legislation had any chance of working its way through the 106th Congress before November, the process had to move quickly in spite of objections still being raised in the San Luis Valley, most notably from the Saguache County commissioners, whose quest to derail the plan continued to focus on lost tax revenues and suspicions that the federal government really just wanted to sell the valley's water, despite repeated assurances and guarantees to the contrary. The commissioners' intransigence eventually inspired a frustrated Scott McInnis to exclaim, "You can meeting this thing to death, at some point you have to cut loose," which is exactly what he did.[12]

On March 28, Representative McInnis officially introduced H.R. 4095, "A Bill to provide for the establishment of the Great Sand Dunes National Park and Great Sand Dunes National Park Preserve in the State of Colorado," in the U.S. House of Representatives.[13] The legislation stipulated that "the Secretary [of the Interior] shall establish the national park as soon as the Secretary determines that sufficient lands, with a sufficient diversity of resources, have been acquired to warrant designation of the lands as a unit of the national park system."[14] In other words, acquisition of the Baca Ranch had to occur before the existing monument could be reclassified as a national park. The bill also required the creation of a Great Sand Dunes National Park Advisory Council to oversee the preparation and implementation of a management plan for the new park, as well as sections on grazing, hunting, wilderness protection, and oversight of the Closed Basin Project, which would continue to be the responsibility of the Bureau of Reclamation. The section dealing with water received perhaps the most scrutiny. Attorney David Robbins, who had been instrumental in drafting this portion of the bill, insisted on language that protected the existing water resources (termed the "Hydrologic Regime") in the new park and preserve by stipulating that "in administering the national park and the park preserve, the Secretary shall protect and maintain the balance in the hydrologic regime necessary for the protection of park resources and park values, while minimizing, to the extent consistent with park protection, adverse impacts on adjacent human and wetland communities."[15]

Water rights and resources associated with the proposed Baca Ranch acquisition received similar treatment, with stipulations requiring that such resources be restricted for use "only within the national park or park preserve or in the immediately surrounding areas of Alamosa or Saguache Counties," which surely must have pleased a certain group of local county commissioners, and "only for purposes that protect park resources and park values, fish and wildlife purposes, and the historical irrigation use."[16] The bill also addressed the issue of future water development schemes in the valley and their potential impact on park resources by requiring that the secretary "shall take such actions as are within the scope of the Secretary's authority to ensure that any new or additional water development in the San Luis Valley aquifers is consistent with the protection and maintenance of the hydrologic balance necessary for preservation of park resources and park values."[17]

As expected, the bill stalled in Hefley's Subcommittee on National Parks, Forests and Public Lands, just as he warned it would. Scott McInnis later recalled, "Joel was very honest with me and told me he was going to kill it."[18] Speculation about the precise reasons for Hefley's opposition, beyond his persistent declarations that he simply didn't feel the dunes were worthy of national park status, soon appeared in local newspapers. Some sources attributed Hefley's opposition to a clash of personal and political styles; one reporter claimed that "Hefley is as laid back as they come in Congress," while describing McInnis as a "former cop . . . aggressive campaigner and fund-raiser who has made no secret of his ambitions for higher office." Those ambitions for higher office apparently included Ben Nighthorse Campbell's Senate seat, which Campbell was expected to vacate if George W. Bush won the presidency in November and offered him a Cabinet post. Hefley was also rumored to have his eye on that same Senate seat, and unidentified sources claimed that both men had expressed their interest to Colorado governor Bill Owens, who would appoint Campbell's replacement if he did in fact join the Bush cabinet.[19] Others suspected that Hefley coveted the groundwater beneath the Baca Ranch for his constituents in Colorado Springs, a charge that he vehemently and repeatedly denied, claiming that it would ruin the economy of the valley and that he would fight any such proposal. He later clarified his position, saying he would "be happy to work with Scott to protect the water resources of the San Luis Valley. If that means having the government purchase the Baca Ranch to do that, let's do it. But I'm not willing to make the dunes into a national park."[20] Hefley also reiterated his considerable expertise in such matters, owing to his years on various committees dealing with public lands, and plainly stated that he didn't want to appear "hypocritical" if he supported the Great Sand Dunes proposal, noting that the "average response is 'that's a great idea.' But I'm the only one who's been working on these issues for years on the parks and lands committee. I've been critical about parks in other states, and I'd be hypocritical to say 'yes' to this just because it's in Colorado."[21]

On April 25, Hefley and McInnis found themselves on opposite sides of the Colorado House of Representatives chambers. The two former state representatives had returned to the state capitol to witness a floor vote on a resolution supporting the Great Sand Dunes National Park bill currently stalled back in Washington, and McInnis took the opportunity to comment on Hefley's opposition to his bill, acknowledging that as a senior member of the committee, "Joel has the power unilaterally to block the national park. Joel has every right to vote his conscience on it. All I want is for him to let the bill get to the House floor. That's a whole different ballgame from disagreeing on the issue."[22] Hefley, who represented Colorado's Fifth Congressional District, left the room just before his wife, Rep. Lynn Hefley (R–Colorado Springs), tried to amend the bill to indicate to Congress that the Colorado legislature supported either the expansion of the monument or its redesignation as national park, which she claimed would give Congress greater flexibility. "This is horrible," lamented McInnis as he conferred with his former colleagues. "She's trying to gut the bill. This is one of the most important issues in my congressional career." In the end, Lynn Hefley's attempted amendment failed miserably and the House resolution passed 57–7, which McInnis said would help move the bill through Congress. Lynn Hefley remained adamant that park status was not the way to go, pointing out that her husband sat on the committee that studies park and public land issues. "I have learned from Congressman Hefley, who has studied national parks for over five years, that this is a façade," she said. "They want the water. They don't want Gary Boyce to sell it." A curious comment indeed, considering her husband's repeated denials about his own rumored interest in exporting the groundwater of the San Luis Valley.[23]

A Magnificent Pile of Sand

On May 11, 2000, Sen. Wayne Allard introduced his version of a Great Sand Dunes National Park bill (S 2547) in the U.S. Senate. Allard's bill, which he called "a major step forward in the protection and preservation of the 750-foot-high dunes and the water that lies beneath them," differed slightly from McInnis's House bill. In describing the difference, Allard aide Sean Conway claimed that the Senate version went further than McInnis's to ensure that the federal government would be unable to bypass Colorado water laws when it came to the subterranean aquifers beneath the valley floor, apparently hoping that the stronger language would make the bill more appealing to Rep. Joel Hefley.[24] At first glance, the change appeared innocuous, but the "stronger language" ended up nearly derailing Allard's Senate bill. As with many conflicts in the San Luis Valley throughout history, the issue boiled down to water rights, as well as variations in the legal interpretations of those rights. The bill stipulated that any application for water rights had to occur "pursuant to State law" and defined the interior secretary's specific authority "to appropriate water under this Act exclusively for the purpose of maintaining ground water levels, surface water levels, and stream flows on, across, and under the national park and the national preserve, in order to accomplish the purposes of the national park and national preserve and to protect park resources and park uses."[25] In late July, Department of the Interior solicitor John Leshy told Allard's staff that the department would oppose the Senate bill as written, primarily over language that seemingly denied the federal government any possibility of acquiring the water rights it had deemed so vital for preserving the ecosystems of the Great Sand Dunes. One section of the bill specifically restricted the federal government's options when it came to water rights by establishing that "except as provided in subsections (c) and (d), no Federal reservation of water may be claimed or established for the national park or the national preserve."[26]

Allard defended the language in the bill, claiming it would make certain "the federal government doesn't take away Colorado's right to manage its own water supplies." According to aides, Allard felt "betrayed" by the threatened opposition to his bill and blasted Interior Secretary Babbitt for reneging on his promise to support the proposal. "I feel sorry for the people of the San Luis Valley," said Allard from Washington. "A lot of them worked hard for many months. And now to have reached a point where they [Interior Department lawyers] have changed their position."[27] Interior Department officials countered that Allard had changed the bill's carefully crafted language and pointed out that the House version of the bill would allow the government to acquire additional federal water rights for the dunes, provided it went through Colorado water courts and explicitly followed Colorado water law.[28] Back in the San Luis Valley, Rio Grande Water Conservation District director Ralph Curtis weighed in on what he termed "political posturing" and provided a dose of common sense. "I don't know if it's as serious as the headlines would say," Curtis cautioned. "Mr. [David] Robbins [the district's lawyer] told me two weeks ago that he had had talks, and that they were trying to get the water language changed around a little bit."[29] Even as the war of words escalated in the press, Allard and the Interior Department continued to express a willingness to find a mutually agreeable resolution to the impasse, just as David Robbins had said they would. With time growing short before the November elections, attorneys and staffers representing both sides hammered out the details, modifying and clarifying the bill's legislative language to produce a suitable compromise.[30]

After several weeks of wrangling, Senator Allard announced on September 19 that the two sides had reached an accord. In essence, the agreement allowed the National Park Service to seek additional water rights under Colorado state law in Colorado water court, but did not explicitly guarantee those rights, and further stipulated that the water could not be used for any other purpose beyond preserving the fragile ecosystems of the proposed Great Sand Dunes National Park and Preserve. A similar provision regarding water rights on the Baca Ranch was also included, if and when the federal government ever acquired it, which would virtually guarantee an end to speculative water development schemes on the property.[31] As Colorado attorney general Ken Salazar later described it, "The bill contains sufficient language to protect existing water rights and provides that the Secretary shall obtain any new water right in accordance with federal and State law."[32] Delighted with the compromise, Allard anticipated little if any opposition in the Senate and expected full approval of the bill before Congress adjourned in October.

Rep. Joel Hefley remained unconvinced. On September 20, the day after Allard announced that his Senate bill had been resuscitated, the *Rocky Mountain News* printed another of Hefley's curiously dismissive descriptions of the Great Sand Dunes: "You go there, you look and you say, 'That's a magnificent pile of sand,' and then you get in your car and drive off," he wrote.[33] Hefley's obstinate refusal to recognize that the whole point of the park proposal was to protect not only the "magnificent pile of sand," but also the incredible alpine watershed, the extensive sand sheet and sabkha, and the precious water that binds these interconnected ecosystems together seems somewhat baffling in retrospect, but he clearly had his reasons and was singularly determined to scuttle the plan if he could. "I don't want it to happen," he explained, "and you have to use whatever tools you can to make things happen or to make things not happen. If I can stop it in committee, I'll stop it in committee."[34]

On October 5, Allard's bill passed the Senate by unanimous consent, a fast-track procedure that required no debate or voting because no senators had opposed the bill. Allard called it "a great victory for the people of the San Luis Valley" and noted that "it's unusual to introduce a piece of legislation and get it passed in the same year it was introduced," no doubt recalling that Sen. Ben Nighthorse Campbell's Black Canyon legislation took over a decade to pass Congress.[35] The Senate's overwhelming approval of the Great Sand Dunes National Park bill thus set the stage for the long-expected showdown in the House of Representatives between Scott McInnis and Joel Hefley.

In the weeks leading up to the Senate passage, Representative McInnis and his staff had been scrambling to figure out a way around Hefley's threat to kill the House bill in committee. Their solution, which McInnis later credited to "some pretty smart people" working for him, as well as his experience serving on the House Rules Committee, involved an obscure, rarely used parliamentary procedure that allowed the Speaker of the House to "hold a bill at the desk" instead of following the normal procedure of sending it on to a particular committee. With the particulars of the unusual procedure in hand, McInnis approached House Speaker Dennis Hastert privately, something he had never done before, and outlined his situation in clear and concise terms. McInnis explained that his Great Sand Dunes bill had widespread, bipartisan support in the San Luis Valley and the state of Colorado, as well as in Congress and the Clinton administration. Next, he reminded Hastert that he had always been a "good foot soldier" for the GOP, recently winning praise from the Speaker himself for raising record amounts of money for the party's congressional campaigns. McInnis then pointed out that he had never asked the Speaker for anything, but in this instance he simply had no choice because Hefley was in a position to kill the bill unilaterally in the House Natural Resources Committee. McInnis concluded by formally requesting that Hastert hold the bill at his desk before sending it to the House floor for a vote. Much to McInnis's surprise and considerable relief, Hastert agreed to honor his request. On October 6, the day after its passage in the Senate, the bill arrived in the House, where Speaker Hastert dutifully held it at his desk as promised. If the bill didn't go to Hefley's committee, Hefley couldn't stop it.[36]

As the drama played out on Capitol Hill, Gary Boyce suddenly reappeared in the local press back in Colorado claiming that the Baca Ranch was not for sale, which if true would have thrown a serious wrench into the legislative works. "No one has said a word to me about it," said Boyce. "But I'll tell you, this ranch isn't for sale."[37] An attorney for Farallon subsequently dismissed Boyce's claim, noting that "Farallon has a controlling interest in the property. Farallon wants to sell and is prepared to direct a sale on terms that are acceptable to the parties, even over Mr. Boyce's objections. Mr. Boyce believes it is practically and legally possible to export that water. Farallon has come to the conclusion it is not."[38] While the owners of the Baca Ranch hashed out their differences in the newspapers, Speaker Hastert placed the bill on the "suspension calendar," clearing the way for its eventual move to the House floor for a vote. McInnis waited patiently as repeated delays in the House left the bill in limbo. He used his time wisely, contacting his colleagues and urging them for their votes in support of the measure, fully aware that Hastert's procedural move to bypass Hefley's committee was technically a "suspension of the rules," which meant the bill needed a two-thirds supermajority vote to pass instead of a simple majority.[39] Joel Hefley also stayed busy lobbying his colleagues, including writing an appeal to his fellow representatives that contained several sentences which seemed to belie his repeated denials of any interest in the valley's groundwater. "What the Sand Dunes are called is unimportant—the key issue is the water underneath them," Hefley wrote. "Anyone familiar with Western water issues knows the treatment of water resources is paramount in a dry climate. And the fact is, many Colorado municipalities have concerns about this proposal."[40]

Finally, on the evening of October 24, 2000, debate on the Great Sand Dunes National Park and Preserve Act of 2000 began on the House floor. Christopher Hatcher, legislative director for Scott McInnis, recalled that the debate had originally been scheduled for the early afternoon but that Hefley had been unavoidably delayed. Rather than proceed without him, all parties agreed to postpone the debate until Hefley could attend, which Hatcher later called "an impressive display" of congressional courtesy and decorum. Perhaps compelled by a similar sense of courtesy to Hefley, Speaker Dennis Hastert chose not to attend the proceedings.[41]

Following a reading of the bill by the Clerk of the House, Scott McInnis stood next to a large photograph of the Great Sand Dunes and briefly described the background of the bill and his reasons for introducing it. Rep. Mark Udall (D-Colo.) spoke next and offered a detailed explanation of his support for the bill, followed by the reading of an extended statement from Colorado attorney general Ken Salazar. When his turn came, Joel Hefley left little doubt about where he stood on the bill:

> I must object to the bill before us. . . . This bill has never been the subject of hearings in the House of Representatives before the Committee on Resources. . . . Now there are a number of questions to be answered. First, most National Park Service regulations say that a park comprises a variety of resources. Now I know the proponents of this would say that there are a variety of resources. There are mountains, there are streams and so forth, but the basic thing is there is a pile of sand, a beautiful pile of sand. But that is the basic resource for this park. . . . I think it is a great national monument, but I do not think it rises to the level of a national park.[42]

Hefley continued by touching on several issues, such as the bill's unusual provisions for land acquisition by the interior secretary and the current disposition of the Baca Ranch purchase, before focusing directly on the San Luis Valley's water and his rumored interest in it. "Lastly, there is a question of water beneath the dunes. One of the main reasons for this bill is to stop the speculation of water in that valley," Hefley declared. "Now, I do not want the water in that valley to come to the front range of Colorado. I do not want it to come to Colorado Springs, Aurora, or anywhere else. I want that water to stay in the valley."[43] To sum up his arguments against a national park designation for the Great Sand Dunes, Hefley finished by accusing McInnis and his supporters of circumventing the customary process of hearings and testimony before the appropriate committees, neglecting to mention that he had refused to grant McInnis's request for a hearing for several months:

> All I am asking is we go through the normal process; we have the hearings, and we make a decision based upon the merit, not based upon who can put the most pressure on the Speaker. This did not come out of committee; this came out of the Speaker's office. He put it on the calendar. I do not know why he put it on the calendar and circumvented the whole

process. I do not think he should have, but this should not be based on that. It should be based on merit.[44]

After a statement of "strong support" for the bill from Rep. Diana DeGette (D-Colo.), McInnis again rose and responded directly to Hefley:

I want the gentleman to know, I have gone to the committee. I have gone to my good colleague, and I say this with all due respect, because our dispute is a professional dispute, not a personal dispute, but I have gone to the gentleman and said, give me a hearing. I want this bill heard on its merits. Let it rise or fall on its merits. . . . I was denied the hearing month after month after month.[45]

McInnis also pointed out that Speaker Hastert's decision to bypass Hefley's committee and take the bill directly to the floor for a vote was perfectly legal and allowable under existing House rules, to which Hefley replied, "I guess I would just close by saying again, yes, this is part of the process; but it is a subversion of the process."[46]

Incredulous, McInnis responded by reminding Hefley of his repeated requests for a hearing on the bill:

Mr. Speaker [pro tempore], again to the colleague, talk about subversion of the process, subversion of the process occurs when you cannot even get a committee hearing. I will not embarrass the gentleman by asking him, but I would if I were in some kind of real knock-down-drag-out, ask the question, did I not in fact request that this go to the committee? Did not the gentleman in fact request that it not go to the committee?[47]

McInnis concluded his allotted time by emphasizing the bill's widespread support across the political spectrum, including unanimous support in the Senate, before submitting a letter from the attorney representing Farallon Capital Management confirming their support of the proposal and willingness to sell the Baca Ranch to the federal government on successful enactment of S. 2547. The contentious debate, which one reporter later described as a "heated exchange," ended with a voice-vote that seemed to indicate unanimous support for McInnis's bill before Representative Hefley, defiant to the end, rose from his seat and demanded a roll-call vote.[48]

The following day, October 25, the House of Representatives voted overwhelmingly in support of the bill, 366 to 34, a tally that, as Sen. Wayne Allard put it, "drives a stake through the heart of any proposals to transfer water out of the San Luis Valley."[49] Not surprisingly, Joel Hefley expressed a different sentiment. "I think it cheapens the title of national park," he said after the vote. "I don't think that national park means the same thing today that it meant yesterday."[50] Hefley's fellow Colorado lawmaker Rep. Bob Schaffer (R-Colo.) also voted no on the bill, largely due to his concerns over federal acquisition of the Baca Ranch. "The notion that somehow the Sand Dunes are going to fall apart or go away unless the government owns another 100,000 more acres is ridiculous," Schaffer claimed. "The Sand Dunes have done just fine under the responsible stewardship of private landowners who have owned the Baca Ranch for decades."[51] Like Hefley, Schaffer believed that the House should have held hearings on the proposal and claimed the legislation had been "railroaded through Congress," despite the fact that it was Hefley himself who had repeatedly refused to convene hearings on the matter.[52]

In the end, the opinions of the small minority in opposition fell on deaf ears; the bill convincingly passed both houses of Congress with overwhelming bipartisan support. The next stop was the White House, which received the Great Sand Dunes legislation on November 14. The Clinton administration, strongly supportive of expanding lands in the public domain, had given every indication that the president would sign the bill, but before he had the chance to do so, White House budget director Jacob Lew received a letter from Secretary of Agriculture Dan Glickman urging the president to veto the bill. Glickman apparently took issue with the bill's requirement that some 44,000 acres of national forest land be transferred to the National Park Service. Glickman believed the bill should allow the transfer, not order it. He declared, "The Department of Agriculture does not support the changes made to the bill and recommends that the president veto the bill. Transferring this middle third (44,000 acres) of the total wilderness area to the National Park Service would not enhance the protective status of the area but would bifurcate management responsibility and add an unnecessary level of bureaucratic complexity."[53]

Spokesmen for McInnis and Allard blasted Glickman's claim. "They were having heartburn over the fact that they would have to give up some of their land to the National Park Service," said McInnis spokesman Josh Penry.[54] Allard spokesman Sean Conway was somewhat less charitable when he described the disagreement as "all about agency envy. It's all about an internal struggle. It's childish, it's immature, it's unprofessional. They are using these very internal concerns to try to deep-six or stop legislation that has great bipartisan support. Quite frankly, they have spent a lot of political capital on nothing."[55]

Faced with such withering criticism, Glickman quickly dispatched another letter to the White House, dated November 16, and formally withdrew his agency's veto request. "The Department of Agriculture has some serious concerns regarding the bill that in other circumstances might warrant a veto," Glickman wrote, "but does not object to the president signing the bill." A spokeswoman for the Agriculture Department was unable to comment on the conflicting messages, and six days later, on November 22, President Bill Clinton signed Public Law 106-530, the Great Sand Dunes National Park and Preserve Act of 2000. The act expanded the boundaries of the existing monument by nearly 70,000 acres and established the Great Sand Dunes National Preserve from roughly 42,000 acres of the Rio Grande National Forest. It also authorized the purchase of the Baca Ranch, an acquisition that would fulfill the bill's requirement that the new park possess "sufficient land having a sufficient diversity of resources." Once the Baca had been acquired by the Department of the Interior, the act mandated the creation of Great Sand Dunes National Park, along with the Baca National Wildlife Refuge.[56] Scott McInnis, who had shepherded the bill since its inception, described himself as "overjoyed" that the legislation had finally become law. Six long years had passed since he raced Superintendent Bill Wellman down that precipitous face of the Great Sand Dunes, and even though the marathon race to establish a national park at the dunes was closer than ever to finishing, McInnis knew there was still a bumpy road ahead, especially where that road intersected with the Baca Ranch.[57]

The Final Piece

With the president's signature came the immediate transfer of authority over U.S. Forest Service lands within the new preserve boundary to the National Park Service, along with an official name change to Great Sand Dunes National Monument and Preserve. The "national park" designation would have to wait for official acquisition of the Baca Ranch. Minutes after the bill was signed into law, Sen. Wayne Allard, who back in December 1999 had written an op-ed piece for the *Denver Post* describing his work with The Nature Conservancy on its efforts to acquire the Baca Ranch, told a reporter during a telephone interview that he and his fellow lawmakers were not involved with the current purchase negotiations. "We're staying out of it," he said. "We don't want to meddle." Following Interior Secretary Babbitt's sage advice to "let the lead horse go out there and pull the wagon," Allard and his colleagues were content to let The Nature Conservancy negotiate with Farallon to purchase the historic ranch property. According to Superintendent Steve Chaney, officials "at the Secretary of the Interior level" were also involved in the process.[58]

On November 24, two days after the president signed the Great Sand Dunes bill, Lexam Explorations, Inc., a Canadian company involved in gas and oil development, announced plans to spend $8 million drilling two deep exploration wells a few miles southeast of Crestone on the Baca Ranch. For many in the valley, the timing of the announcement seemed rather suspicious, as if Lexam were attempting to inflate the value of the Baca Ranch just as negotiations for its sale to the federal government were getting under way in earnest. Lexam had acquired the majority of the mineral, oil, and gas rights under the ranch in a series of deals with AWDI and Newhall Land and Farming Company between 1987 and 1996. The company originally sought gold and other precious minerals on the property, but in the process had discovered strong evidence of oil and gas instead. Lexam claimed that ten years of research indicated the presence of up to

1 *trillion* cubic feet of natural gas beneath the Baca, a figure that surprised many geologists and experts in the energy industry. "Some people say we are only doing this to ratchet up the price," said a Lexam spokesman, "but we've been on this for a long time. We've been exploring there since 1987." Unfortunately, the federal government had no recourse to prevent such drilling, because AWDI had legally separated the mineral rights from the land and water rights on the Baca Ranch back in the 1980s. As Great Sand Dunes superintendent Steve Chaney put it, "If they have the mineral rights, it's theirs to do what they can and want to with. Once we own the property, we ultimately would like to acquire the mineral rights."[59]

The drama surrounding the Baca Ranch took another strange turn in January 2001 when Gary Boyce, acting as president of Stockman's Water Company, filed a lawsuit in U.S. District Court against Farallon Capital Management, charging breach of contract for negotiating with The Nature Conservancy for the sale of the ranch. Boyce sought an injunction against Farallon to prevent it from completing the transaction, claiming that Stockman's had not been notified of the negotiations, nor did it consent to any sale. Although an attorney for Farallon dismissed Boyce's claims as "wholly without merit," the suit did have the potential to cause significant delays and disruptions in the negotiating process. Also in January, the federal government signed an initial agreement to acquire the Baca Ranch from The Nature Conservancy over an extended period of time, provided TNC successfully negotiated the purchase.[60] Meanwhile, Colorado's congressional delegation continued seeking an estimated $25 million to finance the proposed purchase from the administration of newly elected President George W. Bush, with Senator Allard and Representative McInnis taking the lead by lobbying new Interior Secretary Gale Norton, who coincidentally had served as Colorado's attorney general from 1991 to 1999. The lawmakers had an important ally in Sen. Ben Nighthorse Campbell, who served on the powerful Senate Appropriations Committee and remained deeply committed to acquiring the Baca Ranch. Their efforts appeared to be successful in early April 2001 when the White House released a budget that included appropriations totaling some $360 million for the federal Land and Water Conservation Fund, a program administered by the National Park Service that would ultimately provide significant funding for the Baca purchase.[61]

As negotiations continued between Farallon, The Nature Conservancy, and the Interior Department, an unlikely Ivy League connection to the San Luis Valley began to emerge just before Thanksgiving 2001. Christine Canaly, who had joined the San Luis Valley Ecosystem Council in June 2000 after her pioneering work with Citizens for San Luis Valley Water during the 1980s and 1990s, received an anonymous phone call from someone seeking detailed information about Farallon and Stockman's Water Company. Suspicious at first, Canaly finally agreed to the request after the caller convinced her that his intentions were honorable and that he would call back with information that may in fact prove useful to the Baca purchase. As promised, the caller contacted Canaly again in early January 2002 and disclosed that he was a Yale University graduate student, one of many Yale students investigating the university's endowment fund and investment practices on behalf of a coalition of labor unions representing thousands of workers at Yale University and Yale–New Haven Hospital. The results of their inquiries, first posted January 14 on YaleInsider.org, a website operated by the Federation of Hospital and University Employees, revealed that Yale owned 50 percent of Vaca Partners, the for-profit limited partnership formed by Farallon Capital Management to fund the purchase of the Baca Ranch in 1995. Canaly revealed the surprising connection, which she described as "incredibly disappointing," to a board meeting of the Rio Grande Water Conservation District on January 15, noting that the information "means Yale backed those ballot initiatives we had to fight," referring to Gary Boyce's efforts to bankrupt the district back in 1998.[62]

News of Yale's involvement hit the local press and sparked an immediate outcry, not only in the San Luis Valley but also in the offices of Sen. Wayne Allard, who first learned about the controversy in an article published in the *Pueblo Chieftain* and immediately contacted Yale University president Richard Charles Levin seeking an explanation. "Yale University's relatively unknown involvement raises some important issues," Allard wrote, alluding to the repeated efforts to export the valley's groundwater—efforts that, "if successful, potentially would have ravaged both the economy and people of the area." His letter continued with a condemnation of Yale's role in fleecing the public. "I find the fact that Yale University was involved in these attempts extremely disappointing," he admonished. "Simply stated, for a respected institution that participates in advising the federal government on responsible public policy to participate in a business venture that will end in millions of dollars in costs for taxpayers is a great public disservice."[63]

Allard's message clearly hit a nerve with Levin, who responded with a promise to donate all of Yale's profits from the sale of the Baca Ranch to The Nature Conservancy. Allard was quick to praise Levin for his pledge. "Once President Levin learned about the history of the San Luis Valley and the struggle [residents] have gone through to protect the Sand Dunes and their agricultural heritage, he recognized that Yale University should not have been involved in this venture and wanted to make things right," Allard explained. "His prompt response to my request is greatly appreciated and should be praised by all Coloradans."[64] The feelings of goodwill must have been contagious, because on January 30, 2002, The Nature Conservancy announced that it had signed an agreement with Farallon Capital Management to purchase the 97,000-acre Baca Ranch for a reported $31.28 million, making it TNC's most expensive acquisition ever in the Rocky Mountain West. Financing for the deal came from a consortium of private, state, and federal partners and included $10.2 million from the Land and Water Conservation Fund; $7 million in a low-interest loan from the David and Lucille Packard Foundation; $5 million from the Colorado State Land Board; a $3 million loan from the Great Outdoors Colorado Trust Fund; and additional private fundraising coordinated by TNC. Officials cautioned that finalizing the purchase was contingent on successful resolution of several lawsuits between various owners of the Baca currently working their way through state and federal court. They also noted that until additional federal appropriations could be secured to repay The Nature Conservancy and the other lenders, TNC, the Interior Department, and the Colorado State Land Board would hold the land jointly.[65]

Predictably, news of the triumphant purchase agreement touched off impromptu celebrations throughout the San Luis Valley. The *Valley Courier* reported that Superintendent Steve Chaney was "practically turning cartwheels" at the prospect of finally witnessing the creation of Great Sand Dunes National Park.[66] Praise for The Nature Conservancy was understandably widespread and enthusiastic, prompting TNC's Colorado director, Mark Burget, to share the accolades:

> The Nature Conservancy has been working with local communities for years to preserve the magnificent natural wonders of the San Luis Valley. The preservation of the Baca Ranch has long been the community's most important goal. It is a beautiful and remarkable landscape that rightly should be included among America's great national parks. We are honored to be able to play a role in the establishment of a new national park for Colorado, especially with the bipartisan support of so many people in the San Luis Valley and across the state.[67]

Rep. Scott McInnis summed up his sentiments by focusing on the key role the acquisition of the Baca Ranch would play in finally ending the battles over the San Luis Valley's water, stating for the record that the "acquisition of the Baca Ranch was the centerpiece of our efforts to preserve the Great Sand Dunes ecosystem in its totality. With the Baca, the Great Sand Dunes National Park and Preserve will be second to none in natural beauty and ecological diversity. Equally important, the purchase puts an end to the San Luis Valley's water wars once and for all."[68]

Like McInnis, virtually every politician, local official, and common citizen who issued a statement or spoke to a reporter that day referred to the purchase of the Baca Ranch as a foregone conclusion, when in fact the successful completion of the complex deal was anything but certain. Considering how rapidly the legislation authorizing the purchase of the Baca Ranch had made it through Congress, perhaps many in the San Luis Valley were simply expecting a similarly swift resolution to the agreement between The Nature Conservancy and Farallon. In reality, there were still several serious challenges ahead, not the least of which was the issue of securing additional federal funding for the purchase, along with successfully unraveling the tangled web of lawsuits enveloping the current and former owners of the Baca Ranch. In the euphoric aftermath of the historic purchase agreement between TNC and Farallon, few could have guessed that more than two years would pass before the deal would finally be completed.

The legal squabbles involved not only Farallon/Vaca Partners and Cabeza de Vaca Land and Cattle Company (the nominal owner of the Baca Ranch), but also Gary Boyce's Stockman's Water Company, which owned a 37.5 percent interest in the ranch; New York investor Peter Hornick, who owned a 12.5 percent interest in the property; and AWDI, the San Luis Valley's old nemesis, which had sold the Baca Ranch to Cabeza de Vaca in 1995 but retained a 10 percent interest in any future gross revenues realized by the sale of the property's water or water rights.[69] Boyce had already filed suit against Farallon, while Hornick and AWDI were demanding exorbitant compensation for their respective interests in the ranch and sought to block its sale to The Nature Conservancy unless they received a generous return on their investments. The complex series of lawsuits, rulings, and appeals that ensued threatened to delay or even completely derail the fragile purchase agreement. A potential breakthrough came on May 8, 2002, when Saguache County district judge Robert Ogburn, the same judge who had presided over the original AWDI trial back in 1991, ordered the foreclosure sale of the Baca Ranch to satisfy an outstanding $8 million debt that Vaca Partners claimed it was owed by Cabeza de Vaca Land and Cattle Co., a claim that Cabeza did not dispute. Stanley L. Garnett, attorney for Vaca Partners, explained in a brief that the foreclosure action was "related to the continuing effort to sell the Baca Ranch to The Nature Conservancy to create a national park":

> There is litigation pending in state and federal court regarding the various issues involving American Water Development Inc. and (Peter) Hornick (both of whom oppose the plan to create a national park). As a result of that litigation, Cabeza and Vaca Partners agreed with The Nature Conservancy that, as a fall-back position in the event of a foreclosure by Vaca, The Nature Conservancy would purchase the ranch from Vaca rather than Cabeza. This permits the sale to go forward so that the national park can be created while the claims of the various parties are resolved in an orderly fashion in other litigation.[70]

As ordered, on May 30 the Baca Ranch was "sold" at a foreclosure sale to Vaca Partners. Once again, an elegant solution to a seemingly intractable problem had appeared at precisely the right moment, and the ostensibly charmed process of creating a national park at the Great Sand Dunes inched ever closer to its final destination.

With the danger of litigation delaying or even stopping the sale of the Baca Ranch to The Nature Conservancy effectively neutralized, attention turned to securing the massive funding necessary to finalize the deal and fulfill some of the other requirements outlined in the original authorizing legislation. In February 2003, Rep. Scott McInnis announced that the House of Representatives had included $12 million for the Baca purchase in its omnibus Interior Appropriations bill.[71] The following May, Interior Secretary Gale Norton appointed a ten-member Great Sand Dunes National Park Advisory Council, as required by the 2000 legislation. The council's mandate included advising the secretary regarding long-term planning for the new park, as well as assisting in the development of the park's official General Management Plan.[72] In July, Sen. Wayne Allard announced that the Senate Appropriations Subcommittee on the Interior had supported his request for $11 million in funding, followed by an announcement in October that a joint committee of House and Senate members had approved $9 million for the next fiscal year.[73] Then, in January 2004, Superintendent Steve Chaney announced that Yale University, as promised, had donated its $1.5 million share of the profits from the sale of the Baca Ranch to The Nature Conservancy. Chaney also said that TNC would turn over large portions of the Medano Ranch to the NPS for inclusion in the soon-to-be-created Great Sand Dunes National Park.[74] Some grumbling about the precise amount of Yale's donation ensued, mostly among Yale students, who claimed that Yale had originally promised $4 million, a charge that Yale officials denied. "Yale will live up to that agreement, whatever it promised, but it did not promise $4 million," said a university spokesman. "It promised the profits." Regardless, The Nature Conservancy was delighted. "We're not disappointed at all," explained TNC's Colorado associate director, Charles Bedford. "It's a spectacular gift."[75]

Excitement continued to build in March 2004 when $31.28 million, the purchase price of the Baca Ranch, was placed into escrow at a California title company. "Our long journey is nearing its happy destination," said Sen. Wayne Allard in a statement. The federal government still had to come up with approximately $3.5 million in additional funding to reimburse The Nature Conservancy, but officials were confident the money would soon be forthcoming from Congress, clearing the way for the Baca Ranch to be officially transferred in various allotments to the National Park Service, the U.S. Fish and Wildlife Service, and the U.S. Forest Service.[76] Appeals of earlier rulings in the lawsuits filed by Peter Hornick and AWDI against Vaca Partners / Farallon were still pending in court, potentially jeopardizing the federal government's ability to obtain clear title to the property, so in August 2004 Sen. Wayne Allard sent a letter to the U.S. Department of Justice seeking a waiver for the Department of the Interior "so the land transaction that will mark the final step in the creation of our nation's newest park can go forward." Officials with the Justice Department reviewed the facts in the matter, determined that the pending appeals posed little or no threat to the process of obtaining clear title to the Baca Ranch, and granted the waiver as requested, clearing the way for the long-awaited deal to close. The last piece of the complex puzzle of land comprising Great Sand Dunes National Park and Preserve was finally in place.[77]

Monument to Park

On September 10, 2004, much to the delight of the dedicated individuals and organizations that had struggled together to achieve the goal of a new national park at the Great Sand Dunes, Vaca Partners officially transferred ownership of the historic, 97,000-acre Baca Ranch to The Nature Conservancy. As stipulated by the 2000 legislation, "sufficient land having a sufficient diversity of resources" had at last been acquired. Three days later, on September 13, a standing-room-only crowd gathered in the Great Sand Dunes amphitheater under a brilliant late-summer sun to witness Interior Secretary Gale Norton officially designate Great Sand Dunes National Park and Preserve, along with the Baca National Wildlife Refuge:

> Today we dedicate and proclaim a new national park to forever preserve a landscape sculpted by wind and water and we introduce what now becomes the largest national wildlife refuge in Colorado. This area is a haven for wildlife and wonder to modern day visitors. The Great Sand Dunes National Park and Preserve is a living hourglass. The ebb and flow of rushing wind and pulsing water sweeps across the landscape, creating a natural sculpture worthy of permanent preservation.[78]

In addition to Norton, dignitaries in attendance included Sen. Ben Nighthorse Campbell, Rep. Scott McInnis, NPS director Fran Mainella, NPS regional director Steve P. Martin, Great Sand Dunes superintendent Steve Chaney, former Colorado director for The Nature Conservancy Mark Burget, Colorado Department of Resources executive director Russell George, and Mike Blenden of the U.S. Fish and Wildlife Service. Colorado attorney general Ken Salazar and Sen. Wayne Allard both sent representatives with their statements.

Against the stunning backdrop of the Great Sand Dunes, speeches were given, statements were read, and congratulations offered all around to the numerous individuals and organizations most responsible for protecting this unique and incongruous landscape. Senator Campbell commended Allard and McInnis for getting the Great Sand Dunes bill through Congress so quickly. "They got done in one year what it took me fifteen years to get done with the Black Canyon of the Gunnison," he joked, while Mark Burget, who described the process as "the most amazing thing I've been involved with in conservation," thanked the people of the San Luis Valley for "making history today" and emphasized that the partnership and common purpose involved was an example for the rest of the world that The Nature Conservancy would continue to use elsewhere. Senator Allard's statement thanked the local citizenry as well, hailing the dedication as a "tremendous triumph for the people of the San Luis Valley," a sentiment shared by Ken Salazar, who called it "a momentous day for my native San Luis Valley and its people, for the State of Colorado, and indeed the nation."[79]

Figure 8.2 Dignitaries including Interior Secretary Gale Norton (*right, at podium*) address a large crowd gathered in the park amphitheater for the official dedication of Great Sand Dunes National Park and Preserve and the Baca National Wildlife Refuge on September 13, 2004. National Park Service, Great Sand Dunes National Park and Preserve.

With a stroke of her pen, Interior Secretary Gale Norton officially designated some 53,000 acres of meadows, sagebrush, and wetlands for inclusion in the Baca National Wildlife Refuge, bringing its total area to approximately 92,500 acres; transferred roughly 13,000 acres of mountainous terrain, including 14,165-foot Kit Carson Peak, to the Rio Grande National Forest; and protected 31,000 acres of dunes and sand sheet within the boundaries of Great Sand Dunes National Park and Preserve, giving it a total area of 149,512 acres, including over 107,000 acres in the national park and nearly 42,000 acres of the vital Sangre de Cristo watershed contained in the national preserve. Most critically, the surface and subsurface water resources of the Baca Ranch were now protected in perpetuity, ensuring that the entire aeolian system responsible for the creation and continued existence of the Great Sand Dunes would remain intact for generations to come. It also meant that victory could finally be declared in the divisive and contentious San Luis Valley water wars.[80]

On September 15, 2004, Sen. Wayne Allard announced that the Senate Appropriations Committee had included the remaining $3.4 million for the Baca Ranch acquisition in the 2005 appropriations bill, which guaranteed full reimbursement for The Nature Conservancy's significant investment in the property. With the publication of a "Notice of Establishment" in the Federal Register on September 24, establishment of Great Sand Dunes National Park and Preserve became official.[81] Fifteen long years had passed since Scott McInnis and Bob Zimmerman first discussed the idea back in 1989, but the actual legislative and legal process necessary to create the park had in fact been unusually rapid. From the "Summit in the Sand" to the signing of the Great Sand Dunes bill in November 2000 had taken less than a year, and less than four years from that point to the official designation. By every measure, the process of transforming the monument to national park status had been a smashing success, yet administrative conditions on the ground actually changed very little. The NPS assigned an additional backcountry ranger to help deal with new regulations and the expanded law enforcement responsibilities in the National Preserve and along the Medano Pass road, and the NPS also assumed oversight of various outfitters who arranged guided hunts in the area, but much of the sand sheet and sabkha regions of the Medano Ranch to the west of the main dunefield remained under the nominal control of The Nature Conservancy and therefore were off-limits to the majority of park visitors. Herds of bison still roamed throughout the property, posing a potential hazard to curious tourists, and the few primitive roads that did cross the ranch were unsuitable for all but the most rugged four-wheel-drive vehicles, making it unlikely that the NPS would allow unlimited tourist access in the foreseeable future, thus no additional personnel were required for that area of the park. The park's Resource Management Division, which added a trained biologist to the staff, did face some additional challenges with the redesignation, primarily because expansion of the park's boundaries had considerably broadened the field of scientific inquiry to realms far beyond the main dunefield, including an impressive variety of ecosystems from the sabkha

Figure 8.3 North gate of the Baca National Wildlife Refuge, looking east toward the imposing barrier of the Sangre de Cristo Mountains. Photo by the author.

and sand sheet to the crest of the Sangres, with all of the attendant flora, fauna, and archaeological sites they contained. These diverse environments needed to be investigated, catalogued, and analyzed for consideration in the forthcoming General Management Plan that would guide resource management strategies at the park for the next fifteen to twenty years, which understandably increased the workload of the existing resource staff, but did not require hiring legions of additional personnel.[82]

In early October 2004, less than two weeks after the park designation became official, a newly remodeled and expanded visitor center opened to accommodate the expected increase in visitors. As predicted, a slight upsurge in visitation did occur, growing from 267,204 visits in 2004 to 279,589 in 2005, but the numbers continued to fluctuate over the next several years, dropping to less than 259,000 in 2006 before growing to nearly 290,000 in 2009. Despite this gradual increase, the massive gateway development feared by opponents of the new national park simply never materialized. No new hotels or campgrounds were built; no fast-food joints or gas stations or convenience stores suddenly appeared outside the main entrance to mar the stunning visual approaches to the dunes. In fact, the typical visitor experience at the Great Sand Dunes remained virtually unchanged as a result of the redesignation. The overwhelming majority of tourists still confined their activities to the main dunefield, Medano Creek, Pinyon Flats campground, and the rugged trails in the adjacent foothills of the Sangres, just as they had for decades since the creation of the original national monument in 1932. Beyond the new name on the main entrance sign, not much else seemed outwardly different for the casual visitor.[83]

Figure 8.4 New entrance sign at Great Sand Dunes National Park and Preserve. Photo by the author.

What did change as a result of the redesignation was the prevailing perception that the Great Sand Dunes were somehow isolated and disconnected from the landscapes and water resources that surround them and could therefore be managed essentially as an island. With the addition of the Sangre de Cristo watershed, the sand sheet, and the sabkha within the new national park boundaries, the immense sea of sand was finally being managed as part of a much larger ocean of ecosystems, each connected to the others, each dependent on precious water for its continued existence and vitality, each vulnerable to ecological changes. Ironically, the Great Sand Dunes exist as a result of constant, pervasive, inexorable change, literally transforming grain by grain from moment to moment for centuries beyond counting, washed and blown and formed and reformed by the same San Luis Valley winds and waters that caused their existence in the first place. Yet to the naked eye, the dunes also appear to be remarkably static, apparently unchanging from day to day or even year to year. Charged with managing this uniquely enigmatic landscape is the National Park Service, which is required by law to protect and preserve the scenic and historic landscapes of the United States "in such manner and by such means as will leave them unimpaired for the enjoyment of future generations."[84]

In order to fulfill the requirements of this mandate, the NPS relied on the abundant scientific data accumulated during the costly and contentious San Luis Valley water wars of the 1980s and 1990s and concluded that protecting the Great Sand Dunes required expanding the boundaries of the original monument to include the entire aeolian system that created and continues to sustain the dune-building process. It is unlikely, however, that the NPS could have accomplished such an ambitious goal without the unwavering support of the tenacious citizens, businesses, and grassroots organizations of the San Luis Valley, who also recognized the potential threat posed to their beloved dunes by repeated attempts to mine and export the valley's lifeblood of water, and therefore decided *collectively* to agitate for permanent protection by rallying around the concept of a new Great Sand Dunes National Park. With additional assistance from an assortment of local, state, and national politicians who understood how to the manipulate the levers of power in Washington, as well as the incredibly generous and timely financial support of The Nature Conservancy and other unselfish benefactors, the dream of preserving the Great Sand Dunes and the irreplaceable water resources of the San Luis Valley in perpetuity had at long last become a reality.

Ultimately, the successful creation of Great Sand Dunes National Park and Preserve raises the question of whether a new paradigm has emerged for public lands policy in the American West, and perhaps elsewhere in the world. After all, unlike the tortuous, decade-long political process undertaken by Sen. Ben Nighthorse Campbell that created Black Canyon of the Gunnison National Park, the redesignation of Great Sand Dunes National Monument as a national park happened relatively quickly, with (mostly) bipartisan political support at

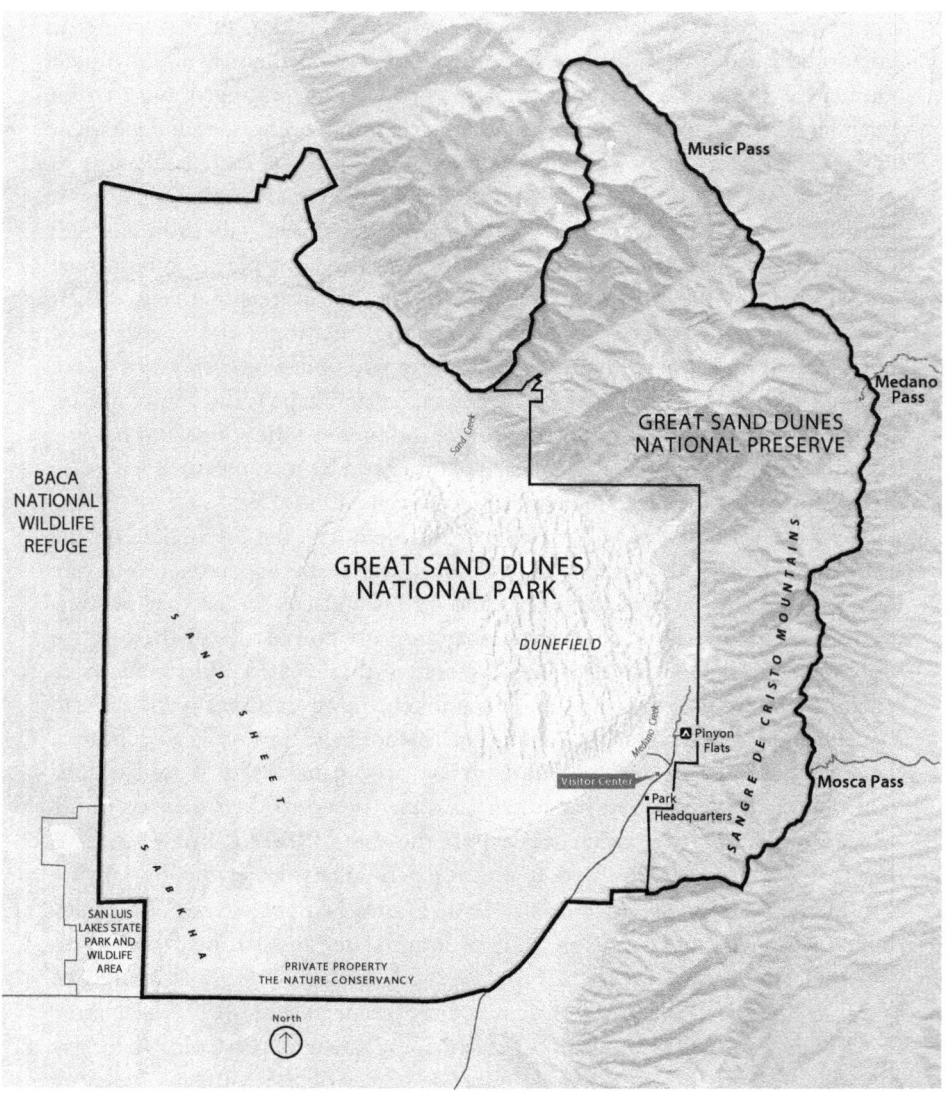

Figure 8.5 Updated map for Great Sand Dunes National Park and Preserve. National Park Service, Great Sand Dunes National Park and Preserve.

both the state and national levels. Coupled with the unprecedented coalition of disparate individuals and organizations, each with their own interests and motives, that came together in common purpose to preserve not only a stunning and spectacular landscape but also a way of life in the San Luis Valley, the Great Sand Dunes example would seem to be an ideal model for land conservation across the planet. Yet the unique conditions and precise opportunities that led to the acquisition of such massive quantities of private property and its subsequent transfer into the public domain are not likely to be repeated in too many other locations around America or the world, especially if precious resources such as abundant fresh water are involved. Thus, rather than a completely new paradigm for conservation, the creation of Great Sand Dunes National Park and Preserve instead demonstrates that effective, commonsense stewardship of public lands first requires a comprehensive understanding of how landscapes and ecosystems interrelate and evolve, as well as how humans influence and interact with the environments around them. That knowledge, along with a recognition that the process of creating and administrating national parks and monuments occurs within a political world, must then be used to influence public policy through grassroots community involvement and an often messy political process, a timeless lesson that can be applied to land conservation anywhere.

The question of why so many different individuals, communities, and organizations cooperated so willingly on the effort to protect the "magnificent pile of sand," as Rep. Joel Hefley had described the Great Sand Dunes, is open to interpretation, but perhaps it's because preserving wild places says something profound about humanity, that we can recognize and appreciate the diverse landscapes in our midst, beyond their obvious value as exploitable commodities, as well as understand the historical connections that inevitably develop between humans and the lands they inhabit. In other words, protecting the Great Sand Dunes gives us the opportunity to imagine generations of humans gazing on this same immense sea of sand, providing us with a crucial link to our own distant past. After all, humans and the Great Sand Dunes share a common history stretching back thousands of years, from the earliest Paleoindian cultures to modern-day tourists who come to gawk and wonder at the sight of gigantic sand dunes piled at the base of soaring mountains. Each successive wave of humanity has imposed its own values, technologies, and cultural beliefs upon the Great Sand Dunes and the San Luis Valley, with wildly differing impacts and outcomes. For example, archaeological evidence indicates that the various Paleoindian cultures that frequented the San Luis Valley in prehistoric times clearly affected mammoth and bison populations in and around the Great Sand Dunes, but these early humans built no fences or roads, constructed no reservoirs or extensive irrigation systems to mark their presence.

Similarly, later Native American cultures certainly impacted local game animals, burned portions of the forests, and scarred the local ponderosa pine trees, but their subsistence activities evidently created few other tangible changes in the landscapes of the Great Sand Dunes or the San Luis Valley. With the arrival of the Europeans, the Native Americans who had been the valley's primary inhabitants for centuries were forced out, their ability to move about the land in search of sustenance severely constrained, their traditions of interacting with the environment no longer tenable. Later Spanish and Anglo-American settlers brought with them their own concepts of market value and scarcity, concepts that had been shaped by the social and ecological conditions in Europe and the eastern United States. They perceived the San Luis Valley as a landscape of great abundance and thus exploited its bounty of animal, timber, mineral, and liquid wealth. With widespread settlement in the late nineteenth century, along with technological advancements that continued well into the twentieth century, the development of large-scale agriculture and irrigation resulted in dramatic changes to the landscape of the valley, creating massive farms and ranches that met not only the needs of local inhabitants, but also the demands of distant markets for cattle, farm produce, and other valuable commodities.[85]

Yet the concept of "value" can also change with surprising rapidity, even within the same culture. For Anglo-American farmers and loggers of the nineteenth-century San Luis Valley, natural vegetation was simply an obstacle to be plowed under, timber a resource to be cut and milled, water a necessity to be exploited for mining operations or irrigation. Now, barely more than a century later, increasing demands for recreation and the "wilderness experience" have conferred a different value upon these same resources, primarily because of their ability to attract tourists and sportsmen, and thus economic prosperity, to the San Luis Valley. Likewise, early settlers initially found few valuable resources beyond irretrievable traces of flour gold in the midst of the Great Sand Dunes, but now the dunes themselves are the valuable resource, attracting waves of visitors who come to relax and recreate in what is perceived to be a unique and thus valuable landscape.

Historic utilization of the San Luis Valley's water resources provides perhaps the best example of this evolving concept of value. To the Tewa Pueblo, the valley's water was the essential source of life itself, a sacred symbol that defined the very beginnings of existence. In contrast, modern water developers simply perceived the valley's water resources as liquid gold, an extractable commodity offering the potential for tremendous profit for a lucky few. The water is the same, but different cultures have wildly divergent concepts of its value and its role in human society. By persistently refusing to allow that precious and valuable water to be exploited, as well as by demanding protection for the uniquely enchanting landscape of the Great Sand Dunes, the citizens of the San Luis Valley have wisely chosen to perpetuate their enduring historic connections to the lands they inhabit, unimpaired for generations to come. That, in and of itself, is a legacy of preservation worthy of acclaim, and justification enough for protecting such a magnificent pile of sand.

Aftermath

On December 30, 2004, the National Park Service filed an application for rights to all unappropriated groundwater in the unconfined aquifer beneath Great Sand Dunes National Park as authorized by the Act of 2000. The nonconsumptive, "in place" water right, which was based on the federal "nonreserved" doctrine that allows the federal government to acquire unappropriated water rights pursuant to state law, did not allow for the withdrawal of any water from the aquifer, but guaranteed rights to maximum groundwater levels sufficient to support and maintain park resources, including the Great Sand Dunes. During the next three years, while the water right application worked its way through district court, Steve Chaney left the Great Sand Dunes to become superintendent of Redwood National and State Parks in California, replaced by Art Hutchinson, a seventeen-year veteran of the National Park Service, who assumed his new post in November 2006.[86]

In July 2007, the National Park Service / Department of the Interior issued its "Record of Decision" regarding the General Management Plan for Great Sand Dunes National Park and Preserve. After several years of careful study that included input from a variety of interested parties and recommendations from the Great Sand Dunes National Park Advisory Council, the NPS "Preferred Alternative" was selected to guide management decisions at the park for the next fifteen to twenty years. Among other issues, the preferred alternative recommended that the main dunefield continue to be the focus of visitor activity, and also presented options for new (limited) access on the northern park boundary and Medano Ranch, along with recommendations and policies concerning horseback and trail access, historic buildings, wilderness areas, future shuttle transportation, dogs, and the disposition of The Nature Conservancy's bison herd, which would be allowed to continue roaming the Medano Ranch for the time being, or at least until the NPS decided on other viable options. As required by the 2000 legislation, the Advisory Council officially disbanded after completion of the General Management Plan.[87]

In August 2008, Division 3 Water Court judge O. John Kuenhold signed a decree approving an in-place groundwater right for the Great Sand Dunes as requested by the Interior Department. Explaining his decision, Judge Kuenhold stated that there "is no question that the United States is entitled to make an appropriation of water in connection with the park. The amount and nature of the right . . . are consistent with maintaining ground water levels, surface water levels and stream flows on, across and under the park, in order to accomplish the purposes of the park and to protect park resources and park uses."[88] Much to the delight of courtroom observers, after signing the decree Judge Kuenhold played the 1942 song made popular by crooner Bing Crosby, "Singing Sands of Alamosa."[89]

Later that year, at the request of Great Sand Dunes staff, the NPS Natural

Resource Program Center deployed an acoustic monitoring system in the northwest corner of the park to measure ambient sound levels. The system collected continuous sound-pressure levels in decibels from September 24 to October 10 and determined, much to the surprise of researchers, that the acoustic environment at the Great Sand Dunes was "exceptionally quiet and revealed natural ambient levels that rivaled those in the Haleakala crater" on Maui, thus ranking the Great Sand Dunes as one of the quietest locations ever measured by the natural sounds program.[90]

In early 2009, Great Sand Dunes resource management specialist Fred Bunch was awarded the 2008 Intermountain Regional Director's Award for Natural Resource Management, a distinction that Superintendent Art Hutchinson called "a major achievement in any career." In presenting the award, NPS regional director Mike Snyder cited Bunch's devotion to science and highlighted his recent work at the dunes "to identify, quantify, assess the condition of, monitor and protect significant and often threatened resources in the park, including ground water, wetlands and other 'vital signs' of park health, ecological and economic sustainability, bison and other ungulates, natural quiet, dark night sky, archaeological resources, and ethnographic resources."[91] A San Luis Valley native, Bunch has served at the Great Sand Dunes since 1988 and helped establish the park's resource management division. He is the first, and to date the only, chief of the division.[92]

Up on the former Baca Ranch, a three-year-old lawsuit involving the U.S. Fish and Wildlife Service, Lexam Explorations, and the San Luis Valley Ecosystem Council finally ended in September 2010 with the signing of a settlement agreement between the parties. In filing the original lawsuit, the Ecosystem Council alleged that the USFWS had failed to comply with the National Environmental Policy Act (NEPA) when it authorized Lexam to drill two exploratory wells on the 92,500-acre Baca National Wildlife Refuge. The agreement nullified the USFWS's prior approval of the project and required another review of the drilling proposal under NEPA guidelines. In light of the settlement, Christine Canaly, director of the Ecosystem Council, vowed to continue with efforts to purchase and retire the mineral rights to the property from Lexam and its partners, a process that as of early 2012 was still in negotiations but appeared to be heading toward a successful resolution, especially considering that Lexam has expressed a willingness to sell its rights for the right price.[93] Also in early 2012, the National Park Service announced that Lisa Carrico, a twenty-eight-year veteran of the NPS, had been named superintendent of Great Sand Dunes National Park and Preserve, the same position her father, Jim Carrico, held at the National Monument from 1969 to 1975. It was definitely a homecoming of sorts for Carrico. When her father was superintendent, she and her family lived in the house that is now the park headquarters, and she expressed amazement that her "new office will be in what was once my family's living room!"[94]

The year 2012 also marked the eightieth anniversary of the establishment of

Great Sand Dunes National Monument in 1932, as well as twelve years since the signing of the original legislation authorizing the creation of a new Great Sand Dunes National Park and Preserve, and eight years since its official redesignation. Finally, in 2014, the U.S. Mint released the Great Sand Dunes National Park Quarter, featuring a depiction of a father and son playing in the sand near Medano Creek, with the Great Sand Dunes and Mount Herard looming in the distance, an indelible image that represents the enduring legacy of a remarkable landscape, where sand and water and humans have coexisted together for years beyond counting.

Yet through all the name changes, expansions and contractions of boundaries, and the evolution of understanding about the colossal forces that created and sustain them, the Great Sand Dunes themselves have remained virtually the same sentinels of sand they've always been, silently bearing witness to the inexorable passage of time and the historic parade of humanity that has passed in their midst. Centuries ago, the Tewa Pueblo perceived the Great Sand Dunes as a mystical place, a paradoxical aberration of sand and water exuding an undeniably supernatural presence that they acknowledged in their oral histories and origin myths. Modern visitors may or may not experience a similar presence, but the creation of Great Sand Dunes National Park and Preserve will at least give them, and their descendants, the opportunity to experience the same sense of mystery and timelessness.

EPILOGUE | **Lessons in the Sand**

A landscape, in sum, is not just a place, it is a story.

Mark Fiege, *Irrigated Eden*

NATIONAL PARK SERVICE resource management specialist Fred Bunch wheeled the truck off County Lane 6N just west of Stewart's Cattle Guard and bounced to a stop in front of a closed gate. The tawny brown mounds of the Great Sand Dunes and the crenellated ramparts of the Sangres loomed above a stark horizon to the northeast, while immediately ahead of us beckoned the immense sprawl of the Medano Ranch. Once through the gate, Fred, his sister Bev and I endured a kidney-rattling, four-wheel-drive ride over what barely qualified as a passable road and arrived at the ranch headquarters, where we climbed down from the truck to begin our journey of discovery. Like Bev, I had never been to the remote western side of the Great Sand Dunes and eagerly agreed when Fred suggested a guided tour of an area of the national park that few visitors ever have the opportunity to see.

Listed on the National Register of Historic Places, the Medano Ranch headquarters, like the Trujillo Homestead and the Zapata Ranch headquarters, is a remnant of the heady days of pioneer cattle ranching in the San Luis Valley, but a quick assessment of site clearly indicated that it was not yet ready to receive many modern visitors. Every structure suffered from the unmistakable effects of years of neglect and exposure to the harsh climate of the San Luis Valley, and while ambitious plans for rehabilitating the property for use as a research facility or interpretive site have been seriously considered, the significant cost of such an effort has understandably delayed any improvements. As we wandered the grounds peering in the windows of the dilapidated buildings, Fred mentioned that The Nature Conservancy was still in the process of turning over most of the ranch and its structures to the National Park Service, and we took it as a good omen when a huge barn owl hooted its apparent approval.

Leaving the ranch headquarters, we set off northeast across the undulating sand sheet toward our rendezvous with Indian (Big) Spring. The primitive road bucked and heaved the truck with nauseating irregularity. At one point I hit my head on the roof of the cab with such force that I actually had to check one of my teeth to determine if the blow had knocked a filling loose. Mercifully, Fred finally stopped the truck as we approached Big Spring Creek, where we got out to photograph several bison grazing placidly on the lush grasses along the bank. Seeing these massive creatures in the bright afternoon sunlight, with the snow-capped summit of Mount Herard and the Great Sand Dunes rising in the radiant distance, immediately conjured images in my mind of Paleoindian hunters stalking similar prey in this very same landscape. The historical connection was inescapable, like looking backward in time across a chasm of centuries. Unbeknownst to me, an actual physical connection to the remote past was waiting just ahead at Indian Spring.

The utter incongruity of encountering a large pool of sparkling fresh water in the middle of the arid sand sheet was startling enough, but what happened next was an experience I shall not soon forget. While Bev waited at the truck and began preparing a light lunch, Fred led me over a small rise adjacent to the spring and pointed to a small pipe and spigot sticking out of the sand. He bent down and fiddled with the spigot, which slowly produced a reluctant trickle

Figure 9.1 Bison grazing on the banks of Big Spring Creek, 2009. Photo by the author.

of water, and then he told me to have a taste. Fred sensed my hesitation and immediately sampled the water to prove that it was safe, after which I did the same. The water was cool, the taste slightly brackish and sulfurous, the scent redolent of decaying organic matter. Fred was smiling now and asked me to guess the age of the water. Remembering that samples taken from monitoring wells at this and other springs on the west side of the park indicated that it had taken at least fifty years for water from Sand and Medano Creeks to work its way through the dunes' complex hydrological system, I fudged a bit and guessed 200 years. "Try 30,000 years," he replied with a chuckle, seeing the obvious look of utter disbelief on my face. When I protested that the water couldn't possibly be that old, Fred explained that Well Y was only one of many monitoring wells that had been drilled around the Great Sand Dunes, but it was the first one where researchers had encountered the "blue clay layer" so close to the dunes, indicating that the well had been drilled in an area that included both the confined and unconfined aquifers. Drilling through the blue clay layer, which separates the shallower unconfined aquifer from the deeper confined aquifer, produced a water sample from the confined aquifer that carbon-14 analysis had reliably dated to 30,000 years old, plus or minus a century or three.[1]

Stunned by the revelation, I bent down and tasted a few more drops, my head reeling with the idea that this ancient water had fallen as rain or snow sometime during the last Great Ice Age, when Columbian mammoths still roamed the San Luis Valley, most likely centuries before the ancestors of modern humans first set foot on the North American continent. Gazing across the sand sheet at the Great Sand Dunes, the flavor of eons still lingering in my mouth, the visceral connection to the distant past was strikingly real and immediate. Aware of my deep and abiding passion for history, Fred had anticipated the profound effect that tasting the water would have on me and chose to remain silent, allowing me a moment of quiet contemplation before wordlessly beckoning me to follow. He had one more surprise in store on the far side of a large dune to the south, where the ceaseless winds had created a "blowout," or deep depression in the sand.

Leaving Well Y behind, we hiked to the crest of the dune and then down the other side into the blowout, where Fred again motioned to the ground. There, sticking out of the gray sand, was the unmistakable shape of a ribcage, the bones bleached a ghostly white by the relentless sun and the passage of years. At first I thought it was an animal of some kind, but when Fred removed his hat, kneeled down next to the bones, and reverently bowed his head, I realized they were human remains. Again my mind reeled. Who was this person? What were they doing in such a lonely place? Had they been ceremoniously buried here by their family or friends, or possibly murdered here by their enemies? Immediately I imagined a Clovis hunter, or a Tewa Pueblo shaman, or a Spanish conquistador, or a long-lost gold miner, but in the end the identity of the person didn't matter much. He (or she?) was a human who at some point had passed through this landscape, probably marveled at its wonders, perhaps wondered at its secrets. I

Figure 9.2 Indian Spring on the west side of the Great Sand Dunes, 2009. Photo by the author.

stood there dumbstruck in the pitiless wind as Fred finished his silent prayer for the soul of the deceased, then fell in behind him as he rose to his feet and began heading back to the truck.

Fred waited until we were well clear of the blowout, almost as if he wanted to be out of sight and earshot of the remains, before he started talking about what we had just experienced. Having served at the Great Sand Dunes for over two decades, Fred was well acquainted with the most remote, hidden reaches of the park and had occasionally come across human bones in the course of his duties, especially in the vicinity of the park's natural springs and wetlands. It was no coincidence that these latest remains had been discovered near Indian Spring. Evidence of historical human activity in this arid climate is typically associated with available water sources. As head of the Resource Management Division at the Great Sand Dunes, it was Fred's responsibility under the provisions of the Native American Graves Protection and Repatriations Act of 1990 (NAGPRA) to notify the various Native American tribes in the region of any positive identification of skeletal remains, if in fact the remains could be identified, usually by an artifact, funerary object, or other "cultural items" discovered in association with them. Unfortunately, these same tribes also prohibit any kind of DNA or other scientific analysis of the remains or artifacts, making positive identification somewhat problematic. Consequently, human remains that cannot be conclusively identified are rarely moved unless they happen to be in a high-traffic area and instead are simply reburied in situ using "culturally sterile sand or soil," locally gathered material that does not contain archaeological or other cultural material. If the remains are discovered in conjunction with identifiable cultural items, NAGPRA requires the National Park Service to inform tribal representatives of the discovery, at which time arrangements are made for appropriate reburial in situ or at a previously agreed-upon secret location, either in the vicinity of the original discovery if appropriate or elsewhere within the park boundaries.[2]

Fred explained that these latest remains discovered near Indian Springs included no readily identifiable cultural items and therefore would likely be reburied in their current location, but he was still awaiting approval from the regional tribes to do so. As we trudged back to the truck, I wondered again about how long those remains had been hidden in the sand near Indian Spring. Decades? Centuries? In this wild and windswept place, a paradoxical landscape that includes an abundant source of freshwater in the very midst of vast stretches of arid sand, the human story of the Great Sand Dunes at last seemed complete. I had dutifully followed former Superintendent Bill Wellman's advice to "follow the water," a journey that had ultimately brought me here, to gaze in wonder at the remains of a long-dead and anonymous human, laid to rest near Indian Springs and a deep well tapping into 30,000-year-old water. Just as the dunes have always depended on water for their continued existence, so too have humans in the San Luis Valley. By choosing to protect that water rather than profit from it, modern-day residents of this remarkable region, along with the National Park Service, have perpetuated the enduring historical connections to the lands they inhabit and manage, a lesson in the sand that hopefully will endure as long as the Great Sand Dunes.

Time Lapse

Prior commitments compelled me to leave the Great Sand Dunes early the next morning. As I broke camp at Pinyon Flats, warm sunlight bathed the entire San Luis Valley in an ethereal, otherworldly glow that revealed every crease and contour of the enigmatic dunes. Off to the west, the canted light of the rising sun pierced the far fringes of perception, revealing the imposing barrier of the rugged San Juans looming high above the staggering immensity of the valley floor. My car packed, I left the campground and stopped briefly at park headquarters to thank my NPS hosts, then pointed my car south, reluctantly leaving the dunes in my wake. Like a far northern outpost of the Land of Enchantment, the San Luis Valley spread out before me, a spellbinding landscape of wonder and mystery shimmering in the brilliant morning sun.

Time and distance soon began to distort in a most peculiar fashion. Covering in minutes what would have taken Pike or Frémont hours if not days, I hurtled through a landscape that encompassed 10,000 years of human history, past places where great herds of bison grazed near Folsom kill sites, past the same sand dunes that mystified the Tewa Pueblo, past the same San Luis Lakes where Father Francisco Torres had likely gasped his last breath ("Sangre de Cristo!") in the early eighteenth century, the same lakes that Don Juan Bautista de Anza may have called a "pleasant little pond" in 1779 and that William Henry Jackson first photographed in 1874. In every direction could be seen traces of the ubiquitous sand that permeates this portion of the valley, stirred by constant gales into churning clouds of airborne dust that rushed inexorably toward the dunes. Continuing west, I sped past enormous center-pivot sprinklers raining water over vast, verdant fields of crops, punctuated by barbed-wire fences dividing the landscape at irregular intervals, confining huge throngs of cattle that grazed among decrepit farmhouses leaning against the restless wind.

Before long I reached the intersection of Highway 17 and turned north, heading for a rendezvous with distant Poncha Pass. To the east, roiling thunderheads already shrouded the serrated summits of the mighty Sangres. As I drove past the sprawling, eighty-acre SunEdison solar farm and the strangely incongruous UFO Watchtower, past endless fields of greasewood and the storied expanse of the massive Baca Grant, it seemed almost inconceivable that the broad sweep of land flying by my windows actually sat atop an estimated 17,000 feet of sand and soil deposits that filled the interior of the immense basin, but such a staggering realization somehow seemed appropriate within the prodigious dimensions of the San Luis Valley. When I finally reached the southern approach to Poncha Pass, it occurred to me that the towering ramparts of the Great Sand Dunes had stubbornly refused to slip from view. Like a time-lapse film, my journey had taken me through at least one hundred centuries of human and natural history, yet still I had not completely separated myself from the graceful bronze contours of the distant dunes.

In much the same way, contemporary residents of the San Luis Valley cannot separate themselves from their environment, nor escape the rich historical legacy of their ancestors. The Great Sand Dunes and the greater San Luis Valley have hosted a remarkably diverse collection of human cultures, each with its own perceptions and values, each inflicting its own transformations upon the land. Manifestations of past land-use decisions are everywhere in the valley. Witness the irrigated fields that carpet the valley floor, or the remnants of old mines and sawmills that dot the rugged slopes of the surrounding mountains, or the decision to protect and preserve the Great Sand Dunes as a national park. In essence, the landscape tells a story, a narrative of past land-use successes and failures that is crucial to our comprehension of humanity's place in the natural world. By exploring and understanding the broad trends of the San Luis Valley's human and natural history, perhaps those who are struggling with similar resource management questions and decisions elsewhere in the American West can benefit from the lessons that a historical narrative has to offer. If they look closely enough, the answers they seek may be revealed somewhere in the landscape, and the history, that surrounds them.

NOTES

PROLOGUE

Epigraph: Richard White, "Trashing the Trails," in *Trails to a New Western History*, ed. Patricia Nelson Limerick, Clyde A. Milner II, and Charles E. Rankin (Lawrence: University Press of Kansas, 1991), 27.

1. Andrew Valdez, interview by author, June 3, 2011. Valdez is the park geologist at Great Sand Dunes National Park and Preserve.
2. Ibid. Other hydrological experts contend that the 2 billion acre-feet figure is based on a flawed U.S. Geological Survey (USGS) study conducted in the early 1970s. The reality is that no one knows precisely how much water the aquifers contain. See Jim Hughes, "Emotions Run Deep in the Valley," *Denver Post*, October 4, 1998, http://extras.denverpost.com/news/water/water12.htm (accessed September 6, 2011).

CHAPTER 1

Epigraph: Dennis Stanford, "A History of Archaeological Research in the San Luis Valley, Colorado," *San Luis Valley Historian* 22, no. 3 (1990): 39.

1. Andrew Valdez, interview by author, June 3, 2011. Estimates of the size of the main dunefield vary from roughly twenty-seven square miles to over forty square miles, depending on how the boundaries of the dunefield are defined. During the interview, park geologist Valdez supplied the "nearly thirty square miles" figure after consulting the latest GIS data on his computer that measures what he called the *active* dunefield; that is, the area of the dunefield where dune formation is actively occurring. Similarly, estimates about the actual size of the San Luis Valley vary widely, primarily due to its ill-defined southern boundary. A common claim is that the valley is "larger than the state of Connecticut," which is 5,544 square miles, but the most common claim for the size of the valley is over 8,000 square miles, in which case many sources claim it is "the largest alpine valley in the world." Others claim that the San Luis Valley is the largest *inhabited* alpine valley in the world. Again, size estimates are relative to the perceived boundaries of the valley. See Virginia McConnell Simmons, *The San Luis Valley: Land of the Six-Armed Cross* (Boulder: Pruett, 1979), 3, endnote 2.

2. Staci Matlock, "Colorado Eruption around 27 Million Years Ago Dwarfs All Others Known Today Worldwide," *Santa Fe New Mexican*, January 15, 2010, www.santafenewmexican.com/Local%20News/16volcano (accessed June 6, 2011).
3. Richard F. Madole et al., "On the Origin and Age of the Great Sand Dunes," *Geomorphology* 99 (2008): 102–103; Valdez interview, June 3, 2011.
4. Michael N. Machette, David W. Marchetti, and Ren A. Thompson, "Ancient Lake Alamosa and the Pliocene to Middle Pleistocene Evolution of the Rio Grande," in Michael N. Machette, Mary-Margaret Coates, and Margo L. Johnson, *2007 Rocky Mountain Section Friends of the Pleistocene Field Trip—Quaternary Geology of the San Luis Basin of Colorado and New Mexico, September 7–9, 2007*, U.S. Geological Survey Open-File Report 2007–1193 (2007): Chapter G, 157, 160, 166, http://pubs.usgs.gov/of/2007/1193/pdf /OF07–1193_ChG.pdf (accessed June 6, 2011); http://pubs.usgs.gov/of/ 2007/1193/ (accessed June 6, 2011). See also National Park Service, U.S. Department of the Interior, Great Sand Dunes National Park and Preserve, "Geology," www.nps.gov/grsa/naturescience/sanddunes.htm (accessed June 6, 2011).
5. U.S. Geological Survey, "Surficial Geology and Geomorphology of the Great Sand Dunes Area, South-Central Colorado" (2008): 2, http://esp.cr.usgs.gov/info/dunes/origin.html (accessed May 14, 2011); Madole et al., "Origin and Age," 100; Ferdinand Vandeveer Hayden for the U.S. Geological Survey, U.S. Department of the Interior, *Geological and Geographical Atlas of Colorado and Portions of Adjacent Territory* (Washington, D.C., 1881): 98, sheet X.
6. Claude E. Siebenthal for the U.S. Geological Survey, U.S. Department of the Interior, *Geology and Water Resources of the San Luis Valley*, Water-Supply Paper 240 (Washington, D.C.: Government Printing Office, 1910), 48.
7. Madole et al., "Origin and Age," 109.
8. U.S. Geological Survey, "Surficial Geology," 2; Roger W. Toll for the National Park Service, U.S. Department of the Interior, *Report on Great Sand Dunes National Monument*, copy of an unpublished report in Great Sand Dunes National Park archives (held at the Visitor Center), 1931, 43.
9. Siebenthal, *Geology and Water Resources*, 52.
10. Ibid.
11. Elliot Coues, ed., *The Journal of Jacob Fowler* (New York: Francis P. Harper, 1898), 112–14.
12. Madole et al., "Origin and Age," 101.
13. Ross B. Johnson, "The Great Sand Dunes of Southern Colorado," *Geological Survey Research 1967*, U.S. Geological Survey Professional Paper 575-C (Washington, D.C.: Government Printing Office, 1967), 177–83; Fred Bunch, telephone interview by author, March 15, 1996. Bunch is a resource management specialist at Great Sand Dunes National Park and Preserve; Madole et al., "Origin and Age," 99.
14. Johnson, "Great Sand Dunes"; Thomas Huber and Robert Larkin, *The San Luis Valley of Colorado: A Geographical Sketch* (Colorado Springs: Hulbert Center Press, 1996), 8.
15. Madole et al., "Origin and Age," 102; Valdez interview, June 3, 2011.
16. Madole et al., "Origin and Age," 99–102, 117; Andrew Valdez, "Separating Fact

from Fiction at Great Sand Dunes," *Outcrop: Newsletter of the Rocky Mountain Association of Geologists* 54, no. 11 (November 2005); Valdez interview, June 3, 2011.

17. Stephen A. Trimble, *Great Sand Dunes: The Shape of the Wind* (Globe, Ariz.: Southwest Parks and Monuments Association, 1978), 8. Trimble's description of the processes that formed the dunes is elegant and straightforward, yet he contends that the Sangre de Cristos actually contribute very little sand to the enormous dunefield. Current research suggests that the Sangres may have contributed as much as 30 percent of the sand in the dunes. See Madole et al., "Origin and Age," 109; Johnson, "Great Sand Dunes," 177–83; and Tom Wolf, *Colorado's Sangre de Cristo Mountains* (Niwot: University Press of Colorado, 1995), 219.

18. Valdez interview, June 3, 2011. See also J. Patrick Wiegand, "Dune Morphology and Sedimentology at Great Sand Dunes National Monument" (master's thesis, Colorado State University, 1977), 1; and National Park Service, U.S. Department of the Interior, Great Sand Dunes National Park and Preserve, "Geology," www.nps.gov/grsa/naturescience/sanddunes.htm (accessed June 10, 2011), 3.

19. Fred Bunch, telephone interview by author, August 20, 1997; Valdez interview, June 3, 2011; Valdez, "Separating Fact from Fiction," 7.

20. Libbie Landreth, "Natural History of the Great Sand Dunes," *San Luis Valley Historian* 22, no. 3 (1990): 5–6; National Park Service, U.S. Department of the Interior, Great Sand Dunes National Park and Preserve, "Insects, Spiders, Centipedes, Millipedes," www.nps.gov/grsa/naturescience/insects.htm (accessed April 13, 2011).

21. National Park Service, U.S. Department of the Interior, Great Sand Dunes National Park and Preserve, "Mammals," www.nps.gov/grsa/naturescience/mammals.htm (accessed April 13, 2011); Landreth, "Natural History," 5–6; See also National Park Service, U.S. Department of the Interior, *Draft Master Plan: Great Sand Dunes National Monument* (Washington, D.C.: Government Printing Office, September 1975), 9.

22. Bruce Bradley and Dennis Stanford, "The North Atlantic Ice-Age Corridor: a Possible Palaeolithic Route to the New World," *World Archaeology* 36, no. 4 (2004): 460; Brian M. Fagan, *Ancient North America: The Archaeology of a Continent* (London: Thames and Hudson Ltd., 1995), 66–71; Maggie Fox, "Ancient Seaweed Chews Confirm Age of Chilean Site," *Reuters*, May 8, 2008, www.reuters.com/article/2008/05/08/ us-humans-chile-idUSN0839099920080508 (accessed April 16, 2011); Eline D. Lorenzen et al., "Species-Specific Responses of Late Quarternary Megafauna to Climate and Humans," *Nature* 479 (November 17, 2011): 359–64.

23. Bradley and Stanford, "North Atlantic," 460, 465.

24. Fox, "Ancient Seaweed."

25. Stanford, "History of Archaeological Research," 33–34; Hillary Mayell, "Bison Kill Site Sheds Light on Ice Age Culture," *National Geographic News*, July 22, 2002, http:// news.nationalgeographic.com/news/2002/07/0722_020722_clovis.html (accessed February 3, 2011).

26. Charles L. Matsch, *North America and the Great Ice Age* (New York: McGraw-Hill, 1976), 83; U.S. Geological Survey, "Surficial Geology," 1.

27. Stanford, "History of Archaeological Research," 33–39; Mayell, "Bison Kill Site."
28. Margaret A. Jodry et al., "Late Quarternary Environments and Human Adaptation in the San Luis Valley, South-Central Colorado," *Water in the Valley* (Lakewood: Colorado Groundwater Association, 1989), 194–95; Jennifer Heinzman, "Archaeologists Uncover SLV 'Ice Age,'" *Alamosa (Colo.) Valley Courier*, July 10, 1993, 2; Mayell, "Bison Kill Site."
29. William K. Stevens, "New Suspect in Ancient Extinctions of the Pleistocene Megafauna: Disease," *New York Times*, April 29, 1997, www.cpluhna.nau.edu/Biota/megafauna_extinctions.htm (accessed February 3, 2011).
30. Paul S. Martin, "Pleistocene Overkill: The Global Model," *Quarternary Extinctions: A Prehistoric Revolution*, ed. Paul S. Martin and Richard G. Klein (Tucson: University of Arizona Press, 1989), 354–403. See also Dan Flores, *Caprock Canyonlands: Journeys into the Heart of the Southern Plains* (Austin: University of Texas Press, 1990), 16; and Fagan, *Ancient North America*, 85–86.
31. Gary Haynes, ed., *American Megafaunal Extinctions at the End of the Pleistocene*, Vertebrate Paleobiology and Paleoanthropology Series 3 (n.p.: Springer, 2009): 1–37; Stevens, "New Suspect."
32. Fagan, *Ancient North America*, 96.
33. Flores, *Caprock Canyonlands*, 16; Matsch, *North America and the Great Ice Age*, 110; Fagan, *Ancient North America*, 85; Stevens, "New Suspect."
34. R. B. Firestone et al., "Evidence of an Extraterrestrial Impact 12,900 Years Ago that Contributed to the Megafaunal Extinctions and the Younger Dryas Cooling," *Proceedings of the National Academy of Sciences* 104, no. 41 (October 9, 2007), www.pnas.org/cgi/doi/10.1073/pnas0706977104 (accessed February 3, 2011); C. Vance Haynes, Jr., "Younger Dryas 'Black Mats' and the Rancholabrean Termination in North America," *Proceedings of the National Academy of Sciences* 105, no. 18 (May 6, 2008); John D. Cox, "Why the Younger Dryas Matters," *Discovery News*, April 2, 2010, http://news.discovery.com/earth/why-the-younger-dryas-matters.html (accessed February 3, 2011); Haynes, *American Megafaunal Extinctions*, 14.
35. Ian Sample, "Sophisticated Hunters Not to Blame for Driving Mammoths to Extinction," *Guardian* (UK), November 19, 2009, www.guardian.co.uk/science/2009/nov/19/hunters-mammoths-extinction (accessed February 3, 2011); *Nature*, November 17, 2011.
36. Fagan, *Ancient North America*, 81; Mayell, "Bison Kill Site"; Stanford, "History of Archaeological Research," 34; *Nature*, November 17, 2011.
37. Stanford, "History of Archaeological Research," 34.
38. K. L. Rogers et al., "Pliocene and Pleistocene Geologic and Climatic Evolution in the San Luis Valley of South-Central Colorado," *Palaeogeography, Palaeoclimatology, Palaeoecology* 94, no. 1 (July 1992): 57; Wolf, *Colorado's Sangre*, 20.
39. Stanford, "History of Archaeological Research," 38.
40. Fagan, *Ancient North America*, 97; Stanford, "History of Archaeological Research," 39.
41. Sloan Emery and Dennis Stanford, "Preliminary Report on Archaeological Investigations at the Cattle Guard Site, Alamosa County, Colorado," *Southwestern Lore* 48, no. 1 (March 1982): 18.

42. William deBuys, *Enchantment and Exploitation: The Life and Hard Times of a New Mexico Mountain Range* (Albuquerque: University of New Mexico Press, 1985), 31.
43. Ibid., 34. See also Stephen J. Pyne, *Fire in America: A Cultural History of Wildland and Rural Fire* (Princeton: Princeton University Press, 1982), 6, 71–83.
44. Mark H. Hunter, "Dwelling May Be 6,000 Years Old," *Denver Post*, August 28, 2000, http://extras.denverpost.com/news/news0828c.htm (accessed February 11, 2011). See also History Colorado, "Great Sand Dunes National Park State Historical Fund Projects," http://www.historycolorado.org/archaeologists/great-sand-dunes-national-park(accessed February 11, 2011).
45. Vince Spero et al. for the U.S. Forest Service, U.S. Department of Agriculture, "An Assessment of the Range of Natural Variability of the Rio Grande National Forest," copy of unpublished internal draft in Great Sand Dunes National Park archives, March 1995, appendix A-7; Stanford, "History of Archaeological Research," 39.
46. Simmons, *San Luis Valley*, 9. See also Edgar L. Hewett and Bertha P. Dutton, eds., *The Pueblo Indian World* (Albuquerque: University of New Mexico Press, 1945), 23.
47. Hewett, *Pueblo Indian World*, 23–24; Simmons, *San Luis Valley*, 9.
48. Simmons, *San Luis Valley*, 9.
49. Yi-Fu Tuan, *Topophilia: A Study of Environmental Perception, Attitudes, and Values* (Englewood Cliffs, N.J.: Prentice-Hall, 1974), 17. See also Alfonso Ortiz, *The Tewa World* (Chicago: University of Chicago Press, 1969), 118–19; and Joseph Campbell, *The Power of Myth* (New York: Doubleday, 1988), 45–50.
50. J. Donald Hughes, *Native Americans in Colorado* (Boulder: Pruett, 1977), 16; Carl Ubbelohde, Maxine Benson, and Duane A. Smith, *A Colorado History* (Boulder: Pruett, 1972), 4–6.
51. Dorothy D. Wilson, "They Came to Hunt: Early Man in the San Luis Valley," in *Guidebook of the San Luis Basin, Colorado*, ed. H. L. James (Socorro: New Mexico Geological Society, 1971), 204. See also Hughes, *Native Americans in Colorado*, 19.
52. Wilson, "They Came to Hunt," 204; Simmons, *San Luis Valley*, 10.
53. Fred Bunch, interview by author, May 14, 2010; Lawrence D. Sundberg, *Dinétah: An Early History of the Navajo People* (Santa Fe: Sunstone Press, 1995), 15; David R. M. White for the National Park Service, U.S. Department of the Interior, *Seinanyédi: An Ethnographic Overview of Great Sand Dunes Park and Preserve* (Santa Fe: Applied Cultural Dynamics, 2005), 164; National Park Service, U.S. Department of the Interior, Great Sand Dunes National Park and Preserve, "Sangre de Cristo National Heritage Area," www.nps.gov/grsa/parknews/sangre-de-cristo-nha.htm (accessed October 17, 2010).
54. Luther E. Bean, *Land of the Blue Sky People* (Monte Vista: The Monte Vista Journal, 1962), 7.
55. Charles S. Marsh, *People of the Shining Mountains* (Boulder: Pruett, 1982), 3, 15. A number of historians, including Marsh, believed the Capote band controlled the San Luis Valley. Other historic sources claim that the Tabeguache/Uncompahgre band controlled the San Luis Valley prior to European contact. Given the potential for intermixing between different bands of the Southern Ute, it is possible that some misidentification occurred in the historical record; Southern Ute Indian

Tribe, "History of the Southern Ute," www.southern-ute.nsn.us/history/ (accessed October 17, 2010); David R. M. White, *Seinanyédi*, 70.

56. William Cronon and Richard White, "Ecological Change and Indian-White Relations," *Handbook of North Native Americans*, ed. William C. Sturtevant (Washington, D.C.: Government Printing Office, 1988), v. 4, *History of Indian-White Relations*, Wilcomb E. Washburn, ed., 417. See also William Cronon and Richard White, "Indians in the Land," *American Heritage* 37 (August-September 1986): 20.
57. Cronon and White, "Ecological Change," 417.
58. Wilson, "They Came to Hunt," 203.
59. Pyne, *Fire in America*, 71–83.
60. Marilyn A. Martorano, "Culturally Peeled Ponderosa Pine Trees," *San Luis Valley Historian* 22, no. 3 (1990): 32.
61. Ibid., 31.

CHAPTER 2

Epigraph 1. George P. Hammond and Agapito Rey, *Don Juan de Oñate: Colonizer of New Mexico 1595–1628* (Albuquerque: University of New Mexico Press, 1953), 335. Another excellent account of Oñate's expedition can be found in Gaspar Pérez De Villagrá, *History of New Mexico*, trans. Gilberto Espinosa (Los Angeles: The Quivira Society, 1933), 125–37.
Epigraph 2. Lawrence D. Sundberg, *Dinétah: An Early History of the Navajo People* (Santa Fe: Sunstone Press, 1995), 17.
1. Virginia McConnell Simmons, *The San Luis Valley: Land of the Six-Armed Cross* (Boulder: Pruett, 1979), 13; David J. Weber, *The Spanish Frontier in North America* (New Haven, Conn.: Yale University Press, 1992), 81; Douglas B. Thomas, *From Fort Massachusetts to the Rio Grande: A History of Southern Colorado and Northern New Mexico from 1850 to 1900* (Washington, D.C.: Thomas International, 2002), 8. See also Herbert E. Bolton, ed., *Spanish Exploration in the Southwest 1542–1706* (New York: Charles Scribner's Sons, 1908), 199–222; and De Villagrá, *History of New Mexico*, 125–37.
2. W. Storrs Lee, ed., *Colorado: A Literary Chronicle* (New York: Funk and Wagnalls, 1970), 16. See also Bolton, *Spanish Exploration*, 199–222; and Weber, *Spanish Frontier*, 83.
3. Bolton, *Spanish Exploration*, 202–204.
4. Alfred W. Crosby, *Ecological Imperialism: The Biological Expansion of Europe 900–1900* (Cambridge: Cambridge University Press, 1986), 2–7.
5. Weber, *Spanish Frontier*, 308–10.
6. Ibid., 27, 206. See also Charles S. Marsh, *People of the Shining Mountains* (Boulder: Pruett, 1982) 15.
7. Ibid., 4, 27–28. See also Crosby, *Ecological Imperialism*, 195–216 and 281–88 for more information on the devastating impact of Old World diseases on New World indigenous populations.
8. Lee, *Colorado*, 17; Bolton, *Spanish Exploration*, 226. There is some discrepancy as to the exact location of this encounter. Lee, citing Bolton as his source, wrote that the encounter with the buffalo "took place in the vicinity of Alamosa, at the foot of the Sangre de Cristo Mountains." Bolton, in a footnote on page 226 of *Spanish*

Exploration, described the encounter taking place while the Oñate expedition was "forty leagues—a hundred miles or more—from Pecos [New Mexico] and the river must have been the Canadian, near Alamosa. It issues from the Sangre de Cristo mountains." However, the headwaters of the Canadian River gather and flow down the east slope of the Sangres, a considerable distance [approx. fifty miles] from Alamosa's location on the western side of the Sangres, in the southern reaches of the San Luis Valley.

9. Lee, *Colorado*, 17, 21.
10. Ibid., 19.
11. Yi-Fu Tuan, *Topophilia: A Study of Environmental Perception, Attitudes, and Values* (Englewood Cliffs: Prentice-Hall Inc., 1974), 67.
12. Ibid.
13. Bolton, *Spanish Exploration*, 230.
14. Weber, *Spanish Frontier*, 86.
15. Ibid., 91.
16. Carroll L. Riley, *Rio del Norte: People of the Upper Rio Grande From Earliest Times to the Pueblo Revolt* (Salt Lake City: University of Utah Press, 1995), 268.
17. J. Manuel Espinosa, "Journal of the Vargas Expedition into Colorado, 1694," *Colorado Magazine* 16, no. 3 (May 1939): 82.
18. Ibid., 88.
19. Simmons, *San Luis Valley*, 15; George Harlan, *Postmarks and Places* (Alamosa: Ye Olde Print Shoppe, 1976), vi.
20. Harlan, *Postmarks*, vi; Simmons, *San Luis Valley*, 15.
21. Luther E. Bean, *Land of the Blue Sky People* (Monte Vista: The Monte Vista Journal, 1962), 94; Harold Schaafsma, "Great Sand Dunes Monument Has Colorful, Exciting Past," *Alamosa Daily Courier*, January 4, 1955, Adams State College Museum Files; Harlan, *Postmarks*, vi; See also Virginia Simmons, "The Mystery of the Valley's Name," *Alamosa Valley Courier*, January 9, 1990, for a detailed examination of the historic origins of the valley's name.
22. Frank C. Spencer, *The Story of the San Luis Valley* (Alamosa: The Alamosa Journal, 1925; reprint, Santa Fe: Sleeping Fox Press, 1975), 20–21 (page references are to original edition); Thomas, *From Fort Massachusetts*, 22. See also Olibama Lopez Tushar, *The People of "El Valle"* (Pueblo: El Escritorio Press, 1992), 19–20; Simmons, *San Luis Valley*, 15; Luther E. Bean, *Land of the Blue Sky People*, 94–95; and Tom Wolf, *Colorado's Sangre de Cristo Mountains* (Niwot: University Press of Colorado, 1995), 64.
23. Stephen M. Voynick, *Colorado Gold: From the Pike's Peak Rush to the Present* (Missoula: Mountain Press, 1992), 152–58; Caroline Bancroft, *Colorado's Lost Gold Mines and Buried Treasure* (Boulder: Johnson Books, 1962), 43; Wolf, *Colorado's Sangre*, 64.
24. Fred Bunch, interview by author, January 30, 1997.
25. Voynick, *Colorado Gold*, 158; Simmons, *San Luis Valley*, 15.
26. Simmons, *San Luis Valley*, 15.
27. Ibid., 16. Pekka Hämäläinen, *The Comanche Empire* (New Haven, Conn.: Yale University Press, 2008), 1–4, 24–25, 27; Ned Blackhawk, *Violence over the Land: Indians and Empires in the Early American West* (Cambridge: Harvard University Press, 2006), 1–15, 18–22.

28. Alfred B. Thomas, *Forgotten Frontiers: A Study of the Spanish Indian Policy of Don Juan Bautista de Anza, Governor of Mexico 1777–1787* (Norman: University of Oklahoma Press, 1932), 127. There is some discrepancy as to the exact location of this "pleasant little pond." On page 67, Thomas describes the pond as a "water hole they named La Ciénega de San Luis." The word *ciénega* (also spelled *ciénaga*) is Spanish for "spring" or "swamp" and often referred to a marshy area. The paragraph from Anza's diary on page 127 that describes the pond has a side note that reads "From Santa Fe to La Cénega [sic] de San Luis, 67 leagues"; John Koshak, telephone interview by author, October 22, 1996. Koshak was chief ranger at San Luis Lakes State Park at the time of interview and told the author that La Ciénega de San Luis is actually a marshy area of interconnected lakes and swamps near Saguache in the northwestern portion of the valley, roughly thirty-five miles from the present-day San Luis Lakes. Such confusion is not uncommon, given the ephemeral nature of lakes and swamps in the Closed Basin region.
29. Ibid., 141; Patricia Nelson Limerick, *Something in the Soil: Legacies and Reckonings in the New West* (New York: W. W. Norton & Company, 2000), 121; Hämäläinen, *Comanche Empire*, pp. 109–10.
30. Weber, *Spanish Frontier*, 10.
31. Forbes Parkhill, "Colorado's Earliest Settlements," *Colorado Magazine* 34, no. 4 (October 1957): 245, 247.
32. Ibid., 247.
33. Jose de Onis, ed., *The Hispanic Contribution to the State of Colorado* (Boulder: Westview Press, 1976), xiii; Wolf, *Colorado's Sangre*, 269; Simmons, *San Luis Valley*, 43–44.
34. William deBuys, *Enchantment and Exploitation: The Life and Hard Times of a New Mexico Mountain Range* (Albuquerque: University of New Mexico Press, 1985), 69; Simmons, *San Luis Valley*, 44.
35. Frances Leon Swadesh, *Los Primeros Pobladores: Hispanic Americans of the Ute Frontier* (South Bend: University of Notre Dame Press, 1974), 72; Simmons, *San Luis Valley*, 43–44.
36. Swadesh, *Los Primeros Pobladores*, 46; Onis, *The Hispanic Contribution*, xv.
37. David Hurst Thomas, ed., *Spanish Borderlands Sourcebooks* 27, *Hispanic Urban Planning in North America*, ed. Daniel J. Garr (New York: Garland, 1991), xv.
38. Marc Simmons, "Settlement Patterns and Village Plans in Colonial New Mexico," in Ibid., 43.
39. Stanley Crawford, *Mayordomo: Chronicle of an Acequia in Northern New Mexico* (Albuquerque: University of New Mexico Press, 1988), xi, xii; Robert W. Ogburn, "A History of the Development of San Luis Valley Water," *San Luis Valley Historian* 28, no. 1 (1996): 7; Simmons, *San Luis Valley*, 48.
40. Swadesh, *Los Primeros Pobladores*, 18.
41. Ibid., 79.

CHAPTER 3

Epigraph 1. Washington Irving, *Astoria, or Anecdotes of an Enterprise Beyond the Rocky Mountains*, ed. Richard Dilworth Rust (1836; reprint, Lincoln: University of Nebraska Press, 1976), 151.

Epigraph 2. William Gilpin, *The Central Gold Region: The Grain, Pastoral, and Gold Regions of North America* (Philadelphia: J. B. Lippincott & Co., 1860), 71.

1. Donald Jackson, *The Journals of Zebulon Montgomery Pike*, vol. 1 (Norman: University of Oklahoma Press, 1966), 286. Wilkinson also instructed Pike to accomplish several other tasks, including the collection and preservation of mineral and botanical specimens, as well as observations of the eclipses of Jupiter's moons. Furthermore, Wilkinson ordered Pike to descend the Red River, presumably by raft, which may explain Pike's search for suitable timber along the Rio Grande (which Pike mistakenly maintained was the Red River) and Conejos Rivers. The fact that Pike later built a stockade on the banks of the Conejos has led some writers to conclude that Pike should have suspected he was on Spanish soil the moment he crossed the Rio Grande (see 375). After being taken into custody by Spanish soldiers and escorted first to Santa Fe and then to Chihuahua, Mexico, Pike's journals and maps of the expedition were confiscated. The first serious effort to locate these documents began in 1906 when Secretary of State Elihu Root asked the U.S. Embassy in Mexico City for help in finding the papers. The investigation met with little success until Secretary Root visited Mexico City on some other official business. There he met Dr. Herbert E. Bolton, an American scholar who was doing research in the Mexican National Archives. Root mentioned the Pike papers to Bolton, who searched the archives on his own and located nineteen of the twenty-one missing documents. The papers were eventually returned to the United States in July 1910 and are currently in the National Archives in Washington, D.C. See W. Eugene Hollon, *The Lost Pathfinder: Zebulon Montgomery Pike* (Norman: University of Oklahoma Press, 1949), 169–70.
2. Ibid. Numerous historians have accused Pike of being involved in a conspiracy allegedly hatched by Aaron Burr and General James Wilkinson. The two men apparently wanted to raise a private army, foment rebellion among the French and Spanish of New Orleans, invade Mexico and establish a new, independent country comprising Mexico and portions of the Louisiana Purchase. No definitive proof of Pike's involvement in such a scheme has ever surfaced, but many believe his expedition to the Southwest was in reality part of a plan to ascertain Spanish military strength along the ill-defined border. See, among others, Hollon, *Lost Pathfinder*, 102, 158–70; Richard White, *"It's Your Misfortune and None of My Own": A New History of the American West* (Norman: University of Oklahoma Press, 1991), 63–64, 121; Elliot Coues, *The Expeditions of Zebulon Montgomery Pike* (New York: Francis P. Harper, 1895), 481; and Jared Orsi, *Citizen Explorer: The Life of Zebulon Pike* (New York: Oxford University Press, 2014).
3. Jack Kyle Cooper, *Zebulon Montgomery Pike's Great Western Adventure, 1806–1807* (Colorado Springs: Clausen Books, 2007), 7; John Steinle, "That Perhaps Necessary Evil: Zebulon Pike and the U.S. Military," *San Luis Valley Historian* 39, no. 3 (2007): 22; National Park Service, U.S. Department of the Interior, Jefferson National Expansion Memorial, *Museum Gazette*, January 1996, 1. There are some translation discrepancies in Pike's letters to Daniel Bissell, so it is possible that Pike actually wrote that his men were a "darn'd set of rascals," but many historians prefer the phrase "dam'd set of rascals," probably for dramatic

effect. See the General Daniel Bissell Papers in the St. Louis Mercantile Library, University of Missouri–St. Louis, www.umsl.edu/mercantile/assets/pdf/special-collections/transcript/M-009_Letters.pdf (accessed November 22, 2011).
4. Jackson, *Journals*, 345.
5. Coues, *Expeditions*, 471.
6. Orsi, *Citizen Explorer*, 159; Hollon, *Lost Pathfinder*, 135. Hollon contends that Humboldt's map of New Spain, compiled from data in Mexico City in 1804, indicates that the Spanish labored under the same error as Pike. Apparently the Spanish believed the headwaters of the Red River were two to three hundred miles northwest of their actual location.
7. Jackson, *Journals*, 373.
8. Cooper, *Pike's Great Western Adventure*, 99–101; Carrol Joe Carter, *Pike in Colorado* (Fort Collins: Old Army Press, 1978), 59–61. Debate over exactly which pass Pike followed through the Sangre de Cristo Mountains in 1807 has continued ever since he entered the San Luis Valley, but currently the general consensus among historians is that he used Medano Pass. Pike historian Carrol Joe Carter argued for Mosca Pass, based on his analysis of specific details gleaned from Pike's journals, including references to snow depth and the location of the Great Sand Dunes. Others, including noted historian Elliot Coues, contend Pike crossed Medano Pass, an assertion supported by the extensive research of the late Jack Kyle Cooper, who argued definitively for Medano Pass. In a conversation between Great Sand Dunes resource management specialist Fred Bunch and this author, Bunch told of a local San Luis Valley man who had found an old knife on Mosca Pass and was attempting to have it accurately dated to determine if it could have belonged to a member of Pike's party. To date, no further information about this alleged knife has surfaced, and the consensus remains that Pike crossed Medano Pass. See also Jack Kyle Cooper, "A Rebuttal Concerning Roads, Trails, Traces and Wagons," *San Luis Valley Historian* 33, no. 3 (2001); Coues, *Expeditions*, 491–92; and Jackson, *Journals*, 373.
9. Jackson, *Journals*, 373–74; Cooper, *Pike's Great Western Adventure*, 99–101.
10. Fred Bunch, telephone interview by author, November 4, 1996.
11. Richard White, *"It's Your Misfortune and None of My Own,"* 121.
12. Coues, *Expeditions*, 525; Orsi, *Citizen Explorer*, 172.
13. Jackson, *Journals*, 376; Cooper, *Pike's Great Western Adventure*, 106–107.
14. National Park Service, U.S. Department of the Interior, Jefferson National Expansion Memorial, *Museum Gazette*, January 1996, 3; National Park Service, U.S. Department of the Interior, *National Register of Historic Places Nomination Form—Pike's Stockade (1961, 1978)*, 1–7; Cooper, *Pike's Great Western Adventure*, 103–107, 136; Orsi, *Citizen Explorer*, 194–203.
15. W. Eugene Hollon, *The Great American Desert, Then and Now* (New York: Oxford University Press, 1966), 65; Cooper, *Pike's Great Western Adventure*, 145.
16. Terry L. Alford, "The West as a Desert in American Thought prior to Long's 1819–1820 Expedition," *Journal of the West* 8, no. 4 (1969): 517; Annie H. Abel, "Trudeau's Description of the Upper Missouri," *Mississippi Valley Historical Review* 8 (1921): 158; Lewis and Clark Trail website, www.Lewisandclarktrail.com/section2/sdcities/FortRandallArea/history2.htm (accessed November 22, 2011).

17. Alford, "West as a Desert," 519.
18. Wallace Stegner, *Beyond the Hundredth Meridian: John Wesley Powell and the Second Opening of the American West* (Boston: Houghton Mifflin, 1954; reprint, New York: Penguin Books, 1992), 215 (page references are to reprint edition).
19. Roger L. Nichols and Patrick L. Halley, *Stephen Long and American Frontier Expansion* (Norman: University of Oklahoma Press, 1995), 137; Stegner, *Beyond the Hundredth Meridian*, 215–16.
20. Alford, "West as a Desert," 519; Coues, *Expeditions*, 525; Stegner, *Beyond the Hundredth Meridian*, 215.
21. William H. Goetzmann, *Exploration and Empire: The Explorer and the Scientist in the Winning of the American West* (New York: W. W. Norton, 1966), 51.
22. David J. Wishart, *The Fur Trade of the American West, 1807–40: A Geographical Synthesis* (Lincoln: University of Nebraska Press, 1979), 22.
23. Harry R. Stevens, "Hugh Glenn," in *The Mountain Men and the Fur Trade of the Far West*, vol. 2, ed. LeRoy R. Hafen (Glendale, Calif.: Arthur H. Clark Company, 1965), 173.
24. David J. Weber, *The Taos Trappers: The Fur Trade in the Far Southwest, 1540–1846* (Norman: University of Oklahoma Press, 1968), preface, viii.
25. Elliot Coues, ed., *The Journal of Jacob Fowler* (New York: Francis P. Harper, 1898), 101, 115. While in the San Luis Valley, Fowler used the remains of Pike's stockade on the banks of the Conejos River as a reference point: "We Came to Slovers party In Camped about two miles up Pikes forke of the Delnort and about three miles below His Block House Wheare He Was taken by the Spanierds [sic]" (114).
26. William deBuys, *Enchantment and Exploitation: The Life and Hard Times of a New Mexico Mountain Range* (Albuquerque: University of New Mexico Press, 1985), 98.
27. Wishart, *Fur Trade*, 31, 205; Tom Wolf, *Colorado's Sangre de Cristo Mountains* (Niwot: University Press of Colorado, 1995), 76.
28. Wishart, *Fur Trade*, 212.
29. Patricia Joy Richmond, "Trail to Disaster," *Monographs in Colorado History*, no. 4 (Denver: Colorado Historical Society, 1989), 7–9; Tom Chaffin, *Pathfinder: John Charles Frémont and the Course of American Empire* (New York: Hill and Wang, 2002), 404.
30. Robert V. Hine, *Edward Kern and American Expansion* (New Haven, Conn.: Yale University Press, 1962), 61.
31. LeRoy R. Hafen and Ann W. Hafen, ed., "Fremont's Fourth Expedition: A Documentary Account of the Disaster of 1848–1849," *The Far West and The Rockies Historical Series 1820–1875*, vol. 11 (Glendale, Calif.: Arthur H. Clark Company, 1960), 241–61. The editors note that expedition artist Richard Kern described the expedition as a "disaster."
32. Ibid., 120; David J. Weber, *Richard H. Kern: Expeditionary Artist in the Far Southwest 1848–1853* (Albuquerque: University of New Mexico Press, 1985), 39.
33. Blanche C. Grant, *When Old Trails Were New: The Story of Taos* (New York: The Press of the Pioneers, 1934), 131.
34. Hine, *Edward Kern*, 195; Richmond, "Trail to Disaster," 9.
35. Robert W. Ogburn, "A History of the Development of San Luis Valley Water," *San Luis Valley Historian* 28, no. 1 (1996): 7.

36. T. Donald Brandes, *Military Posts of Colorado* (Fort Collins: The Old Army Press, 1973), 7–16. In the summer of 1853, Brevet Colonel Joseph K. F. Mansfield, U.S. Army inspector general, visited Fort Massachusetts, which had been named by Boston native Lt. Colonel Edwin Sumner. Mansfield criticized the location of the fort as being indefensible to attack, situated as it was directly beneath the flanks of the Sierra Blanca massif. In addition, although the site had good timber and excellent water, it was too far away from the newly-established settlements in the southern end of the valley to offer much in the way of protection from Indian attacks. Consequently, the army made the decision to establish a new post about six miles south of Fort Massachusetts. The new post was officially opened on June 24, 1858, and named Fort Garland, in honor of Brevet Brigadier General John Garland, commander of the Military Department of New Mexico. The army officially closed the fort on November 30, 1883.
37. Stephen Bonsal, *Edward Fitzgerald Beale: A Pioneer in the Path of Empire, 1822–1903* (New York: G. P. Putnam's Sons, 1912), 66. Heap was a cousin of the commander of the expedition, Lt. Edward F. Beale. Beale had been appointed General Superintendent of Indian Affairs for California and Nevada by President Millard Fillmore in November 1852. In order to reach his new superintendency, Fillmore instructed Beale to proceed via the shortest practicable route to California. He was also instructed to examine the Territories of New Mexico and Utah, and survey the region's suitability for a railroad route to reach what Heap called "our Pacific possessions." Their route led them through the San Luis Valley in 1853. Beale later gained fame for introducing camels to the American West to provide transportation for the U.S. Army. He was convinced that by "the introduction of camels the great desert of Arizona could be robbed of half its terrors." General Beale went on to command the first and last camel corps ever organized in the American West. See 64–68, 199–210.
38. Gwinn Harris Heap, "Central Route to the Pacific," *The Far West and The Rockies Historical Series 1820–1875*, vol. 7, ed. LeRoy R. Hafen and Ann W. Hafen (Glendale, Calif.: Arthur H. Clark Company, 1957), 117–18.
39. Ibid., 118–19.
40. Ibid., 114, 122.
41. Hugh Lovin, "Sage, Jacks, and Snake Plain Pioneers," *Idaho Yesterdays* 22 (Winter 1979): 15.
42. Heap, "Central Route," 123.
43. Dr. James Schiel, "Journey through the Rocky Mountains," *John Williams Gunnison*, ed. Nolie Mumey (Denver: Artcraft Press, 1955), 75.
44. Ibid., 76.
45. Lt. E. G. Beckwith, *Reports of Explorations and Surveys to Ascertain the Most Practicable and Economical Route for a Railroad from the Mississippi River to the Pacific Ocean, 1853–54*, vol. 2, Executive Doc. no. 78, 33rd Congress, 2nd Session (Washington, D.C.: Beverly Tucker, 1855), 44.
46. Ibid.

47. Ibid., 43.
48. John A. Mangimelli, "Climatic Change in the San Luis Valley," *San Luis Valley Historian* 22, no. 3 (1990): 25.
49. James A. Young and B. Abbott Sparks, *Cattle in the Cold Desert* (Logan: Utah State University Press, 1985), 29. See also Stephen A. Trimble, *Great Sand Dunes: The Shape of the Wind* (Globe: Southwest Parks and Monuments Association, 1978), 18. Trimble notes that pronghorn had disappeared from the valley sometime prior to 1900, victims of overhunting and diminished habitat. According to an official press release by Great Sand Dunes National Monument dated February 1, 1963, the Colorado Division of Wildlife reintroduced pronghorn at the Great Sand Dunes in early 1962, and a small herd is currently thriving.
50. Ibid., 20.
51. Beckwith, *Reports*, 45.
52. Virginia McConnell Simmons, *The San Luis Valley: Land of the Six-Armed Cross* (Boulder: Pruett, 1979), 71.
53. Stegner, *Beyond the Hundredth Meridian*, 2–3, 216.
54. Simmons, *San Luis Valley*, 53, 61.
55. Edward R. Crowther, "William Gilpin: Booster and Developer of the San Luis Valley," *San Luis Valley Historian* 26, no. 2 (1994): 38–53; Ed Quillen, "The Baca Ranch: It's Got Quite a Past," *Colorado Central Magazine* (February 2000): 37–38; See also Ed Quillen, "The Baca Ranch—a Legacy from Mexico," http://cozine.com/2000-february/the-baca-ranch-a-legacy-from-mexico/ (accessed June 18, 2011); Virginia Simmons, "Rabbitbrush Rambler: Gilpin, the Promoter, Part 1," *Alamosa Valley Courier*, August 6, 2013, www.alamosanews.com/v2_news_articles.php?heading=0&story_ id=30833&page=74; Robert L. Spude, Regional Historian, Cultural Resources Management, National Park Service Intermountain Region, e-mail correspondence, July 11, 2012.
56. Thomas L. Karnes, *William Gilpin: Western Nationalist* (Austin: University of Texas Press, 1970), 306.
57. Crowther, "William Gilpin," 52; "Nathaniel P. Hill Inspects Colorado," *Colorado Magazine* 33 (October 1956): 259, 268.
58. Mike Foster, *Strange Genius: The Life of Ferdinand Vandeveer Hayden* (Niwot: Roberts Rinehart, 1994), 158; Karnes, *William Gilpin*, 310.
59. Ibid., 159.
60. Karnes, *William Gilpin*, 310.
61. William Gilpin, "Description of the San Luis Park," *Colorado: Its Resources, Parks, and Prospects as a New Field for Emigration; With an Account of the Trenchara and Costilla Estates, in the San Luis Park*, ed. William Blackmore (London: Sampson, Low, Son, and Marston, 1869), 135.
62. Edward Bliss, "A Description of the San Luis Park and the Sangre de Cristo Grant," *Colorado: Its Resources, Parks, and Prospects as a New Field for Emigration; With an Account of the Trenchara and Costilla Estates, in the San Luis Park*, ed. William Blackmore (London: Sampson, Low, Son, and Marston, 1869), 180.
63. William Blackmore, no title, *The Standard* (London), August 19, 1869. Reference from an untitled, undated document in Great Sand Dunes National Park archives.

64. Ferdinand Vandeveer Hayden for the U.S. Geological Survey, U.S. Department of the Interior, "Geological Report for 1869, Embracing Colorado and New Mexico," in *First, Second, and Third Annual Reports of the United States Geological Survey of the Territories for the Years 1867, 1868, and 1869* (Washington, D.C.: Government Printing Office, 1873), n.p., http://esp.cr.usgs.gov/info/dunes/noneolian.html (accessed January 13, 2011).

65. Stegner, *Beyond the Hundredth Meridian*, xix. See also William Cronon, "Landscapes of Scarcity and Abundance," in *The Oxford History of the American West*, ed. Clyde A. Milner, Carol A. O'Conner, and Martha A. Sandweiss (New York: Oxford University Press, 1994), 601–37, for an informative essay on pioneer perceptions of landscapes in the American West that were originally considered to be frontiers of endless abundance that gradually deteriorated into "endangered landscapes of scarcity, so fragile in the face of human destructiveness that only careful management could ensure their survival" (612).

66. Charles Samuel Richardson, "Notebooks 1871–1883," Western History Collection, Denver Public Library, Denver, Colo., notebook no. 15, n.p. Richardson's leather-bound notebooks contain notes from mine surveys, sketches, memoranda, and personal accounts from his travels in Colorado Territory, mostly in Gilpin, Clear Creek, and Boulder counties, with some notes concerning South Park, the San Luis Valley, and Luis Maria Baca Grant no. 4.

67. John Koshak, telephone interview by author, October 22, 1996. Koshak was chief ranger at San Luis Lakes State Park at the time of interview.

68. LeRoy R. Hafen and Ann W. Hafen, ed., "The Diaries of William Henry Jackson, Frontier Photographer," *The Far West and the Rockies Historical Series 1820–1875*, vol. 10 (Glendale, Calif.: Arthur H. Clark Company, 1959), 331.

69. Forbes Parkhill, "Colorado's Earliest Settlements," *Colorado Magazine* 34, no. 4 (October 1957): 249.

70. Ibid.

71. Ibid.

72. John A. Mangimelli, "Climatic Change in the San Luis Valley," *San Luis Valley Historian* 22, no. 3 (1990): 25.

73. Richardson, "Notebooks," no. 15, n.p.

74. Harvey Gardner, telephone interview by author, October 21, 1996. Gardner was working on a biography of Charles Richardson at the time of interview.

CHAPTER 4

Epigraph. William Blackmore, no title, *The Standard* (London), August 19, 1869. Reference from an untitled, undated document in Great Sand Dunes National Park archives.

1. Tom Wolf, *Colorado's Sangre de Cristo Mountains* (Niwot: University Press of Colorado, 1995), 198; Cuvier H. Jones, "The Herard Family," *San Luis Valley Historian* 11, no. 1 (1979): 7.

2. R. W. Shellabarger, "Papers 1901–1941," unpublished papers, Western History Collection, Denver Public Library, Denver, Colo., April 1949, envelope 3, no. 10, 25.

3. Wolf, *Colorado's Sangre*, 238; Virginia McConnell Simmons, *The San Luis Valley: Land of the Six-Armed Cross* (Boulder: Pruett, 1979), 129.

4. George Harlan, *Postmarks and Places* (Alamosa: Ye Olde Print Shoppe, 1976), 1, 9, 154–55.
5. Harlan, *Postmarks*, 154–55; Simmons, *San Luis Valley*, 27, 74, 85; *Montville Trail Guide* (Southwest Parks and Monuments Association, no date), from an undated copy in Great Sand Dunes National Park archives.
6. Donald William Meinig, *The Shaping of America: A Geographical Perspective on 500 Years of History*, vol. 3: *Transcontinental America, 1850–1915* (New Haven, Conn.: Yale University Press, 1998), 120.
7. William deBuys, *Enchantment and Exploitation: The Life and Hard Times of a New Mexico Mountain Range* (Albuquerque: University of New Mexico Press, 1985), 121.
8. Olibama Lopez Tushar, *The People of "El Valle"* (Pueblo: El Escritorio Press, 1992), 35.
9. Jose de Onis, ed., *The Hispanic Contribution to the State of Colorado* (Boulder: Westview Press, 1976), xvi.
10. Richard White, "Animals and Enterprise," in *The Oxford History of the American West*, ed. Clyde A. Milner, Carol A. O'Connor, and Martha A. Sandweiss (New York: Oxford University Press, 1994), 257–58, 269.
11. Vince Spero et al. for the U.S. Forest Service, U.S. Department of Agriculture, "An Assessment of the Range of Natural Variability of the Rio Grande National Forest," copy of an unpublished internal draft in Great Sand Dunes National Park archives, March 1995, A-26; Martha Oliver, "Cattle in the Valley," *San Luis Valley Historian* 17, no. 4 (1985): 2.
12. National Park Service, U.S. Department of the Interior, *National Register of Historic Places Nomination Form—The Trujillo Homestead* (Front Range Research Associates, Inc., 2003), continuation sheets 1–7; "Early Latino Settlement in Colorado Designated National Historic Landmark," National Parks Traveler, January 4, 2012, www.nationalparkstraveler.com/ 2012/01/early-latino-settlement-colorado-designated-national-historic-landmark9248 (accessed January 7, 2012); National Park Service, U.S. Department of the Interior, "Trujillo Homestead," http://nps.gov/nr/feature/hispanic/2004/Trujillo.htm (accessed January 7, 2012); "Medano-Zapata Ranch History," The Nature Conservancy Zapata Ranch website, www.zranch.org/index.cfm?id=df8ef385-d6f6-4393-bca27b2a03flfdba&history-of-medano-zapata-ranch.html (accessed January 7, 2012).
13. Wolf, *Colorado's Sangre*, 237.
14. Spero et al., "Assessment of the Range of Natural Variability," A-27.
15. Ibid.
16. Wolf, *Colorado's Sangre*, 36.
17. Ibid., 198.
18. Oliver, "Cattle," 2–3.
19. Wolf, *Colorado's Sangre*, 198; Jones, "The Herard Family," 7.
20. Jones, "The Herard Family," 7.
21. Simmons, *San Luis Valley*, 129.
22. Herbert E. Bolton, ed., *Spanish Exploration in the Southwest 1542–1706* (New York: Charles Scribner's Sons, 1908), 230; Spero et al., "Assessment of the Range of Natural Variability," A-33–A-35.
23. Wolf, *Colorado's Sangre*, 149; Richard White, "Animals and Enterprise," 270.

24. Wolf, *Colorado's Sangre*, 149.
25. *Trujillo Homestead*, continuation sheet 6.
26. Robert L. Spude, regional historian, Cultural Resources Management, National Park Service Intermountain Region, e-mail correspondence, July 11, 2012; Simmons, *San Luis Valley*, 101–103.
27. Simmons, *San Luis Valley*, 103, 106; Wolf, *Colorado's Sangre*, 149.
28. Muriel Sibell Wolle, *Stampede to Timberline: The Ghost Towns and Mining Camps of Colorado* (Denver: Sage Books, 1949), 308; Harlan, *Postmarks*, 103, 108; Wolf, *Colorado's Sangre*, 231; "Crestone History," Town of Crestone website, http://townofcrestone.org/crestone_history.shtml (accessed January 13, 2012).
29. Noel Harlan, "Broken Dreams: The History of Duncan Mining Camp," *San Luis Valley Historian* 42, no. 3 (2010): 5–7; Harlan, *Postmarks*, 78–80.
30. Harlan, "Broken Dreams," 10; Harlan, *Postmarks*, 84–89; Simmons, *San Luis Valley*, 106.
31. Spude, e-mail correspondence, July 11, 2012; Harlan, "Broken Dreams," 15; Harlan, *Postmarks*, 88.
32. Harlan, *Postmarks*, 141–46.
33. Stephen M. Voynick, *Colorado Gold: From the Pike's Peak Rush to the Present* (Missoula: Mountain Press, 1992), 158.
34. Ibid.
35. Simmons, *San Luis Valley*, 85–100.
36. Spero et al., "Assessment of the Range of Natural Variability," A-21.
37. Wolf, *Colorado's Sangre*, 148.
38. Ibid., 147.
39. Peter Rowlands, "Stand Structure of a Piñon-Juniper Woodland in Great Sand Dunes National Monument Colorado and the Effects of Historical Wood Harvesting," draft copy of report to superintendent of Great Sand Dunes National Monument in Great Sand Dunes National Park archives, November 1996, 15.
40. Paul Reddin, "Sand, Saws, and the Sangre de Cristos: A History of Wood Hauling in the Hooper and Mosca Communities," *San Luis Valley Historian* 16, no. 1 (1980): 2.
41. Spero et al., "Assessment of the Range of Natural Variability," A-13.
42. Wolf, *Colorado's Sangre*, 36, 147. See also Shellabarger, "Papers," envelope 3, no. 10, 25.
43. Reddin, "Sand, Saws," 3.
44. Ibid., 2.
45. Ibid., 2, 5–6.
46. Fred Bunch, interview by author, January 30, 1997.
47. Reddin, "Sand, Saws," 5–6.
48. Ibid., 10.
49. Ibid., 2.
50. Rowlands, "Stand Structure," 3.
51. Ibid., 26.
52. Ibid., 25.
53. Wolf, *Colorado's Sangre*, 149, 150.

54. C. M. King, "The Wild Horse Roundup," *Alamosa Valley Courier*, no date (clipping located in Colorado Springs Pioneer Museum archives, Colorado Springs, Colo.), n.p.; Paul Maddrell, "Mining for Gold in the Sand Dunes," *Del Norte (Colo.) Prospector*, May 19, 1982, 9.
55. Simmons, *San Luis Valley*, 72, 127.
56. Robert W. Ogburn, "A History of the Development of San Luis Valley Water," *San Luis Valley Historian* 28, no. 1 (1996): 9; Simmons, *San Luis Valley*, 74.
57. *San Luis Valley Agriculture* (promotional booklet; Alamosa: Alamosa First National Bank, , no date), 4; Simmons, *San Luis Valley*, 127.
58. *San Luis Valley Agriculture*, 5.
59. Allan G. Bogue, "An Agricultural Empire," in *The Oxford History of the American West*, ed. Clyde A. Milner, Carol A. O'Connor, and Martha A. Sandweiss (New York: Oxford University Press, 1994), 285; Simmons, *San Luis Valley*, 140.
60. Paul Horgan, *Great River: The Rio Grande in North American History*, vol. 2 (New York: Rinehart and Co., Inc., 1954), 889; Ogburn, "History of the Development of San Luis Valley Water," 9; Shellabarger, "Papers," env. 3, no. 10, 10.
61. John Dietz and Albert Larson, "Colorado's San Luis Valley," in *The Mountainous West: Explorations in Historical Geography*, ed. William Wyckoff and Larry M. Dilsaver (Lincoln: University of Nebraska Press, 1995), 359–60.
62. "San Luis Farmer's Big Task: How He Tried Unsuccessfully to Fence the Sand off His Farm," *Denver Times*, November 3, 1898, 8.
63. Simmons, *San Luis Valley*, 140; Ogburn, "History of the Development of San Luis Valley Water," 11.
64. G. E. Radosevich and R. W. Rutz, *San Luis Valley Water Problems: A Legal Perspective* (Fort Collins: Colorado Water Resources Research Institute, January 1979), 20–21; Simmons, *San Luis Valley*, 140; Ogburn, "History of the Development of San Luis Valley Water," 9.
65. Norris Hundley, *The Great Thirst: Californians and Water 1770s–1900s* (Berkeley: University of California Press, 1992), 67–69, 83.
66. Ibid., 73.
67. Ibid., 71.
68. Radosevich and Rutz, *San Luis Valley Water Problems*, 3.
69. Mark H. Hunter, "Water Plan Pitched to State Board," *Alamosa Valley Courier*, September 11, 1996, 1.
70. Wayne Peck, Rio Grande Water Conservation District, telephone interview by author, September 26, 2011; Philip A. Emery, "Water Resources of the San Luis Valley, Colorado," *Guidebook of the San Luis Basin, Colorado*, ed. H. L. James (Socorro: New Mexico Geological Society, 1971), 129; Richard Thomas Raines, "The Effects of Proposed Water Development by AWDI on Great Sand Dunes National Monument and the San Luis Valley in Southern Colorado" (master's thesis, Texas A&M University, 1992), 14; D. H. McFadden, "Aspects of San Luis Valley Water in 1989: Administrative, Investigative, and Litigative," *Water in the Valley* (Lakewood: Colorado Groundwater Association, 1989), 112.
71. Shellabarger, "Papers," env. 3, no. 10, 10; Simmons, *San Luis Valley*, 137.
72. Radosevich and Rutz, *San Luis Valley Water Problems*, 15; Raines, "The Effects of Proposed Water Development," 17.

73. Bureau of Reclamation, U.S. Department of the Interior, *San Luis Valley Project, Closed Basin Division: Facts and Concepts* (Washington, D.C.: Government Printing Office, 1984), 2; Simmons, *San Luis Valley*, 139; Ogburn, "History of the Development of San Luis Valley Water," 11.
74. Bureau of Reclamation, *San Luis Valley Project*, 2; Simmons, *San Luis Valley*, 140; Ogburn, "History of the Development of San Luis Valley Water," 11.
75. Timothy J. Beaton, "Musings on the Conejos River and Ground Water," *Water in the Valley*, 127; Bureau of Reclamation, *San Luis Valley Project*, 6; *San Luis Valley Agriculture*, 4.

CHAPTER 5

Epigraph. Genevieve McLaughlin, "The Sand Dunes," *American Railway Journal*, March 1928: 1–2.
1. Raymond Richards, "Three Lakes Discovered amid Great Sand Dunes of Colorado," *Denver Post*, October 15, 1922, 7(2).
2. Ibid.
3. Ibid.
4. Roger W. Toll for the National Park Service, U.S. Department of the Interior, *Report on Great Sand Dunes National Monument*, copy of an unpublished report in Great Sand Dunes National Park archives, 1931, 47.
5. Ibid., 48.
6. Dietmar Schneider-Hector, "Colorado's Great Sand Dunes: The Making of a National Monument," *Colorado History* 15 (2008): 2–3.
7. Frank J. McEniry, "'Hissing Sands' of Colorado's 'Sahara' One of World's Great Scenic Wonders," *Rocky Mountain News*, July 31, 1921, sec. 2, p. 16; Wallace Hutchinson, "The Desert of Hissing Sands," *Travel*, June 1922, 16–18; Toll, *Report*, 74.
8. McLaughlin, "The Sand Dunes," 1–2.
9. Toll, *Report*, 73.
10. Ibid., 44; Schneider-Hector, "Colorado's Great Sand Dunes," 3.
11. Hal Rothman, *Preserving Different Pasts: The American National Monuments* (Urbana: University of Illinois Press, 1989), xi, 1, 18; Alfred Runte, *National Parks: The American Experience* (Lincoln: University of Nebraska Press, 1987, Second Edition, Revised), 71–74.
12. Rothman, *Preserving Different Pasts*, xii.
13. Toll, *Report*, 44; Schneider-Hector, "Colorado's Great Sand Dunes," 3–4.
14. Toll, *Report*, 48–49.
15. Ibid., 51; Schneider-Hector, "Colorado's Great Sand Dunes," 5.
16. Toll, *Report*, 35; Schneider-Hector, "Colorado's Great Sand Dunes," 4–5.
17. C. Joe Carter, "Great Sand Dunes National Monument: An Administrative History," *San Luis Valley Historian* 19, no. 2 (Spring 1982): 6.
18. Frank Spencer, "The San Luis Valley, a Mecca for Tourists," *Alamosa Journal*, January 27, 1927; Toll, *Report*, 63–64.
19. "Sand Dunes Are Unequalled Attraction for Tourists Who Marvel at Changing Colors," *Alamosa Journal*, January 6, 1928; McLaughlin, "The Sand Dunes," 1–2; Toll, *Report*, 65–71.

20. "Sand Dunes Are Unequalled," *Alamosa Journal*, January 6, 1928; Toll, *Report*, 67.
21. McLaughlin, "The Sand Dunes," 1; Toll, *Report*, 69.
22. Mark H. Hunter, "Dunes Protection Fight Spans 70 Years," *Denver Post*, October 16, 2000, 5B; Stephen M. Voynick, *Colorado Gold: From the Pike's Peak Rush to the Present* (Missoula: Mountain Press, 1992), 158.
23. Jane David, "The Great Sand Dunes of Colorado: A National Monument Thanks to P.E.O.," *P.E.O. Record*, July/August 1991, 6–7; Hunter, "Dunes Protection," 5B; Schneider-Hector, "Colorado's Great Sand Dunes," 6.
24. David, "Thanks to P.E.O.," 7; Hunter, "Dunes Protection," 5B.
25. PEO Chapters V, A.E., and B.H., to State Senator A. Elmer Headlee, June 1930, Chapter V, PEO collection, Monte Vista, Colo.
26. Colorado Legislative Council, "Glossary of Legislative Terms," https://www.colorado.gov/pacific/sites/default/files/Glossary%20of%20Legislative%20Terms_2.pdf (accessed November 13, 2011).
27. Robert L. Schultz for the National Park Service, U.S. Department of the Interior, "Monte Vista P.E.O. Spearheaded Great Sand Dunes Monument Establishment," press release, March 6, 1982; Harold Schaafsma, "Great Sand Dunes Monument Has Colorful, Exciting Past," *Alamosa Daily Courier*, January 5, 1955, Adams State College Museum Files; David, "Thanks to P.E.O.," 7; Schneider-Hector, "Colorado's Great Sand Dunes," 6; Carter, "Great Sand Dunes National Monument," 11.
28. PEO to Headlee, June 1930.
29. Ibid.
30. George M. Corlett to Anna R. Darley, December 24, 1930, Chapter V, PEO collection, Monte Vista, Colo.
31. Senator Lawrence C. Phipps to George Corlett, December 27, 1930, Chapter V, PEO collection, Monte Vista, Colo. See also Great Sand Dunes National Park Archive collections, Box 9.
32. Toll, *Report*, 1.
33. "Senate Joint Memorial," January 16, 1931, newspaper clipping in Chapter V, PEO collection, Monte Vista, Colo.
34. "Our Best Bet," *Alamosa Journal*, February 13, 1931, copy included in Toll, *Report*, 76.
35. Sharon A. Brown, "Roger Wolcott Toll, 1883–1936," in *National Park Service: The First 75 Years*, ed. William H. Sontag (Philadelphia: Eastern National Park and Monument Association, 1990), www.nps.gov/history/ history/online_books/sontag/toll.htm (accessed August 11, 2011); American Academy for Park and Recreation Administration, "Roger Wolcott Toll," http://aapra.org/Pugsley/TollRoger.html (accessed August 11, 2011).
36. Toll, *Report*, 5, 22–23.
37. Ibid., 5–6.
38. "Colorful Sand Dunes of Colorado May Be National Monument," *Denver Post*, February 16, 1931, copy included in Toll, *Report*, 78.
39. "PEO Sponsors Bill for Making Sand Dunes National Monument," *Alamosa Journal*, February 17, 1931, copy included in Toll, *Report*, 77.

40. "Dr. Spencer Urges Creation of National Monument of Sand Dunes," *Alamosa Journal*, February 27, 1931, copy included in Toll, *Report*, 75.
41. Toll, *Report*.
42. Ibid., 32.
43. Ibid., 24.
44. Ibid., 32.
45. Ibid., 33–34.
46. Ibid., 34.
47. Ibid., 3.
48. Ibid.
49. Dietmar Schneider-Hector, "Roger W. Toll, Chief Investigator of Proposed National Parks and Monuments: Setting the Standards for America's National Park System," *Journal of the West* 42, no. 1 (Winter 2003): 89; Schneider-Hector, "Colorado's Great Sand Dunes," 9. Toll's chronological list, included in Schneider-Hector's "Roger W. Toll" on page 89, lists the date of Toll's official submission of the Great Sand Dunes report as April 5, while on page 9 of "Colorado's Great Sand Dunes," Schneider-Hector lists the date as April 3.
50. Representative Guy Hardy to Millie Velhagen, December 19, 1931, Chapter V, PEO collection, Monte Vista, Colo.
51. Chapter V, PEO Committee on Sand Dunes Project to Sen. Edward P. Costigan, December 29, 1931, Chapter V, PEO collection, Monte Vista, Colo.
52. Rep. Guy Hardy to Elizabeth Spencer, January 7, 1932, Chapter V, PEO collection, Monte Vista, Colo.
53. Elizabeth Spencer to Horace M. Albright, January 22, 1932, Chapter V, PEO collection, Monte Vista, Colo.
54. Horace M. Albright to Elizabeth Spencer, January 27, 1932, Chapter V, PEO collection, Monte Vista, Colo.
55. Voynick, *Colorado Gold*, 158.
56. Ibid., 158–59.
57. Rep. Guy Hardy to Elizabeth Spencer, March 14, 1932, Chapter V, PEO collection, Monte Vista, Colo.
58. Carter, "Great Sand Dunes National Monument," 8; Rothman, *Preserving Different Pasts*, xiii; Toll, *Report*, 1; "President Hoover's Proclamation 1932," *Great Sand Dunes Online Curriculum*, www.interactive-earth.com/grsa/resources/curriculum/high/hoover_proclamation.htm (accessed August 14, 2011).
59. Stephen Trimble, "A Sketch on the History of the Establishment of Great Sand Dunes NM," letter to Thea K. Nordling, chief interpreter, Great Sand Dunes National Monument, August 20, 1975, copy in Great Sand Dunes National Park archives; National Park Service, U.S. Department of the Interior, press release, March 22, 1932; Schneider-Hector, "Colorado's Great Sand Dunes," 11.
60. Schneider-Hector, "Roger W. Toll," 87–88.
61. Carter, "Great Sand Dunes National Monument," 6.
62. Dietmar Schneider-Hector, "Great Sand Dunes Wilderness: Creating 'The Very Highest Order of Federal Resource Protection' for North America's Tallest Inland Dunefield," *Colorado Heritage* (September/October 2010): 24; Carter, "Great Sand Dunes National Monument," 10; Schneider-Hector, "Colorado's Great

Sand Dunes," 12; Glen Bean, "Ranger Reminiscing: The Early Days at Great Sand Dunes," *Great Sand Dunes: Celebrating 60 Years 1932–1992*, 3, copy in Great Sand Dunes National Park archives.

63. Paul R. Franke for the National Park Service, U.S. Department of the Interior, *General Report on the Great Sand Dunes National Monument*, May 17, 1934, copy of an unpublished report in Great Sand Dunes National Park archives, 6.

64. Schneider-Hector, "Colorado's Great Sand Dunes," 12; U.S. Department of the Interior, National Park Service, Public Use Statistics Office website, *Great Sand Dunes National Park and Preserve Annual Park Visitation (All Years)*, www.nature.nps.gov/stats/park.cfm?parkid=208 (accessed August 16, 2011).

65. *Alamosa Daily Courier*, January 29, 1940, no title, Chapter V, PEO collection, Monte Vista, Colo.

66. Ibid.

67. Louise Robison and Pearle Hoffman to Sen. Edwin C. Johnson, copy to Rep. Edward T. Taylor, February 6, 1940, Chapter V, PEO collection, Monte Vista, Colo.

68. Sen. Edwin C. Johnson to Louise Robison and Pearle Hoffman, February 10, 1940, Chapter V, PEO collection, Monte Vista, Colo.

69. Carter, "Great Sand Dunes National Monument," 10–11; Glen Bean, "Ranger Reminiscing," 3; Schaafsma, "Great Sand Dunes Monument Has Colorful, Exciting Past."

70. Elizabeth Hall, *Sand Stories: Recollections of Life around the Great Sand Dunes* (Mosca: Friends of the Dunes, 2004), 33; *Buick Magazine* 10, no. 11, no author, no title, May 1949, reference found in National Park Service, U.S. Department of the Interior, *Area Management Study, Great Sand Dunes National Monument*, June 22–23, 1956, Record Group (hereafter RG) 79, Great Sand Dunes National Monument, file A-6423, National Archives and Records Administration (hereafter NARA), Rocky Mountain Region, Denver, Colo.

71. Glen Bean, "PEO Talk," no date, copy in PEO file in Great Sand Dunes National Park archives; Carter, "Great Sand Dunes National Monument," 10–11; Glen Bean, "Ranger Reminiscing," 3; Schaafsma, "Great Sand Dunes Monument Has Colorful, Exciting Past"; Public Use Statistics Office, *Annual Park Visitation*.

72. Bert G. Clarke to George Ziegler, July 30, 1943, RG 79, Great Sand Dunes National Monument, file D-3415, NARA, Rocky Mountain Region, Denver, Colo.

73. Bert G. Clarke to George Ziegler, August 10, 1943, RG 79, Great Sand Dunes National Monument, file D-3415, NARA, Rocky Mountain Region, Denver, Colo.

74. Fred Bunch, telephone interview by author, March 7, 1997.

75. Gary Hasty, "Monte Vista P.E.O. Sisterhood initiated establishment of Sand Dunes National Monument," possibly from the *Monte Vista Journal*, July 24, 1975, copy of newspaper clipping in Great Sand Dunes National Park archives.

76. Glen Bean, "Ranger Reminiscing," 3.

77. Ed Quillen, "How Trinity Didn't Happen at the Sand Dunes," *Colorado Central Magazine*, July 1999, http://cozine.com/1999-july/how-trinity-didnt-happen-at-the-sand-dunes/ (accessed October 18, 2011).

78. Harry S. Truman, "Proclamation 2681—Redefining the Area of Great Sand Dunes National Monument, Colorado," March 12, 1946, available at Gerhard Peters and John T. Woolley, *The American Presidency Project*, www.presidency.ucsb.edu/ws/?pid=87055 (accessed October 18, 2011).
79. "Singing Sands of Alamosa," *Steamboat Pilot*, December 31, 1942; "The Great Sand Dunes: Eight Billion Dollars in Gold," *Folks and Fortunes: Saga of the San Luis Valley in Colorado* 1, no. 1 (Monte Vista: Monte Vista Journal, November 1949): 25.
80. Wilbur B. Foshay to Sen. Edwin C. Johnson, December 1, 1947, RG 79, Great Sand Dunes National Monument, file D-30, NARA, Rocky Mountain Region, Denver, Colo.
81. Wilbur B. Foshay to Sen. Edwin C. Johnson, May 4, 1950, RG 79, Great Sand Dunes National Monument, file K-3819, NARA, Rocky Mountain Region, Denver, Colo.
82. Rep. John Edgar Chenoweth to Conrad L. Worth, August 2, 1952, RG 79, Great Sand Dunes National Monument, file D-30, NARA, Rocky Mountain Region, Denver, Colo.
83. Superintendent Harold Schaafsma to Glen Bean, January 2, 1954, copy of a letter in Great Sand Dunes National Park archives. See also RG 79, Great Sand Dunes National Monument, file L-3019, NARA, Rocky Mountain Region, Denver, Colo.
84. Superintendent Harold Schaafsma to NPS Assistant Regional Director, November 12, 1956, RG 79, Great Sand Dunes National Monument, file D-24, NARA, Rocky Mountain Region, Denver, Colo.
85. National Park Service, U.S. Department of the Interior, "Mission 66 for Great Sand Dunes National Monument," 1956 (date of publication inferred from text), 4, Denver Public Library Government Document I 29.6: M68/G75z; "New Camp Area Ready at 'Dunes,'" possibly from the *Alamosa Valley Courier*, October 14, 1964, newspaper clipping from Great Sand Dunes file in Colorado Room, Nielsen Library, Adams State College, Alamosa, Colo.; Carter, "Great Sand Dunes National Monument," 10–11; Hasty, "Monte Vista P.E.O."; Public Use Statistics Office, *Annual Park Visitation*, 1960–66.
86. Schneider-Hector, "Great Sand Dunes Wilderness," 30; "Leave No Trace," Great Sand Dunes National Park and Preserve website, www.nps.gov/grsa/supportyourpark/leavenotrace.htm (accessed October 24, 2011); Public Use Statistics Office, *Annual Park Visitation*, 1976.

CHAPTER 6

Epigraph. Loren Eiseley, *The Immense Journey* (New York: Franklin Watts, 1957), 15.
1. Bureau of Reclamation, U.S. Department of the Interior, *San Luis Valley Project, Closed Basin Division: Facts and Concepts* (Washington, D.C.: Government Printing Office, 1984), 3.
2. Robert W. Ogburn, "A History of the Development of San Luis Valley Water," *San Luis Valley Historian* 28, no. 1 (1996): 19.
3. Bureau of Reclamation, *San Luis Valley Project*, 4–5.
4. G. E. Radosevich and R. W. Rutz, *San Luis Valley Water Problems: A Legal Perspective* (Fort Collins: Colorado Water Resources Research Institute, January 1979), 24–25.

5. Donald Worster, *Rivers of Empire: Water, Aridity, and the Growth of the American West* (New York: Oxford University Press, 1985), 313; Shelli Mader, "Center Pivot Irrigator Revolutionizes Agriculture," *Fence Post*, May 25, 2010, www.thefencepost.com/article/20100525/NEWS/100529954 (accessed July 15, 2011); "Ag Innovator Frank Zybach," http://livinghistoryfarm.org/farminginthe40s/water_09.html (accessed July 15, 2011); "Center Pivot Irrigator," Asabe Historic Commemoration Event No. 30, www.asabe.org/awards-landmarks/asabe-historic-landmarks/center-pivot-irrigator-30.aspx (accessed July 15, 2011); Randy Alfred, "July 22, 1952: Genuine Crop-Circle Maker Patented," *Wired*, www.wired.com/science/discoveries/news/2008/07/dayintch_0722 (accessed July 15, 2011).
6. Ralph Curtis, telephone interview by author, March 6, 1997. Curtis is the former director of the Rio Grande Water Conservation District; Rio Grande Water Conservation District, telephone interview by author, September 26, 2011; Wayne Peck, Colorado Division of Water Resources, Division 3, telephone interview by author, November 18, 2011.
7. Richard E. Demlo et al., "San Luis Valley Project, Closed Basin Division," in Colorado Groundwater Association, *Water in the Valley* (Lakewood: Colorado Groundwater Association, 1989), 157–58; Radosevich and Rutz, *San Luis Valley Water Problems*, 19.
8. H. N. Dixon, "Relationships of the Greasewood Community to Groundwater in the San Luis Valley," in Colorado Groundwater Association, *Water in the Valley*, 169, 173.
9. Demlo et al., "San Luis Valley Project," 158–61.
10. Bureau of Reclamation, *San Luis Valley Project*, 12–14.
11. Ibid., 14.
12. Ibid., 9; Tom Wolf, *Colorado's Sangre de Cristo Mountains* (Niwot: University Press of Colorado, 1995), 225.
13. Fred Bunch, interview by author, August 26, 1997.
14. Ogburn, "History of the Development of San Luis Valley Water," 28; D. H. McFadden, "Aspects of San Luis Valley Water in 1989: Administrative, Investigative, and Litigative," in Colorado Groundwater Association, *Water in the Valley*, 113; Curtis, telephone interview, March 6, 1997; Peck, telephone interview, November 18, 2011.
15. Julia Rubin, "Battle Brews over SLV Water," *Pueblo Chieftain*, June 12, 1989, 6(A); Helen Martin, "Water Resources and Rights in the Northern San Luis Valley, Part II," *San Luis Valley Historian* 37, no. 4 (2005): 13; Ogburn, "History of the Development of San Luis Valley Water," 30; American Water Development Inc. v. City of Alamosa (Colorado Supreme Court, May 9, 1994), http://co.findacase.com/research/wfrmDocViewer.aspx/xq/fac.%5CSAC%5CCO%5C1994%5C19940509_0040691.CO.htm/qx (accessed August 9, 2011).
16. Malcom Ebright, *Land Grants and Lawsuits in Northern New Mexico* (Albuquerque: University of New Mexico Press, 1994), 174–78; David Remley, *Bell Ranch: Cattle Ranching in the Southwest, 1824–1947*, rev. ed. (Las Cruces: Yucca Tree Press, 2000), 38, 48; J. J. Bowden, "Luis Maria Cabeza de Baca Grant," New Mexico Office of the State Historian website, www.newmexicohistory.org/

filedetails.php?fileID=24833 (accessed June 18, 2011); Kurt F. Anschuetz and Thomas Merlan for the Rocky Mountain Research Station, U.S. Forest Service, *More Than a Scenic Mountain Landscape: Valles Caldera National Preserve Land Use History*. General Technical Report RMRS-GTR-196 (September 2007): 32–33; Fred Roeder, "The Baca Floats," The American Surveyor, April 11, 2009, www.amerisurv.com/content/view/6083/ (accessed June 18, 2011); Mary Abdoo, "Baca Grande History," Baca Grande Property Owners Association website, www.bacapoa.org/Baca-Grande-History-120603-14064.htm (accessed June 18, 2011); Virginia McConnell Simmons, *The San Luis Valley: Land of the Six-Armed Cross* (Boulder: Pruett, 1979), 81.

17. Ed Quillen, "The Baca Ranch: It's Got Quite a Past," *Colorado Central Magazine* (February 2000): 37–38; See also Ed Quillen, "The Baca Ranch—a Legacy from Mexico," *Colorado Central Magazine* (February 2000), http://cozine.com/2000-february/the-baca-ranch-a-legacy-from-mexico/ (accessed June 18, 2011); Ebright, *Land Grants*, 177–78; Remley, *Bell Ranch*, 67–68, 74; Anschuetz and Merlan, *More Than a Scenic Mountain Landscape*, 32; Simmons, *San Luis Valley*, 81; Bowden, "Luis Maria Cabeza de Baca"; Roeder, "Baca Floats"; Abdoo, "Baca Grande."

18. Martin, "Water Resources and Rights," 10–11.

19. Sam Bingham, *The Last Ranch: A Colorado Community and the Coming Desert* (New York: Pantheon Books, 1996), 138. Bingham writes that the U.S. government gave the land to Baca's heirs in 1863, but San Luis Valley historian Virginia Simmons and several other sources give 1860 as the date for the grant. See Simmons, *San Luis Valley*, 81; Jacqui Ainley-Conley, "Timeline of Baca Grant No. 4," July 10, 2010, unpublished document, copy on file at Great Sand Dunes National Park archives; Martin, "Water Resources and Rights," 9–12; Quillen, "The Baca Ranch—a Legacy from Mexico"; Anschuetz and Merlan, *More Than a Scenic Mountain Landscape*, 32; Bowden, "Luis Maria Cabeza de Baca"; Roeder, "Baca Floats"; Abdoo, "Baca Grande"; "Factual Background," American Water Development Inc. v. City of Alamosa, 1994.

20. Bingham, *Last Ranch*, 138.

21. Rubin, "Battle Brews," 6(A); "Water Facts," Citizens for San Luis Valley Water, copy of press release, n.d., n.p. Located in AWDI newspaper clippings file, Colorado Room, Nielsen Library, Adams State College, Alamosa, Colo.; Martin, "Water Resources and Rights," 13.

22. "A Rape in the Making," *Pueblo Chieftain*, November 26, 1989, 2(E); Diane Bairstow, "The Water War with AWDI Begins," *Crestone (Colo.) Eagle*, January 2009, 21; Rick Boychuk, "'Division of Interest' Causes Maurice Strong to Quit AWDI," *Crestone Eagle*, January 1990, 1, 3, www.scribd. com/doc/40083195/Big-Bad-Bank (accessed June 18, 2011); Martin, "Water Resources and Rights," 15.

23. Marc Reisner, *Cadillac Desert: The American West and Its Disappearing Water* (New York: Penguin Books, 1986), 55–103; Ogburn, "History of the Development of San Luis Valley Water," 31; Bingham, *Last Ranch*, 139; Rubin, "Battle Brews," 6(A).

24. Richard D. Lamm, "Water Project May Be as Good as It's Going to Get," op-ed column, *Rocky Mountain News*, January 30, 1991; Bingham, *Last Ranch*, 275.

25. Bingham, *Last Ranch*, 144–45.

26. Ibid.
27. Ibid.
28. Fred Bunch, interview by author, August 26, 1997.
29. Dick Foster, "Study: Water Project Will Imperil Dunes," *Rocky Mountain News*, February 24, 1991, 49.
30. Fred Bunch, interview by author, August 26, 1997; Andrew Valdez, interview by author, June 3, 2011; Erin Smith, "Dune Mystery Drilled," *Pueblo Chieftain*, October 12, 1990. Although Stewart's data initially seemed conclusive, the reality is that at the time the connection between the dunes and the underground aquifer remained unclear at best, and the NPS thought it prudent to base its legal arguments on the proven hydrologic connection discovered during studies of Sand Creek. By 2011, Stewart's theory had generally been dismissed as unreliable.
31. Erin Smith, "Judge Unveils Written Report on AWDI Decision," *Pueblo Chieftain*, February 12, 1992, 1–2(A); Todd Hartman, "Capitalist Cowboy Wants to Tap San Luis Aquifer," *Coloradoan* (Fort Collins, Colo.), March 17, 1997, 5(A); Alex Prud'homme, *The Ripple Effect: The Fate of Fresh Water in the Twenty-First Century* (New York: Scribner, 2007), 147. "Procedural History and Issues Presented," American Water Development Inc. v. City of Alamosa, 1994; Bingham, *Last Ranch*, 143, 168.
32. Bingham, *Last Ranch*, 139; Bairstow, "Water War," 21; Boychuk, "Division of Interest," 1, 3.
33. National Park Service, U.S. Department of the Interior, *Resource Management Strategy, Great Sand Dunes National Monument* (March 1994), 4; Fred Bunch, interview by author, October 12, 2009; Steve Chaney, telephone interview by author, September 22, 2011; Andrew Valdez, e-mail correspondence with author, December 12, 2011; Martin, "Water Resources and Rights," 19.
34. Fred Bunch, interview by author, October 12, 2009; Martin, "Water Resources and Rights," 19.
35. Erin Smith, "Sand Dunes Encompass 9 Ecosystems," *Pueblo Chieftain*, April 27, 2000, 1A, 3A; Andrew Valdez, interview by author, June 3, 2011; Andrew Valdez, "The Role of Streams in the Development of the Great Sand Dunes and Their Connection with the Hydrologic Cycle," Great Sand Dunes National Park and Preserve website, www.nps.gov/grsa/ naturescience/upload/Trp2029.pdf (accessed August 23, 2011).
36. Bunch interview, October 12, 2009; Erin Smith, "Sand Dunes Encompass 9 Ecosystems"; Valdez interview, June 3, 2011.
37. "Hydrology," formerly titled "Mysterious Waters of the Dunes," Great Sand Dunes National Park and Preserve website, www.nps.gov/gras/index.htm (accessed August 23, 2011); Michael G. Rupert and L. Niel Plummer for the U.S. Geological Survey, U.S. Department of the Interior, *Ground-Water Age and Flow at Great Sand Dunes National Monument, South- Central Colorado*, Fact Sheet 2004–3051 (July 2004); Bunch interview, October 12, 2009; Valdez interview, June 3, 2011.
38. "Hydrology"; Bunch interview, October 12, 2009; Valdez interview, June 3, 2011.
39. Ibid.
40. Ibid.

41. Ed Quillen, "Appearing on Your November Ballot: A Water War," *Colorado Central Magazine*, no. 56 (October 1998): n.p.; David Robbins, telephone interview by author, December 7, 2011. Robbins is an attorney with Hill and Robbins PC in Denver and has served as general counsel for the Rio Grande Water Conservation District since 1981. Bingham, *Last Ranch*, 171.
42. Mark Burget, interview by author, November 19, 2009. Burget is the former Colorado state director for The Nature Conservancy; The Nature Conservancy, "Our History," http://www.nature.org/about-us/vision-mission/history/index.htm?intc=nature.tnav.about.list (accessed June 19, 2011).
43. Gary Gerhardt, "Written in Sand," *Rocky Mountain News*, October 15, 2000, 7A, 18A; Bingham, *Last Ranch*, 234.
44. Bingham, *Last Ranch*, 219, 221, 238.
45. Ibid., 219–38.
46. David Nicholas, "Water and Sand: On the Congressional Fast Track toward a National Park," *Crestone Eagle*, February 2000; Katharine Q. Seelye, "Complex Deal Is First Step to Create New National Park," *New York Times*, January 31, 2002; Erin Smith, "Baca Ranch Sold in Foreclosure, Speeding Sand Dunes Park," *Pueblo Chieftain*, May 31, 2002; William Baue, "Private Equity Investment Puts Yale in Deep Water," *Social Funds*, February 8, 2002, www.socialfundds.com/news/save.cgi?sfArticleId=774 (accessed September 6, 2011); Loch Adamson, "Steyer Power," *Institutional Investor*, February 23, 2005, http://www.institutionalinvestor.com/article/1026224/steyer-power.html#.VZw6i0XlY7A (accessed September 6, 2011); The Responsible Endowment Project, "Baca Ranch," no date, www.responsibleendowment.com/baca-ranch.html (accessed September 6, 2011).
47. Gary Boyce, "Tap Bounty without Building Dam," *Denver Post*, August 30, 1998, http://extras.denverpost.com/opinion/boyce0830.htm (accessed September 6, 2011).
48. Jim Hughes, "Emotions Run Deep in the Valley," *Denver Post*, October 4, 1998, http://extras.denverpost.com/news/water/water12.htm (accessed September 6, 2011).
49. Ed Quillen, "San Luis Valley Water War Goes Statewide," *Colorado Central Magazine*, November 1998, http://cozine.com/1998-november/san-luis-valley-water-war-goes-statewide/ (accessed September 6, 2011); Patrick O'Driscoll, "Wanted: San Luis Water," *Denver Post*, April 7, 1996, 1, 7(B); Mark H. Hunter, "Municipal League Talks Water," *Alamosa Valley Courier*, September 21, 1996, 1, 3; Tom Wolf, "Longevity of Ranches Linked to Water Sale," *Coloradoan* (Fort Collins, Colo.), February 16, 1997, 3(E).
50. O'Driscoll, "Wanted," 7(B).
51. Quillen, "San Luis Valley Water War Goes Statewide."
52. Marty Jones, "High and Dry," *Westword*, November 12, 1998, http://www.westword.com/1998-11-12/news/high-and-dry/ (accessed September 7, 2011).
53. Ed Quillen, "A Water Baron Takes On the Establishment," *High Country News*, October 26, 1998, http://hcn.org/issues/141/4560 (accessed September 7, 2011); Colorado Legislative Council, "Historical Ballot Information," www.colorado.gov/cs/Satellite/CGA-LegislativeCouncil/ CLC/1200536136114 (accessed September 7, 2011); Adamson, "Steyer Power."

54. Quillen, "San Luis Valley Water War Goes Statewide."
55. Ibid.
56. Quillen, "San Luis Valley Water War Goes Statewide"; Quillen, "Water Baron"; Colorado Legislative Council, "Historical Ballot Information."
57. Bill Martin, e-mail correspondence with author, December 27, 2011. Martin was the real estate GIS planner for the Colorado State Land Trust at the time of correspondence. See also Colorado State Land Board website, http://trustlands.state.co.us/Pages/SLB.aspx (accessed December 22, 2011).
58. "Colorado Trust Lands and Education Funding," Lincoln Institute of Land Planning website, www.lincolninst.edu/subcenters/managing-state-trust-lands/state/ed-funding-co.pdf (accessed December 22, 2011).
59. Christine Canaly, telephone interview by author, December 8, 2011. Canaly was the director of the San Luis Valley Ecosystem Council at the time of interview. Quillen, "San Luis Valley Water War Goes Statewide."
60. Jones, "High and Dry."
61. Canaly interview, December 8, 2011; Adamson, "Steyer Power."
62. Brendan Smith, "Great Sand Dunes National Park and Preserve/Baca Ranch Purchase," September 1, 2005, Red Lodge Clearinghouse website, www.rlch.org/stories/great-sand-dunes-national-park-and-preservebaca-ranch-purchase (accessed December 8, 2011).
63. Erin Smith, "Colorado Voters Uphold Water Management System in San Luis Valley," *Pueblo Chieftain*, November 5, 1998; Jones, "High and Dry."
64. Erin Smith, "Colorado Voters."
65. Jones, "High and Dry."
66. Ibid.

CHAPTER 7

Epigraph. National Park Service, "Famous Quotes Concerning the National Parks," www.cr.nps.gov/history/hisnps/NPSThinking/famousquotes.htm (accessed July 2, 2012).

1. Scott McInnis, interview by author, July 7, 2011. McInnis is a former U.S. representative for Colorado's Third Congressional District. Fred Bunch, e-mail correspondence with author, August 31, 2011.
2. Ruth Heide, "McInnis Salutes 'Champions of the Dunes,'" *Valley Courier*, April 1, 2004; Fred Bunch, interview by author, June 2, 2011; Robert Zimmerman, telephone interview by author, August 23, 2011. Zimmerman is a former Alamosa County commissioner.
3. Heide, "McInnis Salutes 'Champions.'"
4. Robert W. Righter, "National Monuments to National Parks: The Use of the Antiquities Act of 1906," *Western Historical Quarterly* 20, no. 3 (August 1989): 281–301; Fred Bunch, interview by author, October 12, 2009.
5. "Difference between a National Park and a National Monument," National Park Service website, www.nps.gov/history/history/online_books/portfoli00b.htm (accessed July 11, 2011); "A National Monument, Memorial, Park . . . What's the Difference?," National Atlas website, http://nationalatlas.gov/articles/government/a_nationalparks.html (accessed July 11, 2011); Gerald W. Williams, "National Monuments and the Forest Service," National Park Service, U.S.

Department of the Interior, www.nps.gov/history/history/online_books/fs/monuments.htm (accessed July 11, 2011).

6. Stephan Weiler, Andrew Seidl, and Erich Weeks, "What's in a Name? Economic Impact of National Park Designation on the Great Sand Dunes and the San Luis Valley," *Agricultural and Resource Policy Report*, no. 3 (June 2001), Colorado State University, Department of Agricultural and Resource Economics, Fort Collins.

7. Ibid.

8. Steve Chaney, telephone interview by author, September 22, 2011. Chaney is the former superintendent of Great Sand Dunes National Park and Preserve. Bunch interview, October 12, 2009; McInnis interview, July 7, 2011.

9. Kurt Repanshek, "National Park Service Opposes Redesignation of Pinnacles National Monument as a National Park," *National Parks Traveler*, November 18, 2009, www.nationalparkstraveler.com/2009/11/national-park-service-opposes-redesignation-pinnacles-national-monument-national-park4951 (accessed July 11, 2001).

10. Christine Canaly, interview by author, December 8, 2011. Canaly was the director of the San Luis Valley Ecosystem Council at the time of interview. Chaney interview, September 22, 2011.

11. Erin Smith, "Incoming Superintendent Sees Need to Keep Protecting Great Sand Dunes," *Pueblo Chieftain*, November 14, 1997.

12. Andrew Valdez, interview by author, June 3, 2011.

13. Ibid.

14. Mark Burget, interview by author, November 19, 2009. Burget is the former Colorado state director for The Nature Conservancy. Paul Robertson, interview by author, October 14, 2009. Robertson was The Nature Conservancy's San Luis Valley Project director at the time of interview. "Nature Conservancy Signs Agreement to Buy 100,000-Acre Zapata/Medano Ranch," The Nature Conservancy (TNC), press release, January 21, 1999; "Medano-Zapata Ranch History," TNC Zapata Ranch website, www.zranch.org/index.cfm?id=df8ef385-d6f6-4393-bca27b2a03f1fdba&history-of-medano-zapata-ranch.html (accessed October 11, 2011).

15. Jack Kyle Cooper, *Zebulon Montgomery Pike's Great Western Adventure, 1806–1807* (Colorado Springs: Clausen Books, 2007), 101; "Medano-Zapata Ranch History," TNC Zapata Ranch website.

16. William Blackmore, *Colorado: Its Resources, Parks, and Prospects as a New Field for Emigration; With an Account of the Trenchara and Costilla Estates, in the San Luis Park*, (London: Sampson, Low, Son, and Marston, 1869), 199; Ora Brooks Peake, *The Colorado Range Cattle Industry* (Glendale, Calif.: Arthur H. Clark Company, 1937), 61; Thomas L. Karnes, *William Gilpin: Western Nationalist* (Austin: University of Texas Press, 1970), 310.

17. National Park Service, U.S. Department of the Interior, *National Register of Historic Places (NRHP) Registration Form—Medano Ranch Headquarters* (Front Range Research Associates, Inc., 2003), continuation sheet 12.

18. Peake, Colorado Range Cattle, 63; National Park Service, *NRHP Registration Form—Medano Ranch Headquarters*, continuation sheets 13–15.

19. National Park Service, *NRHP Registration Form—Medano Ranch Headquarters*, continuation sheets 15–16. According to Bob Linger, whose family later owned

and operated the Medano-Zapata, Adee was a heavy drinker who committed suicide after a serious bender. An article in the Rocky Mountain News, dated March 21, 1887, claimed that Adee had "seemed humiliated" by his relationship with a Pueblo Indian woman who had cheated him out of a large sum of money, a claim that was discounted three days later by the *Alamosa Independent* newspaper.
20. Ibid., continuation sheets 16–19.
21. Ibid., continuation sheets 21–22; Treasure Bruce, Perspectives in Sand: Township 40N Range 12E of the NM Meridian in the San Luis Valley of Colorado, University of Denver College of Law (November 2003), 17, Appendices P and Q. Copy on file at Great Sand Dunes National Park and Preserve, Resource Management Office; Ruth Heide, "Zapata Course Gets Conditional OK," Valley Courier, August 24, 1988, 1, 11; Fred Bunch, interview by author, October 13, 2009.
22. "Conservancy Buys 2 Ranches in Colorado," Rocky Mountain News, July 11, 1999; Bunch interview, October 13, 2009.
23. Sanjay Advani, "Nature Conservancy Shows Off Medano Zapata Ranches," Del Norte (Colo.) Prospector, September 16, 1999, 10A; TNC press release, January 21, 1999; Bunch interview, October 13, 2009.
24. Mark H. Hunter, "Group Buys Zapata Ranch," Denver Post, July 10, 1999; "Conservancy Acquires Spectacular Medano and Zapata Ranches," TNC, press release, July 1, 1999; TNC press release, January 21, 1999; Bunch interview, October 13, 2009.
25. Erin Smith, "Nature Conservancy Closes Dunes Golf Course," Pueblo Chieftain, March 31, 2000; "Medano-Zapata Ranch History," TNC Zapata Ranch website; Bunch interview, October 13, 2009.
26. Gary Taylor, "The Sand Dunes as a National Park," *Valley Courier*, May 6, 1999; Chaney interview, September 22, 2011.
27. "Background, History and Status Great Sand Dunes National Park and Preserve Act of 2000," March 1, 2001, unpublished internal document provided by former Great Sand Dunes National Park superintendent Steve Chaney; "Great Sand Dunes National Park Timeline," n.d., unpublished internal document provided by former Great Sand Dunes National Park superintendent Steve Chaney; Chaney interview, September 22, 2011.
28. Ibid.
29. Ibid.
30. Erin Smith, telephone interview by author, September 30, 2011. Smith covered the San Luis Valley for the *Pueblo Chieftain* during a long and distinguished career in journalism that began in 1972 and ended with her semi-retirement in 2009. Erin Smith, "Campbell Backs Sand Dunes Park," *Pueblo Chieftain*, June 2, 1999.
31. Michael Hesse, telephone interview by author, October 11, 2011. Hesse is the former chief of staff for U.S. representative Scott McInnis. Erin Smith, "McInnis Leaning toward Great Sand Dunes Park," *Pueblo Chieftain*, August 12, 1999.
32. Erin Smith, "Great Sand Dunes Park Idea Backed by Allard," *Pueblo Chieftain*, August 21, 1999; Wayne Allard, "Finally—It's a DUNES DEAL," *Pueblo Chieftain*, September 19, 2004, www.chieftain.com/editorial/finally-it-s-a-dunes-deal/article_c200ce82-95f9-5a3a-ad25-0914cbde51e7.html (accessed August 5, 2011).

33. Dick Foster, "Move Afoot to Upgrade Dunes to National Park," *Denver Post*, November 12, 1999.
34. Bill McAllister, "Babbitt to Back Dunes Park Plan," *Denver Post*, November 30, 1999, 1B, 8B; McInnis interview by author, July 7, 2011.
35. "Protecting a Sandy Treasure," *Denver Post*, November 26, 1999, 14B; McAllister, "Babbitt to Back."
36. Bill McAllister, "National Park Push Grows for Dunes," *Denver Post*, December 2, 1999, 5B; Tania Anderson, "Dunes Tour Draws Big Interest," *Pueblo Chieftain*, December 2, 1999; Erin Smith, "Happy Destination," *Pueblo Chieftain*, March 28, 2004, www.chieftain.com/metro/happy-destination/article_3a9705e4-f0ea-5f62–8f69–5133bfe2c540.html (accessed August 5, 2011).
37. Ruth Heide, "Baca Part of Sand Dunes Park Plan," *Valley Courier*, December 2, 1999, 1, 3; Yvonne Smith, "Saguache Board on Record in Opposition to Baca Sale," *Valley Courier*, December 9, 1999, 1, 3; Ruth Heide, "Resident Resistance Rises: Dunes National Park Proposal Revives Area Tax, Water Worries," *Valley Courier*, December 9, 1999, 1, 3.
38. Nelda Curtiss, "Winds of Change at Sand Dunes," *Valley Courier*, December 17, 1999, 1, 7.
39. Andrew Valdez, interview by author, June 3, 2011.
40. Steve Chaney, e-mail correspondence with author, January 13, 2012; "Draft Itinerary, Secretary Babbitt's Visit to Great Sand Dunes National Monument," December 18, 1999. Copy of unpublished internal document provided by former Great Sand Dunes National Park superintendent Steve Chaney; Andrew Valdez, e-mail correspondence with author, December 12, 2011; Valdez interview, June 3, 2011.
41. Bunch interview, June 2, 2011; Chaney interview, September 22, 2011; McInnis interview, July 7, 2011; Valdez interview, June 3, 2011.
42. Nelda Curtiss, "Dunes Deal Said Do-Able," *Valley Courier*, December 21, 1999, 1, 5; Erin Smith, "Sand Dunes Park Plan Gets Boost from Babbitt," *Pueblo Chieftain*, December 19, 1999, 1A, 2A; Electa Draper, "Interior Chief, Lawmakers Back New National Park," *Denver Post*, December 19, 1999, 1B, 5B.
43. David Nichols, "Nature Conservancy, Federal and State Gov't. Collaborating to Purchase the Baca Grant; National Park in the Forming," *Crestone Eagle*, January 2000, 9.
44. Curtiss, "Dunes Deal"; Erin Smith, "Sand Dunes Park Plan"; Draper, "Interior Chief."
45. Nichols, "Nature Conservancy," 8.
46. Curtiss, "Dunes Deal," 5.
47. Chaney interview, September 22, 2011.

CHAPTER 8

Epigraph 1. National Park Service Organic Act, August 25, 1916.
Epigraph 2. "Hefley Opposes Efforts to Reclassify Great Sand Dunes National Monument as a National Park," Office of Congressman Joel Hefley, press release, December 20, 1999.

1. "Sand Dunes National Park?," *Rocky Mountain News*, December 26, 1999, 2B; "Getting Burned by Steamtown," *Reading (Pa.) Eagle*, July 2, 1995, A16, http://news.google.com/newpapers?id=GVExAAAAIBAJ&sjid=AKIFAAAAIBAJ&pg=1554,2402937dq=steamtown+bankrupt&hl=en (accessed October 2, 2011).
2. Hefley press release, December 20, 1999.
3. Bill McAllister, "Hefley Opposes Park Status for Sand Dunes," *Denver Post*, December 21, 1999, 5B, 7B.
4. Barney Mcmanigal, "State Delegation Puts Sand Dunes at Top of Agenda," *Pueblo Chieftain*, January 16, 2000; Mike Soraghan, "Dunes Move toward U.S. Park Status," *Denver Post*, October 7, 2000, 1B, 6B.
5. Mcmanigal, "State Delegation."
6. Cong. Rec. 146, no. 134 (October 24, 2000): H10714.
7. David Robbins, telephone interview by author, December 7, 2011. Robbins is an attorney with Hill and Robbins PC in Denver and has served as general counsel for the Rio Grande Water Conservation District since 1981. Charlie Meyers, "Bill No Walk in the Park for Hunters," *Denver Post*, April 4, 2000, www.denverpost.com/rec/hunt0404.htm (accessed December 7, 2011); Erin Smith, "Sand Dunes Park Called Key to Water Future," *Pueblo Chieftain*, February 17, 2000, www.chieftain.com/metro/sand-dunes-park-called-key-to-water-future/article_ad6ef659-f052-5c11-8bfd-f522d71b5556.html (accessed December 7, 2011).
8. Mary Jean Porter, "Key Issues Laid Out in Dunes Plan," *Pueblo Chieftain*, February 23, 2000.
9. Erin Smith, "Method Proposed to Smooth Loss of Dunes Taxes," *Pueblo Chieftain*, January 12, 2000, 1B, 2B; Gary Taylor, "Park Idea Could Help Saguache," *Valley Courier*, January 14, 2000, 1, 3; Dianne James, "McInnis Proposes Recouping Baca Taxes," *Valley Courier*, January 19, 2000; Ruth Heide, "Refuge Kicks In for Dunes Park," *Valley Courier*, June 8, 2000, 1, 3; Steve Chaney, telephone interview by author, September 22, 2011; Fred Bunch, interview by author, June 2, 2011; Robbins interview, December 7, 2011; "Refuge Revenue Sharing," U.S. Fish and Wildlife Service website, www.fws.gov/refuges/ realty/rrs.html (accessed December 7, 2011).
10. Erin Smith, "Dunes Park Gains Backing," *Pueblo Chieftain*, February 3, 2000, 1A, 2A.
11. "A New National Park?," *Colorado Central Magazine*, February 2000, cover; "How Many Gates to the Great Sand Dunes?," *Colorado Central Magazine*, February 2000, 43.
12. David Nicholas, "Dunes Park Bill HR 4095 Submitted to Congress," *Crestone Eagle*, April 2000, 1, 5.
13. Cong. Rec. 146, no. 36 (March 28, 2000): H1476.
14. *Great Sand Dunes National Park Act of 2000*, H.R. 4095, 106th Cong., 2nd sess., section 3, page 6.
15. Ibid., sec. 4, p. 10; Robbins interview, December 7, 2011.
16. Ibid., sec. 6, p. 17.
17. Ibid.
18. Scott McInnis, interview by author, July 7, 2011.

19. Mary Boyle, "Tension in Congress," *Colorado Springs Gazette*, April 9, 2000.
20. Peter Roper, "Hefley Defends Dunes Stance," *Pueblo Chieftain*, July 8, 2000, 1A, 2A.
21. Eric Gorski, Mary Boyle and Todd Hartman, "Political Quicksand," *Colorado Springs Gazette*, April 9, 2000, A10.
22. Tom Mcavoy, "After Feud, Lawmakers Back Park," *Pueblo Chieftain*, April 26, 2000.
23. Lynn Bartels, "McInnis, Hefley Argue over Dunes Proposal," *Rocky Mountain News*, April 26, 2000; Mcavoy, "After Feud."
24. Mary Boyle, "New Push for Sand Dunes/Allard Introduces Senate Bill Seeking National Park Status," *Colorado Springs Gazette*, May 12, 2000, www.highbeam.com/doc/1P2-5949609.html (accessed June 14, 2011).
25. Great Sand Dunes National Park and Preserve Act of 2000, S. 2547, 106th Cong., 2nd sess., sec. 9, p. 7.
26. Craig Linder, "Attorney Wants Changes in Dunes Bill," *Pueblo Chieftain*, July 28, 2000, 1A, 2A; Mary Boyle, "Water a New Snag in Dunes Park Plan," *Colorado Springs Gazette*, August 3, 2000; S. 2547, sec. 9, p. 7.
27. Mike Soraghan, "Clinton Drops Support for Dunes Park," *Denver Post*, July 27, 2000; Gary Taylor, "Dunes Bill Demand a Shock," *Valley Courier*, July 28, 2000, 1, 3; Michael Romano, "Allard Blames Babbitt for Delay on Sand Dunes," *Rocky Mountain News*, July 28, 2000, www.rockymountainnews.com/news/0728dune6.shtml (accessed June 14, 2011).
28. Mike Soraghan, "Water Words Snag Dunes Park," *Denver Post*, July 28, 2000, www.denverpost.com/news/news0728.htm (accessed June 14, 2011); Mike Soraghan, "Dunes Park Back on Track," *Denver Post*, September 20, 2000.
29. Taylor, "Dunes Bill Demand."
30. Linder, "Attorney Wants Changes"; Robbins interview, December 7, 2011.
31. Kevin Flynn, "Dunes Closer to Becoming Park," *Rocky Mountain News*, September 20, 2000, 7A, 21A; Craig Linder, "Allard, Interior Closer on Dunes Park," *Pueblo Chieftain*, September 20, 2000, 1A, 2A; Cong. Rec. 146, no. 134, H10714.
32. Cong. Rec. 146, no. 134, H10714.
33. Flynn, "Dunes Closer."
34. Linder, "Allard, Interior."
35. Ruth Heide, "Dunes Park Is Passed," *Valley Courier*, October 6, 2000, 1, 3; Craig Linder, "Sand Dunes Park Proposal Clears Senate," *Pueblo Chieftain*, October 6, 2000, www.chieftain.com/friday/news/index.htm?article=3 (accessed June 14, 2011).
36. Scott McInnis, "Paying Tribute to the History of the Great Sand Dunes National Park and Preserve," Cong. Rec. 150, no. 55 (April 27, 2004): E666; Christopher Hatcher, telephone interview by author, November 22, 2011. Hatcher is the former legislative director for Rep. Scott McInnis; Soraghan, "Dunes Move Toward"; McInnis interview, July 7, 2011.
37. "Boyce Says Baca Ranch Not for Sale," *Valley Courier*, October 11, 2000, 1, 3.
38. Mike Soraghan, "Sand Dunes' Outcome Shifting," *Denver Post*, October 16, 2000, 1B, 5B.
39. Peter Roper, "House of Representatives Leaves Dunes Bill in Limbo," *Pueblo Chieftain*, October 12, 2000; McInnis interview, July 7, 2011.

40. Craig Linder, "National Park Issue Debated in House," *Pueblo Chieftain*, October 25, 2000, 1A, 2A; Gary Taylor, "Dunes Vote Postponed," *Valley Courier*, October 26, 2000, 1, 3.
41. Hatcher interview, November 22, 2011.
42. Cong. Rec. 146, no. 134, H10714.
43. Ibid.
44. Ibid.
45. Ibid., H10715.
46. Ibid., H10716.
47. Ibid.
48. Mike Soraghan, "Dunes OK Expected Today," *Denver Post*, October 25, 2000; Taylor, "Dunes Vote."
49. Gary Taylor, "Valley Wins Its National Park," *Valley Courier*, October 26, 2000, 1, 3.
50. Craig Linder, "House Strongly Approves Sand Dunes Park," *Pueblo Chieftain*, October 26, 2000, 1A, 2A.
51. Ibid.
52. Ibid.
53. Craig Linder, "Dunes Bill Awaiting Clinton's Signature," *Pueblo Chieftain*, November 22, 2000, 1A, 2A; Mike Soraghan, "Clinton to Sign Bill Elevating Great Sand Dunes," *Denver Post*, November 22, 2000, 6B.
54. Linder, "Dunes Bill."
55. Ibid.
56. Mike Soraghan, "Clinton Signs Sand Dunes Bill," *Denver Post*, November 23, 2000; National Park Service, U.S. Department of the Interior, "Secretary Norton Creates Great Sand Dunes National Park and Preserve and Baca National Wildlife Refuge," press release, September 13, 2004; Linder, "Dunes Bill."
57. McInnis, "Paying Tribute," E666.
58. Wayne Allard, "Dunes Deserve Park Status," *Denver Post*, December 18, 1999, 7B; Gary Taylor, "Dunes Bill Signed," *Valley Courier*, November 27, 2000, 1, 3; Ruth Heide, "Dunes-Baca Buy On," *Valley Courier*, December 14, 2000, 1, 3.
59. Deborah Frazier, "Lexam to Drill Near Dunes," *Rocky Mountain News*, November 24, 2000, 7A, 18A; Teresa L. Benns, "Chaney Denies Threat of Wells," *Valley Courier*, October 28, 2000, 1, 3; Mike Soraghan, "Company Files Plans to Drill Near Dunes," *Denver Post*, October 24, 2000, 1B, 3B; Christine Canaly, telephone interview by author, December 8, 2011. Canaly was the director of the San Luis Valley Ecosystem Council at the time of interview.
60. David Nichols, "Boyce Files Suit against Farallon in an Attempt to Block Sale of Baca Ranch," *Crestone Eagle*, February 2001, 1, 20; Mike Soraghan, "Lawsuit Complicates Dunes Transition to Park Status," *Denver Post*, January 23, 2001; "A History of Great Sand Dunes National Park and Preserve," U.S. Department of the Interior website, http://doi.net/news/04_News_Releases/greatsanddunes.pdf (accessed August 24, 2011).
61. Sylvia Lobato, "Bush Budget Funds Baca Buy," *Valley Courier*, April 11, 2001, 1, 3; Megan Scully, "Push for Baca Bucks in Full Gear," *Pueblo Chieftain*, March 3, 2001, www.chieftain.com/metro/push-for-baca-bucks-in-full-gear/article_b022d 4ae-50a3–5das-b6b4–38d2edf9a3e7.html (accessed August 24, 2011).

62. Ruth Heide, "Farallon Roots Examined, Baca's Bedfellows Seen," *Valley Courier*, January 16, 2002, http://hillandrobbins.com/pdf/AVC_Bacas_01–16–02.pdf (accessed August 24, 2011); Erin Smith, "Yale University Silent Partner in Baca Dealings," *Pueblo Chieftain*, January 20, 2002, www.chieftain.com/metro/yale-university-silent-partner-in-baca-dealings/article_fe422319–3c0f-5d5a-a781–1cec21908794.html (accessed August 24, 2011); Erin Smith, "Yale to Donate Profits to Nature Conservancy on the Sale of Baca Ranch," *Pueblo Chieftain*, January 26, 2002, www.chieftain.com/metro/yale-to-donate-profits-to-nature-conservancy-on-sale-of/ article_6cfdd4a7–5050–59df-bbac-7fe5d0ce8a9f.html (accessed August 24, 2011); William Baue, "Private Equity Investment Puts Yale in Deep Water," *Social Funds*, February 8, 2002, www.socialfunds.com/news/print.cgi?sfArticleId=774 (accessed August 24, 2011); Canaly interview, December 8, 2011.
63. Erin Smith, "Allard Thrilled with Results of Yale-Baca Talks," *Pueblo Chieftain*, January 27, 2002, 1A, 2A.
64. Ibid.
65. Erin Smith, "The Nature Conservancy Agrees to Buy Baca Ranch," *Pueblo Chieftain*, January 31, 2002, www.chieftain.com/metro/the-nature-conservancy-agrees-to-buy-baca-ranch/article_764cdae1–672d-52ac-8a66-e49a8231654e.html (accessed November 5, 2011).
66. Tim Flowers and Ruth Heide, "Locals Elated," *Valley Courier*, January 31, 2002, 1, 3.
67. Tim Flowers, "Baca Deal Inked," *Valley Courier*, January 31, 2002, 1, 3.
68. Tim Flowers, "Official Reaction Swift, Positive," *Valley Courier*, January 31, 2002, 1, 3.
69. David Nicholas, "Water and Sand: On the Congressional Fast Track toward a National Park," *Crestone Eagle*, February 2000; Ruth Heide, "Court Cases Cloud Dunes Deal," *Valley Courier*, July 28, 2004, 1, 3; Ed Quillen, "Baca Foreclosure May Clear Way for National Park," *Colorado Central Magazine* June 2002, http://cozine.com/2002-june/baca-foreclosure-may-clear-way-for-national-park/ (accessed November 5, 2011).
70. Erin Smith, "Judge Orders Baca Ranch Foreclosure," *Pueblo Chieftain*, May 10, 2002, http://chieftain.com/metro/judge-orders-baca-ranch-foreclosure/article_c6d622cl-8b73–50c6–94d6-e21ae1567f52.html (accessed November 5, 2011).
71. "McInnis Secures Funding to Purchase Baca Ranch," *Pueblo Chieftain*, February 14, 2003, www.chieftain.com/metro/mcinnis-secures-funding-to-purchase-baca-ranch/article_46757421-e688–57e1-b691–8bacaf8cf95ee.html (accessed November 5, 2011).
72. Erin Smith, "10 Named to Grand [sic] Sand Dunes National Park Advisory Council," *Pueblo Chieftain*, May 28, 2003, www.chieftain.com/metro/named-to-grand-sand-dunes-national-park-advisory-council/article_4e925a94-ac21–5b35–9f5c-15abafdc43a7.html (accessed November 5, 2011).
73. "Committee Backs Dunes Purchase," *Valley Courier*, October 29, 2003, 1A.
74. Erin Smith, "Yale Donates $1.5 Million to Sand Dunes," *Pueblo Chieftain*, January 24, 2004, www.chieftain.com/metro/named-to-grand-sand-dunes-national-park-advisory-council/article_4e925a94-ac21–5b35–9f5c-15abafdc43a7.html (accessed November 5, 2011).

75. Erin Smith, "Yale Subsidy Urged for Dunes Park," *Pueblo Chieftain*, February 12, 2004, 5A; Kim Martineau, "Protesters Will Hit Yale Investments," *Hartford Courant*, March 3, 2004, http://articles.courant.com/2004-03-03/news/0403030797_1_hedge-students-from-stanford-university-yale-s-endowment (accessed November 5, 2011).
76. Julie Poppen, "Great Sand Dunes Closer to Becoming National Park," *Rocky Mountain News*, March 13, 2004, 28A.
77. Glenn Maffei, "Allard Tries to Fit Final Piece into Sand Dunes Park Puzzle," *Pueblo Chieftain*, August 6, 2004; Erin Smith, "Park Includes Nation's Largest Wildlife Refuge," *Pueblo Chieftain*, September 14, 2004, www.chieftain.com/metro/park-includes-nation-s-largest-wildlife-refuge/article_65eb4ae6-4a0a-59a7-af70-de8bc30e89a4.html (accessed November 5, 2011); David Nicholas, "Great Sand Dunes Becomes a National Park: History Made on September 13, 2004," *Crestone Eagle*, October 2004, http://crestoneeagle.com/archives2004/oct04_a1.html (accessed November 6, 2011).
78. "Secretary Norton Creates," NPS press release, September 13, 2004.
79. "Dunes Become U.S. 58th National Park," *Center (Colo.) Post-Dispatch*, September 16, 2004, 1A, 18A; Jillian Lloyd, "Tallest US Dunes Become Newest Park," *Christian Science Monitor*, September 14, 2004, www.csmonitor.com/2004/0914/p02s02-usgn.html (accessed November 6, 2011); Smith, "Park Includes"; "Secretary Norton Creates," NPS press release, September 13, 2004.
80. "Secretary Norton Creates," NPS press release, September 13, 2004.
81. "Allard: Dunes Funding Completed in New Bill," *Pueblo Chieftain*, September 16, 2004, www.chieftain.com/metro/allard-dunes-funding-completed-in-new-bill/article_ab686b5e-716b-56bb-be06-96328c827004.html (accessed November 6, 2011); Office of the Secretary, U.S. Department of the Interior, "Notice of Establishment," Fed. Reg. 69, no. 185 (September 24, 2004): 57355–56.
82. Fred Bunch, interview by author, June 2, 2011; Fred Bunch, e-mail correspondence with author, February 10, 2012.
83. Erin Smith, "Great Sand Dunes Visitor Center Opens," *Pueblo Chieftain*, October 3, 2004, www.chieftain.com/metro/great-sand-dunes-visitor-center-opens/article_955701fb-d415-5074-a4d9-2375aa00fb90.html (accessed November 7, 2011); National Park Service, U.S. Department of the Interior, Public Use Statistics Office website, *Great Sand Dunes National Park and Preserve Annual Park Visitation (All Years)*, www.nature.nps.gov/stats/park.cfm?parkid=208 (accessed November 7, 2011).
84. Act of August 25, 1916 (National Park Service Organic Act), PL 64–235, 16 USC §1 et seq. as amended.
85. William Cronon, *Changes in the Land: Indians, Colonists, and the Ecology of New England* (New York: Hill and Wang, 1983), 159.
86. "Federal 'Non-Reserved' Water Rights: The Great Sand Dunes National Park Application for Absolute Groundwater Rights," *University of Denver Water Law Review* 427 (Spring 2007): 427–32; Erin Smith, "Park Service Files for Dunes' Water," *Pueblo Chieftain*, December 31, 2004, www.chieftain.com/metro/park-service-files-for-dunes-water/article_b5cc4dd3-2911-5a47-a69e-00f036258383.html (accessed November 7, 2011); Matt Hildner, "Judge Signs Off on Dunes Water Rights," *Pueblo Chieftain*, August 5, 2008, www.chieftain.

com/news/local/judge-signs-off-on-dunes-water-rights/article_b35c6515–2619–535b-b8d9–9cbfc02eb308.html (accessed November 7, 2011); "Sand Dunes Gets New Superintendent," *Pueblo Chieftain*, October 7, 2006, www.chieftain.com/metro/sand-dunes-gets-new-superintendent/article_1b90ec79–71e4–5ab3–9faa-8d5b6df6a06d.html (accessed November 7, 2011).

87. National Park Service, U.S. Department of the Interior, "Record of Decision," *General Management Plan/Wilderness Study/Environmental Impact Statement*, July 2007.

88. Ruth Heide, "Kuenhold Signs Dunes Water Decree," *Valley Courier*, August 5, 2008, www.alamosanews.com/V2_news_articles.php?heading=0&page=72&story_id=8611 (accessed November 7, 2011).

89. Hildner, "Judge Signs"; Heide, "Kuenhold Signs."

90. Emma Lynch and Frank Turina, "Monitoring Reveals Park Enjoys Exceptional Quiet," *InsideNPS*, October 29, 2008, copy in Great Sand Dunes National Park archives; Bunch interview, June 2, 2011.

91. "Fred Bunch Receives 2008 Regional Director's Award," *Valley Courier*, January 19, 2009, htpp://www.alamosanews.com/V2_news_articles.php?heading=0&story_id=11220&page=75 (accessed November 7, 2011).

92. Ibid.

93. Matt Hildner, "Settlement Ends Lawsuit over Drilling at Baca Refuge," *Pueblo Chieftain*, September 25, 2010, www.chieftain.com/news/local/settlement-ends-lawsuit-over-drilling-at-baca-refuge/article_f3c293e8-c878–11df-9228–001cc4c002e0.html (accessed November 8, 2011); Christine Canaly, e-mail correspondence with author, February 3, 2012.

94. "Lisa Carrico Named Superintendent of Great Sand Dunes," *NPS Digest*, January 4, 2012, www.nps.gov/applications/digest/headline.cfm?type=PeopleNews&id=3400 (accessed February 3, 2012).

EPILOGUE

Epigraph. Mark Fiege, *Irrigated Eden: The Making of an Agricultural Landscape in the American West* (Seattle: University of Washington Press, 1999), 202.

1. Michael G. Rupert and L. Niel Plummer for the U.S. Geological Survey, U.S. Department of the Interior, "Ground-Water Age and Flow at Great Sand Dunes National Monument, South-Central Colorado," Fact Sheet 2004–3051 (July 2004); Fred Bunch, e-mail correspondence with author, October 21, 2011, January 5, 2012; Andrew Valdez, interview by author, June 3, 2011.

2. Bunch interview, June 2, 2011.

BIBLIOGRAPHY

GOVERNMENT DOCUMENTS

Anschuetz, Kurt F., and Thomas Merlan for the Rocky Mountain Research Station, U.S. Forest Service. *More Than a Scenic Mountain Landscape: Valles Caldera National Preserve Land Use History*. General Technical Report RMRS-GTR-196, September 2007.

"Background, History and Status, Great Sand Dunes National Park and Preserve Act of 2000." March 1, 2001. Copy of unpublished internal document provided by former Great Sand Dunes National Park Superintendent Steve Chaney.

Beckwith, Lt. E. G. *Reports of Explorations and Surveys to Ascertain the Most Practicable and Economical Route for a Railroad from the Mississippi River to the Pacific Ocean, 1853–54*. Vol. 2, Executive Doc. No. 78, 33rd Cong., 2nd sess. Washington, D.C.: Beverly Tucker, 1855.

Bureau of Reclamation, U.S. Department of the Interior. *San Luis Valley Project, Closed Basin Division: Facts and Concepts*. Washington, D.C.: Government Printing Office, 1984.

"Draft Itinerary, Secretary Babbitt's Visit to Great Sand Dunes National Monument." December 18, 1999. Copy of unpublished internal document provided by former Great Sand Dunes National Park superintendent Steve Chaney.

Franke, Paul R., for the National Park Service, U.S. Department of the Interior. *General Report on the Great Sand Dunes National Monument*. Copy of an unpublished report in Great Sand Dunes National Park archives, May 17, 1934.

"Great Sand Dunes National Park Timeline." N.d. Copy of unpublished internal document provided by former Great Sand Dunes National Park superintendent Steve Chaney.

Hayden, Ferdinand Vandeveer, for the U.S. Geological Survey, U.S. Department of the Interior. *Geological and Geographical Atlas of Colorado and Portions of Adjacent Territory*. Washington, D.C.: Government Printing Office, 1881.

———. "Geological Report for 1869, Embracing Colorado and New Mexico." In *First, Second, and Third Annual Reports of the U.S. Geological Survey of the Territories for the Years 1867, 1868, and 1869* (Washington, D.C.: Government Printing Office, 1873). http://esp.cr.usgs.gov/info/dunes/noreolian.html (accessed January 13, 2011).

"Hefley Opposes Efforts to Reclassify Great Sand Dunes National Monument as a National Park." Office of Congressman Joel Hefley, press release, December 20, 1999.

Johnson, Ross B. "The Great Sand Dunes of Southern Colorado." *Geological Survey Research 1967.* U.S. Geological Survey Professional Paper 575-C. Washington, D.C.: Government Printing Office, 1967.

Machette, Michael N., David W. Marchetti, and Ren A. Thompson. "Ancient Lake Alamosa and the Pliocene to Middle Pleistocene Evolution of the Rio Grande." In Michael N. Machette, Mary-Margaret Coates, and Margo L. Johnson, *2007 Rocky Mountain Section Friends of the Pleistocene Field Trip—Quaternary Geology of the San Luis Basin of Colorado and New Mexico, September 7–9, 2007:* U.S. Geological Survey Open-File Report 2007-1193 (2007). http://pubs.usgs.gov/of/2007/1193/pdf /OF07-1193_ChG.pdf (accessed June 6, 2011).

McInnis, Scott. "Paying Tribute to the History of the Great Sand Dunes National Park and Preserve." *Congressional Record* 150, no. 55 (April 27, 2004): E664–E666.

National Park Service, U.S. Department of the Interior. *Draft Master Plan: Great Sand Dunes National Monument.* Washington, D.C.: Government Printing Office, September 1975.

———. *Great Sand Dunes National Monument: Statement for Management.* Washington, D.C.: Government Printing Office, 1988.

———. "Jefferson National Expansion Memorial." *Museum Gazette,* January 1996.

———. "Mission 66 for Great Sand Dunes National Monument," 1956 (date of publication inferred from text). Denver Public Library Government Document I 29.6: M68/G75z.

———. *National Register of Historic Places (NRHP) Registration Form—Medano Ranch Headquarters.* Denver: Front Range Research Associates, 2003.

———. *National Register of Historic Places (NRHP) Nomination Form—Pike's Stockade, 1961, 1978.*

———. *National Register of Historic Places (NRHP) Nomination Form—The Trujillo Homestead.* Denver: Front Range Research Associates, 2003.

———. Public Use Statistics Office website. *Great Sand Dunes National Park and Preserve Annual Park Visitation (All Years).* http://www.nature.nps.gov/stats/park.cfm?parkid=208 (accessed August 16, 2011).

———. "Record of Decision." *General Management Plan/Wilderness Study/Environmental Impact Statement,* July 2007.

———. *Resource Management Strategy, Great Sand Dunes National Monument,* March 1994.

———. "Secretary Norton Creates Great Sand Dunes National Park and Preserve and Baca National Wildlife Refuge." Press release, September 13, 2004.

Office of the Secretary, U.S. Department of the Interior. "Notice of Establishment." Federal Register 69, no. 185 (Washington, D.C.: Government Printing Office, September 24, 2004): 57355–56.

Rowlands, Peter. "Stand Structure of a Piñon-Juniper Woodland in Great Sand Dunes National Monument Colorado and the Effects of Historical Wood Harvesting." Draft copy of report to superintendent of Great Sand Dunes National Monument, 1996. Copy at Great Sand Dunes National Park archives.

Rupert, Michael G., and L. Niel Plummer for the U.S. Geological Survey, U.S. Department of the Interior. *Ground-Water Age and Flow at Great Sand Dunes National Monument, South-Central Colorado*, Fact Sheet 2004-3051, July 2004.

Schaafsma, Harold, superintendent of Great Sand National Monument, to Glen Bean, former superintendent, January 2, 1954. Copy of letter in Great Sand Dunes National Park archives.

Schultz, Robert L., for the National Park Service, U.S. Department of the Interior. "Monte Vista P.E.O. Spearheaded Great Sand Dunes Monument Establishment." Press release, March 6, 1982.

Siebenthal, Claude E., for the U.S. Geological Survey, U.S. Department of the Interior. *Geology and Water Resources of the San Luis Valley*. Water-Supply Paper 240. Washington, D.C.: Government Printing Office, 1910.

Spero, Vince, et al., for the U.S. Forest Service. "An Assessment of the Range of Natural Variability of the Rio Grande National Forest." copy of internal draft in Great Sand Dunes National Park archives, March 1995.

Toll, Roger W., for the National Park Service, U.S. Department of the Interior. *Report on Great Sand Dunes National Monument*, copy of an unpublished report in Great Sand Dunes National Park archives, 1931.

U.S. Congress. Cong. Rec. 146, no. 36 (March 28, 2000).

U.S. Congress. Cong. Rec. 146, no. 134 (October 24, 2000).

U.S. Congress. House. Great Sand Dunes National Park Act of 2000, H. R. 4095, 106th Cong., 2nd sess.

U.S. Congress. Senate. Great Sand Dunes National Park and Preserve Act of 2000, S. 2547, 106th Cong., 2nd sess.

U.S. Geological Survey, U.S. Department of the Interior. "Surficial Geology and Geomorphology of the Great Sand Dunes Area, South-Central Colorado." June 3, 2008. http://esp.cr.usgs.gov/info/dunes/origin.html (accessed May 14, 2011).

White, David R. M., for the National Park Service, U.S. Department of the Interior. *Seinanyédi: An Ethnographic Overview of Great Sand Dunes Park and Preserve*. Santa Fe: Applied Cultural Dynamics, 2005.

BOOKS, ARTICLES, AND THESES

Abel, Annie H. "Trudeau's Description of the Upper Missouri." *Mississippi Valley Historical Review* 8 (1921).

Alford, Terry L. "The West as a Desert in American Thought prior to Long's 1819–1820 Expedition." *Journal of the West* 8, no. 4 (1969): 515–25.

Bancroft, Caroline. *Colorado's Lost Gold Mines and Buried Treasure*. Boulder: Johnson Books, 1962.

Bean, Luther E. *Land of the Blue Sky People*. Monte Vista: Monte Vista Journal, 1962.

Bingham, Sam. *The Last Ranch: A Colorado Community and the Coming Desert*. New York: Pantheon Books, 1996.

Blackhawk, Ned. *Violence over the Land: Indians and Empires in the Early American West*. Cambridge, Mass.: Harvard University Press, 2006.

Blackmore, William, ed. *Colorado: Its Resources, Parks, and Prospects as a New Field for Emigration; with an Account of the Trenchara and Costilla Estates, in the San Luis Park*. London: Sampson, Low, Son, and Marston, 1869.

Bolton, Herbert E., ed. *Spanish Exploration in the Southwest, 1542–1706*. New York: Charles Scribner's Sons, 1908.

Bonsal, Stephen. *Edward Fitzgerald Beale: A Pioneer in the Path of Empire, 1822–1903*. New York: G. P. Putnam's Sons, 1912.

Bradley, Bruce, and Dennis Stanford. "The North Atlantic Ice-Age Corridor: A Possible Palaeolithic Route to the New World." *World Archaeology* 36, no. 4 (2004): 459–78.

Brandes, T. Donald. *Military Posts of Colorado*. Fort Collins, Colo.: Old Army Press, 1973.

Brooks, James F. *Captives and Cousins: Slavery, Kinship, and Community in the Southwest Borderlands*. Chapel Hill: University of North Carolina Press, 2002.

Bruce, Treasure. "Perspectives in Sand: Township 40N Range 12E of the NM Meridian in the San Luis Valley of Colorado." *University of Denver College of Law* (November 2003). Copy on file at Great Sand Dunes National Park and Preserve, Resource Management Office.

Butler, Mike. *Images of America: Great Sand Dunes National Park*. Charleston: Arcadia, 2013.

Campbell, Joseph. *The Power of Myth*. New York: Doubleday, 1988.

Carter, C. Joe. "Cultural History of the Great Sand Dunes." *San Luis Valley Historian* 22, no. 3 (1990): 6–16.

———. "Great Sand Dunes National Monument: An Administrative History." *San Luis Valley Historian* 19, no. 2 (Spring 1982): 2–12.

———. *Pike in Colorado*. Fort Collins, Colo.: Old Army Press, 1978.

Catton, Theodore. *National Park, City Playground: Mount Rainier in the Twentieth Century*. Seattle: University of Washington Press, 2006.

Chaffin, Tom. *Pathfinder: John Charles Frémont and the Course of American Empire*. New York: Hill and Wang, 2002.

Chambers, Frank. *Hayden and His Men: Being a Selection of 108 Photographs by William Henry Jackson of the United States Geological and Geographical Survey of the Territories for the Years 1870–1878, Ferdinand V. Hayden, Geologist in Charge*. Dillsburg, Pa.: Francis Paul Geoscience Literature, 1988.

Colorado Groundwater Association. *Water in the Valley*. Lakewood: Colorado Groundwater Association, 1989.

Cooper, Jack Kyle. "A Rebuttal Concerning Roads, Trails, Traces and Wagons." *San Luis Valley Historian* 33, no. 3 (2001): 4–41.

———. *Zebulon Montgomery Pike's Great Western Adventure 1806–1807*. Colorado Springs: Clausen Books, 2007.

Coues, Elliot. *The Expeditions of Zebulon Montgomery Pike*. New York: Francis P. Harper, 1895.

———, ed. *The Journal of Jacob Fowler*. New York: Francis P. Harper, 1898.

Cox, John D. "Why the Younger Dryas Matters." *Discovery News*, April 2, 2010. http://news.discovery.com/earth/why-the-younger-dryas-matters.html (accessed February 3, 2011).

Crawford, Stanley. *Mayordomo: Chronicle of an Acequia in Northern New Mexico*. Albuquerque: University of New Mexico Press, 1988.

Cronon, William. *Changes in the Land: Indians, Colonists, and the Ecology of New England*. New York: Hill and Wang, 1983.

———. "The Uses of Environmental History." *Environmental History Review* 17, no. 3 (Fall 1993): 1–22.

———. "A Place for Stories: Nature, History, and Narrative." *Journal of American History* 78 (March 1992): 1347–76.

Cronon, William, and Richard White. "Indians in the Land." *American Heritage* 37 (August–September 1986): 19–25.

Crosby, Alfred W. *Ecological Imperialism: The Biological Expansion of Europe, 900–1900.* Cambridge: Cambridge University Press, 1986.

Crowther, Edward R. "William Gilpin: Booster and Developer of the San Luis Valley." *San Luis Valley Historian* 26, no. 2 (1994): 35–56.

David, Jane. "The Great Sand Dunes of Colorado: A National Monument Thanks to P.E.O." *P.E.O. Record*, July/August 1991, 6–7.

deBuys, William. *Enchantment and Exploitation: The Life and Hard Times of a New Mexico Mountain Range.* Albuquerque: University of New Mexico Press, 1985.

de Onis, Jose, ed. *The Hispanic Contribution to the State of Colorado.* Boulder: Westview Press, 1976.

De Villagrá, Gaspar Pérez. *History of New Mexico.* Translated by Gilberto Espinosa. Los Angeles: Quivira Society, 1933.

"Early Latino Settlement in Colorado Designated National Historic Landmark." *National Parks Traveler*, January 4, 2012. http://www.nationalparkstraveler.com/2012/01/early-latino-settlement-colorado-designated-national-historic-landmark9248 (accessed January 7, 2012).

Ebright, Malcolm. *Land Grants and Lawsuits in Northern New Mexico.* Albuquerque: University of New Mexico Press, 1994.

Eiseley, Loren. *The Immense Journey.* New York: Franklin Watts, 1957.

Emery, Sloan, and Dennis Stanford. "Preliminary Report on Archaeological Investigations at the Cattle Guard Site, Alamosa County, Colorado." *Southwestern Lore* 48, no. 1 (March 1982): 10–20.

Espinosa, J. Manuel. "Journal of the Vargas Expedition into Colorado, 1694." *Colorado Magazine* 16, no. 3 (May 1939): 81–90.

Fagan, Brian M. *Ancient North America: The Archaeology of a Continent.* London: Thames and Hudson Ltd., 1995.

"Federal 'Non-Reserved' Water Rights: The Great Sand Dunes National Park Application for Absolute Groundwater Rights." *University of Denver Water Law Review* 427 (Spring 2007): 427–32.

Fiege, Mark. *Irrigated Eden: The Making of an Agricultural Landscape in the American West.* Seattle: University of Washington Press, 1999.

———. *The Republic of Nature: An Environmental History of the United States.* Seattle: University of Washington Press, 2012.

Firestone, R. B., et al. "Evidence of an Extraterrestrial Impact 12,900 Years Ago That Contributed to the Megafaunal Extinctions and the Younger Dryas Cooling." *Proceedings of the National Academy of Sciences* 104, no. 41 (October 9, 2007). http://www.pnas.org/cgi/doi/10.1073/pnas 0706977104 (accessed February 3, 2011).

Flores, Dan. *Caprock Canyonlands: Journeys into the Heart of the Southern Plains.* Austin: University of Texas Press, 1990.

Foster, Mike. *Strange Genius: The Life of Ferdinand Vandeveer Hayden.* Niwot, Colo.: Roberts Rinehart, 1994.

Frank, Jerry J. *Making Rocky Mountain National Park: The Environmental History of an American Treasure*. Lawrence: University Press of Kansas, 2013.

Gilpin, William. *The Central Gold Region: The Grain, Pastoral, and Gold Regions of North America*. Philadelphia: J. B. Lippincott & Co., 1860.

———. *Guide to the Kansas Gold Mines at Pike's Peak, from Notes of Captain J. W. Gunnison 1859*. Denver: Nolie Mumey, 1952.

Goetzmann, William H. *Exploration and Empire: The Explorer and the Scientist in the Winning of the American West*. New York: W. W. Norton & Company, 1966.

Grant, Blanche C. *When Old Trails Were New: The Story of Taos*. New York: Press of the Pioneers, 1934.

"The Great Sand Dunes: Eight Billion Dollars in Gold." *Folks and Fortunes: Saga of the San Luis Valley in Colorado* 1, no. 1 (Monte Vista: Monte Vista Journal, November 1949): 25.

Hafen, LeRoy R., ed. *The Mountain Men and the Fur Trade of the Far West*. Vol. 2. Glendale, Calif.: Arthur H. Clark Company, 1965.

Hafen, LeRoy R., and Ann W. Hafen. "The Diaries of William Henry Jackson, Frontier Photographer." Vol. 10 of *The Far West and the Rockies Historical Series, 1820–1875*. Glendale, Calif.: Arthur H. Clark Company, 1959.

———, eds. "Fremont's Fourth Expedition: A Documentary Account of the Disaster of 1848–1849." Vol. 11 of *The Far West and the Rockies Historical Series, 1820–1875*. Glendale, Calif.: Arthur H. Clark Company, 1960.

Hafen, LeRoy R., and Harlin M. Fuller, eds. "The Journal of Captain John R. Bell." Vol. 5 of *The Far West and the Rockies Historical Series, 1820–1875*. Glendale, Calif.: Arthur H. Clark Company, 1957.

Hall, Elizabeth. *Sand Stories: Recollections of Life around the Great Sand Dunes*. Mosca, Colo.: Friends of the Dunes, 2004.

Hämäläinen, Pekka. *The Comanche Empire*. New Haven, Conn.: Yale University Press, 2008.

Hammond, George P., and Agapito Rey. *Don Juan de Oñate: Colonizer of New Mexico 1595–1628*. Albuquerque: University of New Mexico Press, 1953.

Harlan, George. *Postmarks and Places*. Alamosa, Colo.: Ye Olde Print Shoppe, 1976.

Harlan, Noel. "Broken Dreams: The History of Duncan Mining Camp." *San Luis Valley Historian* 42, no. 3 (2010): 5–7.

Harvey, Thomas J. *Rainbow Bridge to Monument Valley: Making the Modern Old West*. Norman: University of Oklahoma Press, 2011.

Haynes, C. Vance, Jr. "Younger Dryas 'Black Mats' and the Rancholabrean Termination in North America." *Proceedings of the National Academy of Sciences* 105, no. 18 (May 6, 2008): 6520–25.

Haynes, Gary, ed. *American Megafaunal Extinctions at the End of the Pleistocene*. Vertebrate Paleobiology and Paleoanthropology Series 3. N.p.: Springer, 2009.

Heap, Gwinn Harris. "Central Route to the Pacific." Vol. 7 of *The Far West and the Rockies Historical Series, 1820–1875*, edited by LeRoy R. and Ann W. Hafen. Glendale, Calif.: Arthur H. Clark Company, 1957.

Hewett, Edgar L., and Bertha P. Dutton, eds. *The Pueblo Indian World*. Albuquerque: University of New Mexico Press, 1945.

Hine, Robert V. *Edward Kern and American Expansion*. New Haven, Conn.: Yale University Press, 1962.

Hollon, W. Eugene. *The Great American Desert, Then and Now*. New York: Oxford University Press, 1966.

———. *The Lost Pathfinder: Zebulon Montgomery Pike*. Norman: University of Oklahoma Press, 1949.

Horgan, Paul. *Great River: The Rio Grande in North American History*. Vol. 2. New York: Rinehart and Co., 1954.

"How Many Gates to the Great Sand Dunes?" *Colorado Central Magazine* 72 (February 2000): 43.

Huber, Thomas, and Robert Larkin. *The San Luis Valley of Colorado: A Geographical Sketch*. Colorado Springs: Hulbert Center Press, 1996.

Hughes, J. Donald. *Native Americans in Colorado*. Boulder: Pruett, 1977.

Hundley, Norris. *The Great Thirst: Californians and Water, 1770s–1900s*. Berkeley: University of California Press, 1992.

Hutchinson, Wallace. "The Desert of Hissing Sands." *Travel*, June 1922, 16–18.

Irving, Washington. *Astoria, or Anecdotes of an Enterprise beyond the Rocky Mountains*. Edited by Richard Dilworth Rust. Philadelphia: Carey, Lea, and Blanchard, 1836; reprint, Lincoln: University of Nebraska Press, 1976.

Jackson, Donald. *The Journals of Zebulon Montgomery Pike*. Volume 1. Norman: University of Oklahoma Press, 1966.

James, H. L., ed. *Guidebook of the San Luis Basin, Colorado*. Socorro: New Mexico Geological Society, 1971.

Jones, Cuvier H. "The Herard Family." *San Luis Valley Historian* 11, no. 1 (1979): 7–11.

Karnes, Thomas L. *William Gilpin: Western Nationalist*. Austin: University of Texas Press, 1970.

Landreth, Libbie. "Natural History of the Great Sand Dunes." *San Luis Valley Historian* 22, no. 3 (1990): 4–6.

Lee, W. Storrs, ed. *Colorado: A Literary Chronicle*. New York: Funk and Wagnalls, 1970.

Leopold, Aldo. *A Sand County Almanac*. New York: Oxford University Press, 1949.

Leopold, Estella B., and Herbert W. Meyer. *Saved in Time: The Fight to Establish Florissant Fossil Beds National Monument, Colorado*. Albuquerque: University of New Mexico Press, 2012.

Limerick, Patricia Nelson. *Something in the Soil: Legacies and Reckonings in the New West*. New York: W. W. Norton & Company, 2000.

Limerick, Patricia Nelson, Clyde A. Milner II, and Charles E. Rankin, eds. *Trails to a New Western History*. Lawrence: University Press of Kansas, 1991.

Lorenzen, Eline D. et al., "Species-Specific Responses of Late Quaternary Megafauna to Climate and Humans." *Nature* 479 (November 17, 2011): 359–64.

Louter, David. *Windshield Wilderness: Cars, Roads, and Nature in Washington's National Parks*. Seattle: University of Washington Press, 2006.

Lovin, Hugh. "Sage, Jacks, and Snake Plain Pioneers." *Idaho Yesterdays* 22 (Winter 1979): 13–24.

Madole, Richard, et al. "On the Origin and Age of the Great Sand Dunes." *Geomorphology* 99 (2008): 102–103.

Mangimelli, John A. "Climatic Change in the San Luis Valley." *San Luis Valley Historian* 22, no. 3 (1990): 21–26.

Martin, Helen. "Water Resources and Rights in the Northern San Luis Valley, Part 2." *San Luis Valley Historian* 37, no. 4 (2005): 5–43.
Marsh, Charles S. *People of the Shining Mountains.* Boulder: Pruett, 1982.
Marsh, George Perkins. *Man and Nature.* Cambridge, Mass.: Harvard University Press, 1864.
Martin, Paul S., and Richard G. Klein, eds. *Quaternary Extinctions: A Prehistoric Revolution.* Tucson: University of Arizona Press, 1989.
Martorano, Marilyn A. "Culturally Peeled Ponderosa Pine Trees." *San Luis Valley Historian* 22, no. 3 (1990): 26–32.
Matsch, Charles L. *North America and the Great Ice Age.* New York: McGraw-Hill, 1976.
Mayell, Hillary. "Bison Kill Site Sheds Light on Ice Age Culture." *National Geographic News,* July 22, 2002. http://news.nationalgeographic.com/news/2002/07/ 0722_020722_clovis.html (accessed February 3, 2011).
McLaughlin, Genevieve. "The Sand Dunes." *American Railway Journal,* March 1928, 1–3.
Meinig, Donald William. *The Shaping of America: A Geographical Perspective on 500 Years of History,* Vol. 2: *Continental America, 1800–1867.* New Haven, Conn.: Yale University Press, 1993.
———. *The Shaping of America: A Geographical Perspective on 500 Years of History,* Vol. 3: *Transcontinental America, 1850–1915.* New Haven, Conn.: Yale University Press, 1998.
Miles, John C. *Wilderness in National Parks: Playground or Preserve.* Seattle: University of Washington Press, 2009.
Milner, Clyde A., Carol A. O'Connor, and Martha A. Sandweiss, eds. *The Oxford History of the American West.* New York: Oxford University Press, 1994.
Montoya, María E. *Translating Property: The Maxwell Land Grant and the Conflict over Land in the American West, 1840–1900.* Berkeley: University of California Press, 2002.
Mumey, Nolie, ed. *John Williams Gunnison.* Denver: Artcraft Press, 1955.
Nash, Roderick. *Wilderness and the American Mind.* 3rd ed. New Haven, Conn.: Yale University Press, 1983.
"Nathaniel P. Hill Inspects Colorado." *Colorado Magazine* 33 (October 1956).
"A New National Park?" *Colorado Central Magazine* 72 (February 2000).
Nichols, Roger L., and Patrick L. Halley. *Stephen Long and American Frontier Expansion.* Norman: University of Oklahoma Press, 1995.
Ogburn, Robert W. "A History of the Development of San Luis Valley Water." *San Luis Valley Historian* 28, no. 1 (1996): 5–35.
Oliver, Martha. "Cattle in the Valley." *San Luis Valley Historian* 17, no. 4 (1985): 1–7.
Orsi, Jared. *Citizen Explorer: The Life of Zebulon Pike.* New York: Oxford University Press, 2014.
Ortiz, Alfonso. *The Tewa World.* Chicago: University of Chicago Press, 1969.
Parkhill, Forbes. "Colorado's Earliest Settlements." *Colorado Magazine* 34, no. 4 (October 1957): 241–53.
Peake, Ora Brooks. *The Colorado Range Cattle Industry.* Glendale, Calif.: Arthur H. Clark Company, 1937.
Prud'homme, Alex. *The Ripple Effect: The Fate of Fresh Water in the Twenty-First Century.* New York: Scribner, 2007.

Pyne, Stephen J. *Fire in America: A Cultural History of Wildland and Rural Fire*. Princeton, N.J.: Princeton University Press, 1982.

Quillen, Ed. "Appearing on Your November Ballot: A Water War." *Colorado Central Magazine* no. 56 (October 1998).

———. "Baca Foreclosure May Clear Way for National Park." *Colorado Central Magazine* June 2002. http://cozine.com/2002-june/baca-foreclosure-may-clear-way-for-national-park/ (accessed November 5, 2011).

———. "The Baca Ranch: It's Got Quite a Past." *Colorado Central Magazine* (February 2000): 37–39.

———. "How Trinity Didn't Happen at the Sand Dunes." *Colorado Central Magazine*, July 1999. http://cozine.com/1999-july/how-trinity-didnt-happen-at-the-sand-dunes/ (accessed October 18, 2011).

———. "San Luis Valley Water War Goes Statewide." *Colorado Central Magazine*, November 1998. http://cozine.com/1998-november/san-luis-valley-water-war-goes-statewide/ (accessed September 6, 2011).

Radosevich, G. E., and R. W. Rutz. *San Luis Valley Water Problems: A Legal Perspective*. Fort Collins: Colorado Water Resources Research Institute, 1979.

Raines, Richard Thomas. "The Effects of Proposed Water Development by AWDI on Great Sand Dunes National Monument and the San Luis Valley in Southern Colorado." Master's thesis, Texas A&M University, 1992.

Reddin, Paul. "Sand, Saws, and the Sangre de Cristos: A History of Wood Hauling in the Hooper and Mosca Communities." *San Luis Valley Historian* 16, no. 1 (1980): 2–13.

Reisner, Marc. *Cadillac Desert: The American West and Its Disappearing Water*. New York: Penguin Books, 1986.

Remley, David. *Bell Ranch: Cattle Ranching in the Southwest, 1824–1947*. Rev. ed. Las Cruces, N.M.: Yucca Tree Press, 2000.

Repanshek, Kurt. "National Park Service Opposes Redesignation of Pinnacles National Monument as a National Park." *National Parks Traveler*, November 18, 2009. http://www.nationalparkstraveler.com/2009/11/national-park-service-opposes-redesignation-pinnacles-national-monument-national-park4951 (accessed July 11, 2001).

Richmond, Patricia Joy. "Trail to Disaster." *Monographs in Colorado History 4*. Denver: Colorado Historical Society, 1989.

Righter, Robert W. "National Monuments to National Parks: The Use of the Antiquities Act of 1906." *Western Historical Quarterly* 20, no. 3 (August 1989): 281–301.

Riley, Carroll L. *Rio del Norte: People of the Upper Rio Grande from Earliest Times to the Pueblo Revolt*. Salt Lake City: University of Utah Press, 1995.

Robbins, William G. *Colony and Empire: The Capitalist Transformation of the American West*. Lawrence: University Press of Kansas, 1994.

Rogers, K. L., et al. "Pliocene and Pleistocene Geologic and Climatic Evolution in the San Luis Valley of South-Central Colorado." *Palaeogeography, Palaeoclimatology, Palaeoecology* 94, no. 1 (July 1992): 55–86.

Rothman, Hal. *Preserving Different Pasts: The American National Monuments*. Urbana: University of Illinois Press, 1989.

———. *Devil's Bargains: Tourism in the Twentieth-Century American West*. Lawrence: University Press of Kansas, 1998.

Rothman, Hal, and Char Miller. *Death Valley National Park: A History*. Reno: University of Nevada Press, 2013.

Runte, Alfred. *National Parks: The American Experience*. 2nd ed., rev. Lincoln: University of Nebraska Press, 1987.

Schneider-Hector, Dietmar. "Colorado's Great Sand Dunes: The Making of a National Monument." *Colorado History* 15 (2008).

———. "Great Sand Dunes Wilderness: Creating 'the Very Highest Order of Federal Resource Protection' for North America's Tallest Inland Dunefield." *Colorado Heritage* (September/October 2010): 22–31.

———. "Roger W. Toll, Chief Investigator of Proposed National Parks and Monuments: Setting the Standards for America's National Park System." *Journal of the West* 42, no. 1 (Winter 2003): 82–90.

Sellars, Richard West. *Preserving Nature in the National Parks: A History*. New Haven, Conn.: Yale University Press, 1997.

Shaffer, Marguerite S. *See America First: Tourism and National Identity, 1880–1940*. Washington, D.C.: Smithsonian Institution Press, 2001.

Simmons, Virginia McConnell. *The San Luis Valley: Land of the Six-Armed Cross*. Boulder: Pruett, 1979.

———. *The Ute Indians of Utah, Colorado, and New Mexico*. Boulder: University Press of Colorado, 2000.

Sontag, William H., ed. *National Park Service: The First 75 Years*. Philadelphia: Eastern National Park and Monument Association, 1990.

Spencer, Frank C. *The Story of the San Luis Valley*. Alamosa: Alamosa Journal, 1925; reprint, Santa Fe: Sleeping Fox Press, 1975.

Stanford, Dennis. "A History of Archaeological Research in the San Luis Valley, Colorado." *San Luis Valley Historian* 22, no. 3 (1990): 33–39.

Stegner, Wallace. *Beyond the Hundredth Meridian: John Wesley Powell and the Second Opening of the American West*. Introduction by Bernard De Voto. Boston: Houghton Mifflin, 1954; reprint, New York: Penguin Books, 1992.

Steinle, John. "That Perhaps Necessary Evil: Zebulon Pike and the U.S. Military." *San Luis Valley Historian* 39, no. 3 (2007).

Sturtevant, William C., ed. *Handbook of North Native Americans*. Vol. 4 of *History of Indian-White Relations*, edited by Wilcomb E. Washburn. Washington, D.C.: Government Printing Office, 1988.

Sundberg, Lawrence D. *Dinétah: An Early History of the Navajo People*. Santa Fe: Sunstone Press, 1995.

Sutter, Paul S. *Driven Wild: How the Fight against Automobiles Launched the Modern Wilderness Movement*. Seattle: University of Washington Press, 2002.

Swadesh, Frances Leon. *Los Primeros Pobladores: Hispanic Americans of the Ute Frontier*. South Bend: University of Notre Dame Press, 1974.

Thomas, Alfred B. *Forgotten Frontiers: A Study of the Spanish Indian Policy of Don Juan Bautista de Anza, Governor of Mexico 1777–1787*. Norman: University of Oklahoma Press, 1932.

Thomas, David Hurst, ed. *Spanish Borderlands Sourcebooks*. Vol. 27 of *Hispanic Urban Planning in North America*, edited by Daniel J. Garr. New York: Garland, 1991.

Thomas, Douglas B. *From Fort Massachusetts to the Rio Grande: A History of Southern Colorado and Northern New Mexico from 1850 to 1900*. Washington, D.C.: Thomas International, 2002.

Thoreau, Henry David. *Walden; or, Life in the Woods*. Mount Vernon, N.Y.: Peter Pauper Press, 1966. First published 1854.

Trimble, Stephen A. *Great Sand Dunes: The Shape of the Wind*. Globe, Ariz.: Southwest Parks and Monuments Association 1978.

Tuan, Yi-Fu. *Topophilia: A Study of Environmental Perception, Attitudes, and Values*. Englewood Cliffs, N.J.: Prentice-Hall, 1974.

Turner, James Morton. *The Promise of Wilderness: American Environmental Politics since 1964*. Seattle: University of Washington Press, 2012.

Tushar, Olibama Lopez. *The People of "El Valle."* Pueblo: El Escritorio Press, 1992.

Ubbelohde, Carl, Maxine Benson, and Duane A. Smith. *A Colorado History*. Boulder: Pruett, 1972.

Valdez, Andrew. "Separating Fact from Fiction at Great Sand Dunes." *Outcrop: Newsletter of the Rocky Mountain Association of Geologists* 54, no. 11 (November 2005).

Voynick, Stephen M. *Colorado Gold: From the Pike's Peak Rush to the Present*. Missoula: Mountain Press, 1992.

Weber, David J. *Richard H. Kern: Expeditionary Artist in the Far Southwest, 1848–1853*. Albuquerque: University of New Mexico Press, 1985.

———. *The Spanish Frontier in North America*. New Haven, Conn.: Yale University Press, 1992.

———. *The Taos Trappers: The Fur Trade in the Far Southwest, 1540–1846*. Norman: University of Oklahoma Press, 1968.

Weiler, Stephan, Andrew Seidl, and Erich Weeks. "What's in a Name? Economic Impact of National Park Designation on the Great Sand Dunes and the San Luis Valley." *Agricultural and Resource Policy Report*, no. 3 (June 2001), Colorado State University, Department of Agricultural and Resource Economics, Fort Collins.

White, Richard. *"It's Your Misfortune and None of My Own": A New History of the American West*. Norman: University of Oklahoma Press, 1991.

———. *Land Use, Environment, and Social Change: The Shaping of Island County, Washington*. Seattle: University of Washington Press, 1980.

Wiegand, J. Patrick. "Dune Morphology and Sedimentology at Great Sand Dunes National Monument." Master's thesis, Colorado State University, 1977.

Wilkinson, Charles F. *Crossing the Next Meridian: Land, Water, and the Future of the American West*. Washington, D.C.: Island Press, 1992.

Wishart, David J. *The Fur Trade of the American West, 1807–40: A Geographical Synthesis*. Lincoln: University of Nebraska Press, 1979.

Wittmann, Kelly. *Explorers of the American West*. Philadelphia: Mason Crest, 2003.

Wolf, Tom. *Colorado's Sangre de Cristo Mountains*. Niwot: University Press of Colorado, 1995.

Wolle, Muriel Sibell. *Stampede to Timberline: The Ghost Towns and Mining Camps of Colorado*. Denver: Sage Books, 1949.

Worster, Donald. *Rivers of Empire: Water, Aridity, and the Growth of the American West*. New York: Oxford University Press, 1985.

Wyckoff, William, and Larry M. Dilsaver, eds. *The Mountainous West: Explorations in Historical Geography*. Lincoln: University of Nebraska Press, 1995.

Yochim, Michael. *Protecting Yellowstone: Science and the Politics of National Park Management*. Albuquerque: University of New Mexico Press, 2013.

Young, James A., and B. Abbott Sparks. *Cattle in the Cold Desert*. Logan: Utah State University Press, 1985.

NEWSPAPER ARTICLES

Advani, Sanjay. "Nature Conservancy Shows Off Medano Zapata Ranches." *Del Norte (Colo.) Prospector*, September 16, 1999.

"Allard: Dunes Funding Completed in New Bill." *Pueblo (Colo.) Chieftain*, September 16, 2004. http://www.chieftain.com/metro/allard-dunes-funding-completed-in-new-bill/article_ab686b5e-716b-56bb-be06-96328c827004.html (accessed November 6, 2011).

Allard, Wayne. "Dunes Deserve Park Status." *Denver Post*, December 18, 1999.

———. "Finally—It's a DUNES DEAL." *Pueblo Chieftain*, September 19, 2004. http://www.chieftain.com/editorial/ finally-it-s-a-dunes-deal/article_c200ce82-95f9-5a3a-ad25-0914cbde51e7.html (accessed August 5, 2011).

Anderson, Tania. "Dunes Tour Draws Big Interest." *Pueblo Chieftain*, December 2, 1999.

Bairstow, Diane. "The Water War with AWDI Begins." *Crestone (Colo.) Eagle*, January 2009.

Bartels, Lynn. "McInnis, Hefley Argue over Dunes Proposal." *Rocky Mountain News*, April 26, 2000.

Bean, Glen. "Ranger Reminiscing: The Early Days at Great Sand Dunes." *Great Sand Dunes: Celebrating 60 Years 1932–1992*. Copy in Great Sand Dunes National Park archives.

Benns, Teresa L. "Chaney Denies Threat of Wells." *Alamosa (Colo.) Valley Courier*, October 28, 2000.

Blackmore, William. No title. *The Standard* (London), August 19, 1869.

Boyce, Gary. "Tap Bounty without Building Dam." *Denver Post*, August 30, 1998. http://extras.denverpost.com/opinion/boyce0830.htm (accessed September 6, 2011).

"Boyce Says Baca Ranch Not for Sale." *Valley Courier*, October 11, 2000.

Boychuk, Rick. "'Division of Interest' Causes Maurice Strong to Quit AWDI." *Crestone Eagle*, January 1990. http://www.scribd.com/doc/40083195/Big-Bad-Bank (accessed June 18, 2011).

Boyle, Mary. "New Push for Sand Dunes; Allard Introduces Senate Bill Seeking National Park Status." *Colorado Springs Gazette*, May 12, 2000. http://www.highbeam.com/doc/1P2-5949609.html (accessed June 14, 2011).

———. "Tension in Congress." *Colorado Springs Gazette*, April 9, 2000.

———. "Water a New Snag in Dunes Park Plan." *Colorado Springs Gazette*, August 3, 2000.

"Colorful Sand Dunes of Colorado May Be National Monument." *Denver Post*, February 16, 1931.

"Committee Backs Dunes Purchase." *Valley Courier*, October 29, 2003.

"Conservancy Buys 2 Ranches in Colorado." *Rocky Mountain News*, July 11, 1999.

Curtiss, Nelda. "Dunes Deal Said Do-Able." *Valley Courier*, December 21, 1999.

———. "Winds of Change at Sand Dunes." *Valley Courier*, December 17, 1999.

"Dr. Spencer Urges Creation of National Monument of Sand Dunes," *Alamosa (Colo.) Journal*, February 27, 1931.

Draper, Electa. "Interior Chief, Lawmakers Back New National Park." *Denver Post*, December 19, 1999.

"Dunes Become U.S. 58th National Park." *Center (Colo.) Post-Dispatch*, September 16, 2004.

Flores, Dan. "The West That Was, and the West That Can Be." *High Country News*, August 18, 1997.

Flowers, Tim. "Baca Deal Inked." *Valley Courier*, January 31, 2002.

———. "Official Reaction Swift, Positive." *Valley Courier*, January 31, 2002.

Flowers, Tim, and Ruth Heide. "Locals Elated." *Valley Courier*, January 31, 2002.

Flynn, Kevin. "Dunes Closer to Becoming Park." *Rocky Mountain News*, September 20, 2000.

Foster, Dick. "Move Afoot to Upgrade Dunes to National Park." *Denver Post*, November 12, 1999.

———. "Study: Water Project Will Imperil Dunes." *Rocky Mountain News*, February 24, 1991.

Fox, Maggie. "Ancient Seaweed Chews Confirm Age of Chilean Site." *Reuters*, May 8, 2008. http://www.reuters.com/article/2008/05/08/us-humans-chile-idUSN08390999 20080508 (accessed April 16, 2011).

Frazier, Deborah. "Lexam to Drill Near Dunes." *Rocky Mountain News*, November 24, 2000.

"Fred Bunch Receives 2008 Regional Director's Award." *Valley Courier*, January 19, 2009. htpp://www.alamosanews.com/V2_news_articles.php?heading= 0&story_id= 11220&page=75 (accessed November 7, 2011).

Gerhardt, Gary. "Written in Sand." *Rocky Mountain News*, October 15, 2000.

"Getting Burned by Steamtown." *Reading (Pa.) Eagle*, July 2, 1995. http://news.google.com/newpapers?id=GVExAAAAIBAJ&sjid=AKIFAAAAIBAJ&pg=1554,2402937dq=steamtown+bankrupt&hl=en (accessed October 2, 2011).

Gorski, Eric, Mary Boyle, and Todd Hartman. "Political Quicksand." *Colorado Springs Gazette*, April 9, 2000.

Hartman, Todd. "Capitalist Cowboy Wants to Tap San Luis Aquifer." *Coloradoan* (Fort Collins, Colo.), March 17, 1997.

Heide, Ruth. "Baca Part of Sand Dunes Park Plan." *Valley Courier*, December 2, 1999.

———. "Court Cases Cloud Dunes Deal." *Valley Courier*, July 28, 2004.

———. "Dunes-Baca Buy On." *Valley Courier*, December 14, 2000.

———. "Dunes Park Is Passed." *Valley Courier*, October 6, 2000.

———. "Farallon Roots Examined, Baca's Bedfellows Seen." *Valley Courier*, January 16, 2002. http://hillandrobbins.com/pdf/AVC_Bacas_01-16-02.pdf (accessed August 24, 2011).

———. "Kuenhold Signs Dunes Water Decree." *Valley Courier*, August 5, 2008. http://www.alamosanews.com/V2_news_articles.php?heading=0&page=72&story_id=8611 (accessed November 7, 2011).

———. "McInnis Salutes 'Champions of the Dunes.'" *Valley Courier*, April 1, 2004.

———. "Refuge Kicks In for Dunes Park." *Valley Courier*, June 8, 2000.

———. "Resident Resistance Rises: Dunes National Park Proposal Revives Area Tax, Water Worries." *Valley Courier*, December 9, 1999.

———. "Zapata Course Gets Conditional OK." *Valley Courier*, August 24, 1988.

Heinzman, Jennifer. "Archaeologists Uncover SLV 'Ice Age.'" *Valley Courier*, July 10, 1993.

Hildner, Matt. "Judge Signs Off on Dunes Water Rights." *Pueblo Chieftain*, August 5, 2008. http://www.chieftain.com/news/local/judge-signs-off-on-dunes-water-rights/article_b35c6515-2619-535b-b8d9-9cbfc02eb308.html (accessed November 7, 2011).

———. "Settlement Ends Lawsuit over Drilling at Baca Refuge." *Pueblo Chieftain*, September 25, 2010. http://www.chieftain.com/news/local/settlement-ends-lawsuit-over-drilling-at-baca-refuge/article_f3c293e8-c878-11df-9228-001cc4c002e0.html (accessed November 8, 2011).

Hughes, Jim. "Emotions Run Deep in the Valley." *Denver Post*, October 4, 1998. http://extras.denverpost.com/news/water/water12.htm (accessed September 6, 2011).

Hunter, Mark H. "Dunes Protection Fight Spans 70 Years." *Denver Post*, October 16, 2000.

———. "Dwelling May Be 6,000 Years Old." *Denver Post*, August 28, 2000. http://extras.denverpost.com/news/news0828c.htm (accessed February 11, 2011).

———. "Group Buys Zapata Ranch." *Denver Post*, July 10, 1999.

———. "Municipal League Talks Water." *Valley Courier*, September 21, 1996.

———. "Water Plan Pitched to State Board." *Valley Courier*, September 11, 1996.

James, Dianne. "McInnis Proposes Recouping Baca Taxes." *Valley Courier*, January 19, 2000.

Jones, Marty. "High and Dry." *Westword*, November 12, 1998. http://www.westword.com/1998-11-12/news/high-and-dry/ (accessed September 7, 2011).

King, C. M. "The Wild Horse Roundup." *Valley Courier*, no date (clipping located in Colorado Springs Pioneer Museum archives, Colorado Springs, Colo.).

Lamm, Richard D. "Water Project May Be as Good as It's Going to Get." *Rocky Mountain News*, January 30, 1991.

Linder, Craig. "Allard, Interior Closer on Dunes Park." *Pueblo Chieftain*, September 20, 2000.

———. "Attorney Wants Changes in Dunes Bill." *Pueblo Chieftain*, July 28, 2000.

———. "House Strongly Approves Sand Dunes Park." *Pueblo Chieftain*, October 26, 2000.

———. "National Park Issue Debated in House." *Pueblo Chieftain*, October 25, 2000.

———. "Sand Dunes Park Proposal Clears Senate." *Pueblo Chieftain*, October 6, 2000. http://www.chieftain.com/friday/news/index.htm?article=3 (accessed June 14, 2011).

Lloyd, Jillian. "Tallest US Dunes Become Newest Park." *Christian Science Monitor*, September 14, 2004. http://www.csmonitor.com/2004/0914/p02s02-usgn.html (accessed November 6, 2011).

Lobato, Sylvia. "Bush Budget Funds Baca Buy." *Valley Courier*, April 11, 2001.

Maddrell, Paul. "Mining for Gold in the Sand Dunes." *Del Norte (Colo.) Prospector*, May 19, 1982.

Maffei, Glenn. "Allard Tries to Fit Final Piece into Sand Dunes Park Puzzle." *Pueblo Chieftain*, August 6, 2004.

Martineau, Kim. "Protesters Will Hit Yale Investments." *Hartford (Conn.) Courant*, March 3, 2004. http://articles.courant.com/2004-03-03/news/0403030797_1_hedge-students-from-stanford-university-yale-s-endowment (accessed November 5, 2011).

Matlock, Staci. "Colorado Eruption around 27 Million Years Ago Dwarfs All Others Known Today Worldwide." *Santa Fe New Mexican*, January 15, 2010.

McAllister, Bill. "Babbitt to Back Dunes Park Plan." *Denver Post*, November 30, 1999.

———. "Hefley Opposes Park Status for Sand Dunes." *Denver Post*, December 21, 1999.

———. "National Park Push Grows for Dunes." *Denver Post*, December 2, 1999.

Mcavoy, Tom. "After Feud, Lawmakers Back Park." *Pueblo Chieftain*, April 26, 2000.

McEniry, Frank J. "'Hissing Sands' of Colorado's 'Sahara' One of World's Great Scenic Wonders." *Rocky Mountain News*, July 31, 1921.

"McInnis Secures Funding to Purchase Baca Ranch." *Pueblo Chieftain*, February 14, 2003. http://www.chieftain.com/metro/mcinnis-secures-funding-to-purchase-baca-ranch/article_46757421-e688-57e1-b691-8bacaf8cf95ee.html (accessed November 5, 2011).

Mcmanigal, Barney. "State Delegation Puts Sand Dunes at Top of Agenda." *Pueblo Chieftain*, January 16, 2000.

Meyers, Charlie. "Bill No Walk in the Park for Hunters." *Denver Post*, April 4, 2000. http://www.denverpost.com/rec/hunt0404.htm (accessed December 7, 2011).

Nicholas, David. "Boyce Files Suit against Farallon in an Attempt to Block Sale of Baca Ranch." *Crestone Eagle*, February 2001.

———. "Dunes Park Bill HR 4095 Submitted to Congress." *Crestone Eagle*, April 2000.

———. "Great Sand Dunes Becomes a National Park: History Made on September 13, 2004." *Crestone Eagle*, October 2004. http://crestoneeagle.com/archives 2004/oct04_a1.html (accessed November 6, 2011).

———. "Nature Conservancy, Federal and State Gov't. Collaborating to Purchase the Baca Grant; National Park in the Forming." *Crestone Eagle*, January 2000.

———. "Water and Sand: On the Congressional Fast Track toward a National Park." *Crestone Eagle*, February 2000.

O'Driscoll, Patrick. "Wanted: San Luis Water." *Denver Post*, April 7, 1996.

"Our Best Bet." *Alamosa (Colo.) Journal*, February 13, 1931.

"PEO Sponsors Bill for Making Sand Dunes National Monument." *Alamosa Journal*, February 17, 1931.

Poppen, Julie. "Great Sand Dunes Closer to Becoming National Park." *Rocky Mountain News*, March 13, 2004.

Porter, Mary Jean. "Key Issues Laid Out in Dunes Plan." *Pueblo Chieftain*, February 23, 2000.

"Protecting a Sandy Treasure." *Denver Post*, November 26, 1999.

Quillen, Ed. "A Water Baron Takes on the Establishment." *High Country News*, October 26, 1998. http://hcn.org/issues/141/4560 (accessed September 7, 2011).

"A Rape in the Making." *Pueblo Chieftain*, November 26, 1989.

Richards, Raymond. "Three Lakes Discovered amid Great Sand Dunes of Colorado." *Denver Post*, October 15, 1922.

Romano, Michael. "Allard Blames Babbitt for Delay on Sand Dunes." *Rocky Mountain News*, July 28, 2000. http://www.rockymountainnews.com/news/ 0728dune6.shtml (accessed June 14, 2011).

Roper, Peter. "Hefley Defends Dunes Stance." *Pueblo Chieftain*, July 8, 2000.

———. "House of Representatives Leaves Dunes Bill in Limbo." *Pueblo Chieftain*, October 12, 2000.

Rubin, Julia. "Battle Brews over SLV Water." *Pueblo Chieftain*, June 12, 1989.

Sample, Ian. "Sophisticated Hunters Not to Blame for Driving Mammoths to Extinction." *Guardian* (UK), November 19, 2009. http://www.guardian.co.uk/science/2009/nov/19/hunters-mammoths-extinction (accessed February 3, 2011).

"San Luis Farmer's Big Task: How He Tried Unsuccessfully to Fence the Sand off His Farm." *Denver Times*, November 3, 1898.

"Sand Dunes Are Unequalled Attraction for Tourists Who Marvel at Changing Colors." *Alamosa Journal*, January 6, 1928.

"Sand Dunes Gets New Superintendent." *Pueblo Chieftain*, October 7, 2006. http://www.chieftain.com/metro/sand-dunes-gets-new-superintendent/article_1b90ec79-71e4-5ab3-9faa-8d5b6df6a06d.html (accessed November 7, 2011).

"Sand Dunes National Park?" *Rocky Mountain News*, December 26, 1999.

Schaafsma, Harold. "Great Sand Dunes Monument Has Colorful, Exciting Past." *Alamosa Daily Courier*, January 4, 1955.

Scully, Megan. "Push for Baca Bucks in Full Gear." *Pueblo Chieftain*, March 3, 2001. http://www.chieftain.com/metro/push-for-baca-bucks-in-full-gear/article_b022d4ae-50a3-5das-b6b4-38d2edf9a3e7.html (accessed August 24, 2011).

Seelye, Katharine Q. "Complex Deal Is First Step to Create New National Park." *New York Times*, January 31, 2002.

Simmons, Virginia. "The Mystery of the Valley's Name." *Valley Courier*, January 9, 1990.

———. "Rabbitbrush Rambler: Gilpin, the Promoter, Part 1." *Valley Courier*, August 6, 2013. http://www.alamosanews.com/v2_news_articles.php?heading=0&story_id=30833&page=74 (accessed July 23, 2012).

"Singing Sands of Alamosa." *Steamboat (Colo.) Pilot*, December 31, 1942.

Smith, Erin. "10 Named to Grand [sic] Sand Dunes National Park Advisory Council." *Pueblo Chieftain*, May 28, 2003. http://www.chieftain.com/metro/named-to-grand-sand-dunes-national-park-advisory-council/article_4e925a94-ac21-5b35-9f5c-15abafdc43a7.html (accessed November 5, 2011).

———. "Allard Thrilled with Results of Yale-Baca Talks." *Pueblo Chieftain*, January 27, 2002.

———. "Baca Ranch Sold in Foreclosure, Speeding Sand Dunes Park." *Pueblo Chieftain*, May 31, 2002.

———. "Campbell Backs Sand Dunes Park." *Pueblo Chieftain*, June 2, 1999.

———. "Colorado Voters Uphold Water Management System in San Luis Valley." *Pueblo Chieftain*, November 5, 1998.

———. "Dune Mystery Drilled." *Pueblo Chieftain*, October 12, 1990.

———. "Dunes Park Gains Backing." *Pueblo Chieftain*, February 3, 2000.

———. "Great Sand Dunes Park Idea Backed by Allard." *Pueblo Chieftain*, August 21, 1999.

———. "Great Sand Dunes Visitor Center Opens." *Pueblo Chieftain*, October 3, 2004. http://www.chieftain.com/metro/great-sand-dunes-visitor-center-opens/article_955701fb-d415-5074-a4d9-2375aa00fb90.html (accessed November 7, 2011).

———. "Happy Destination." *Pueblo Chieftain*, March 28, 2004. http://www.chieftain.com/metro/happy-destination/article_3a9705e4-f0ea-5f62-8f69-5133bfe2c540.html (accessed August 5, 2011).

———. "Incoming Superintendent Sees Need to Keep Protecting Great Sand Dunes." *Pueblo Chieftain*, November 14, 1997.

———. "Judge Orders Baca Ranch Foreclosure." *Pueblo Chieftain*, May 10, 2002. http://chieftain.com/metro/judge-orders-baca-ranch-foreclosure/article_c6d622cl-8b73-50c6-94d6-e21ae1567f52.html (accessed November 5, 2011).

———. "Judge Unveils Written Report on AWDI Decision." *Pueblo Chieftain*, February 12, 1992.

———. "McInnis Leaning toward Great Sand Dunes Park." *Pueblo Chieftain*, August 12, 1999.

———. "Method Proposed to Smooth Loss of Dunes Taxes." *Pueblo Chieftain*, January 12, 2000.

———. "The Nature Conservancy Agrees to Buy Baca Ranch." *Pueblo Chieftain*, January 31, 2002. http://www.chieftain.com/metro/the-nature-conservancy-agrees-to-buy-baca-ranch/article_764cdae1-672d-52ac-8a66-e49a8231654e.html (accessed November 5, 2011).

———. "Nature Conservancy Closes Dunes Golf Course." *Pueblo Chieftain*, March 31, 2000.

———. "Park Includes Nation's Largest Wildlife Refuge." *Pueblo Chieftain*, September 14, 2004. http://www.chieftain.com/metro/park-includes-nation-s-largest-wildlife-refuge/article_65eb4ae6-4a0a-59a7-af70-de8bc30e89a4.html (accessed November 5, 2011).

———. "Park Service Files for Dunes' Water." *Pueblo Chieftain*, December 31, 2004. http://www.chieftain.com/metro/park-service-files-for-dunes-water/article_b5cc4dd3-2911-5a47-a69e-00f036258383.html (accessed November 7, 2011).

———. "Sand Dunes Encompass 9 Ecosystems." *Pueblo Chieftain*, April 27, 2000.

———. "Sand Dunes Park Called Key to Water Future." *Pueblo Chieftain*, February 17, 2000. http://www.chieftain.com/metro/sand-dunes-park-called-key-to-water-future/article_ad6ef659-f052-5c11-8bfd-f522d71b5556.html (accessed December 7, 2011).

———. "Sand Dunes Park Plan Gets Boost from Babbitt." *Pueblo Chieftain*, December 19, 1999.

———. "Yale Donates $1.5 Million to Sand Dunes." *Pueblo Chieftain*, January 24, 2004. http://www.chieftain.com/metro/named-to-grand-sand-dunes-national-park-advisory-council/article_4e925a94-ac21-5b35-9f5c-15abafdc43a7.html (accessed November 5, 2011).

———. "Yale Subsidy Urged for Dunes Park." *Pueblo Chieftain*, February 12, 2004.

———. "Yale to Donate Profits to Nature Conservancy on the Sale of Baca Ranch." *Pueblo Chieftain*, January 26, 2002. http://www.chieftain.com/metro/yale-to-donate-profits-to-nature-conservancy-on-sale-of/article_6cfdd4a7-5050-59df-bbac-7fe5d0ce8a9f.html (accessed August 24, 2011).

———. "Yale University Silent Partner in Baca Dealings." *Pueblo Chieftain*, January 20, 2002. http://www.chieftain.com/metro/yale-university-silent-partner-in-baca-dealings/article_fe422319-3c0f-5d5a-a781-1cec21908794.html (accessed August 24, 2011).

Smith, Yvonne. "Saguache Board on Record in Opposition to Baca Sale." *Valley Courier*, December 9, 1999.

Soraghan, Mike. "Clinton Drops Support for Dunes Park." *Denver Post*, July 27, 2000.

———. "Clinton Signs Sand Dunes Bill." *Denver Post*, November 23, 2000.

———. "Clinton to Sign Bill Elevating Great Sand Dunes." *Denver Post*, November 22, 2000.

———. "Company Files Plans to Drill Near Dunes." *Denver Post*, October 24, 2000.

———. "Dunes Move toward U.S. Park Status." *Denver Post*, October 7, 2000.

———. "Dunes OK Expected Today." *Denver Post*, October 25, 2000.

———. "Dunes Park Back on Track." *Denver Post*, September 20, 2000.

———. "Lawsuit Complicates Dunes Transition to Park Status." *Denver Post*, January 23, 2001.

———. "Sand Dunes' Outcome Shifting." *Denver Post*, October 16, 2000.

———. "Water Words Snag Dunes Park." *Denver Post*, July 28, 2000. http://www.denverpost.com/news/news0728.htm (accessed June 14, 2011).

Spencer, Frank. "The San Luis Valley, a Mecca for Tourists." *Alamosa Journal*, January 27, 1927.

Stevens, William K. "New Suspect in Ancient Extinctions of the Pleistocene Megafauna: Disease." *New York Times*, April 29, 1997. http://www.cpluhna.nau.edu/Biota/megafauna_extinctions.htm (accessed February 3, 2011).

Taylor, Gary. "Dunes Bill Demand a Shock." *Valley Courier*, July 28, 2000.

———. "Dunes Bill Signed." *Valley Courier*, November 27, 2000.

———. "Dunes Vote Postponed." *Valley Courier*, October 26, 2000.

———. "Park Idea Could Help Saguache." *Valley Courier*, January 14, 2000.

———. "The Sand Dunes as a National Park." *Valley Courier*, May 6, 1999.

———. "Valley Wins Its National Park." *Valley Courier*, October 26, 2000.

Wolf, Tom. "Longevity of Ranches Linked to Water Sale." *Coloradoan* (Fort Collins, Colo.), February 16, 1997.

MISCELLANEOUS

Abdoo, Mary. "Baca Grande History." Baca Grande Property Owners Association website. http://www.bacapoa.org/Baca-Grande-History-120603-14064.htm (accessed June 18, 2011).

Adamson, Loch. "Steyer Power." February 23, 2005. http://www.Institutionalinvestor.com.Popups/PrintArticle.aspx?ArticleID=1024622 (accessed September 6, 2011).

"Ag Innovator Frank Zybach." http://livinghistoryfarm.org/farminginthe40s/water_09.html (accessed July 15, 2011).

Alfred, Randy. "July 22, 1952: Genuine Crop-Circle Maker Patented." http://www.wired.com/science/discoveries/news/2008/07/dayintch_0722 (accessed July 15, 2011).

American Water Development Inc. v. City of Alamosa (Colorado Supreme Court, May 9,

1994). http://co.findacase.com/research/wfrmDocViewer.aspx/xq/ac.%5CSAC%5CCO%5C1994%5C19940509_0040691.CO.htm/qx (accessed August 9, 2011).

"Baca Ranch." http://www.responsibleendowment.com/baca-ranch.html (accessed September 6, 2011).

Baue, William. "Private Equity Investment Puts Yale in Deep Water." *Social Funds*, February 8, 2002. http://www.socialfundds.com/news/save.cgi?sfArticleId=774 (accessed September 6, 2011).

Bean, Glen. "P.E.O. Talk." no date, copy in PEO file in Great Sand Dunes National Park archives.

Bowden, J. J. "Luis Maria Cabeza de Baca Grant." New Mexico Office of the State Historian website. http://www.newmexicohistory.org/filedetails.php?fileID=24833 (accessed June 18, 2011).

Buick Magazine 10, no. 11, no author, no title, May 1949. Reference found in National Park Service, U.S. Department of the Interior. *Area Management Study, Great Sand Dunes National Monument*, June 22–23, 1956. Record Group 79, Great Sand Dunes National Monument, file A-6423, National Archives and Records Administration, Rocky Mountain Region, Denver, Colo.

"Center Pivot Irrigator." Asabe Historic Commemoration Event No. 30. http://www.asabe.org/awards-landmarks/asabe-historic-landmarks/center-pivot-irrigator-30.aspx (accessed July 15, 2011).

Chapter V, PEO collection, Monte Vista, Colo. Colorado Legislative Council, "Historical Ballot Information." http://www.colorado.gov/cs/Satellite/CGALegislativeCouncil/CLC/1200536136114 (accessed September 7, 2011).

"Colorado Trust Lands and Education Funding." Lincoln Institute of Land Planning website. http://www.lincolninst.ecu/subcenters/managing-state-trust-lands/state/ed-funding-co.pdf (accessed December 22, 2011).

"Conservancy Acquires Spectacular Medano and Zapata Ranches." The Nature Conservancy, press release, July 1, 1999.

"Crestone History." Town of Crestone website. http://townofcrestone.org/crestone_history.shtml (accessed January 13, 2012).

"Difference between a National Park and a National Monument." National Park Service website. http://www.nps.gov/history/history/online_books/portfoli00b.htm (accessed July 11, 2011).

General Daniel Bissell Papers. St. Louis Mercantile Library, University of Missouri–St. Louis. http://www.umsl.edu/mercantile/assets/pdf/special-collections/transcript/M-009_Letters.pdf (accessed November 22, 2011).

"Great Sand Dunes National Park State Historical Fund Projects." http://coloradohistory-oahp.org/programareas/shf/projects/2005/sand.htm (accessed February 11, 2011).

Hasty, Gary. "Monte Vista P.E.O. Sisterhood initiated establishment of Sand Dunes National Monument." Possibly from the *Monte Vista (Colo.) Journal*, July 24, 1975. Copy of newspaper clipping on file in Great Sand Dunes National Park Archive collections.

"A History of Great Sand Dunes National Park and Preserve." U.S. Department of the Interior website. http://doi.net/news/04_News_Releases/greatsanddunes.pdf (accessed August 24, 2011).

"Hydrology." Formerly titled "Mysterious Waters of the Dunes." Great Sand Dunes National Park and Preserve website. http://www.nps.gov/ gras/index.htm (accessed August 23, 2011).

"Leave No Trace." Great Sand Dunes National Park and Preserve website, http://www.nps.gov/grsa/supportyourpark/leavenotrace.htm (accessed October 24, 2011).

"Legislative Glossary of Terms." http://www.leg.state.or.us/glossary.html (accessed November 13, 2011).

Lewis and Clark Trail website. http://www.Lewisandclarktrail.com/section2/sdcities/FortRandallArea/history2.htm (accessed November 22, 2011).

"Lisa Carrico Named Superintendent of Great Sand Dunes." *NPS Digest*, January 4, 2012. http://www.nps.gov/applications/digest/headline.cfm?type=PeopleNews&id=3400 (accessed February 3, 2012).

Lynch, Emma, and Frank Turina. "Monitoring Reveals Park Enjoys Exceptional Quiet." *InsideNPS*, October 29, 2008. Copy in Great Sand Dunes National Park archives.

Mader, Shelli. "Center Pivot Irrigator Revolutionizes Agriculture." *Fence Post*, May 25, 2010, http://www.thefencepost.com/article/20100525/NEWS/100529954 (accessed July 15, 2011).

"Medano-Zapata Ranch History." The Nature Conservancy Zapata Ranch website. http://www.zranch.org/index.cfm?id=df8ef385-d6f6-4393-bca27b2a03flfdba&history-of-medano-zapata-ranch.html (accessed January 7, 2012).

Montville Trail Guide. N.p.: Southwest Parks and Monuments Association, n.d. Copy in Great Sand Dunes National Park archives.

"A National Monument, Memorial, Park . . . What's the Difference?" National Atlas website. http://nationalatlas.gov/articles/government/a_nationalparks.html (accessed July 11, 2011).

"Nature Conservancy Signs Agreement to Buy 100,000-Acre Zapata/Medano Ranch." The Nature Conservancy, press release, January 21, 1999.

"New Camp Area Ready at 'Dunes.'" Possibly from the *Alamosa Valley Courier*, October 14, 1964. Newspaper clipping from Great Sand Dunes file in Colorado Room, Nielsen Library, Adams State College, Alamosa, Colo.

"President Hoover's Proclamation 1932." *Great Sand Dunes Online Curriculum*, http://www.interactive-earth.com/grsa/resources/curriculum/high/hoover_proclamation.htm (accessed August 14, 2011).

"Refuge Revenue Sharing." U.S. Fish and Wildlife Service website. http://www.fws.gov/refuges/ realty/rrs.html (accessed December 7, 2011).

Richardson, Charles Samuel. "Notebooks 1871–1883." Unpublished manuscript, Western History Collection, Denver Public Library, Denver, Colo. Notebook no. 15, n.p.

Roeder, Fred. "The Baca Floats." *American Surveyor*, April 11, 2009. http://www.amerisurv.com/content/view/6083/ (accessed June 18, 2011).

San Luis Valley Agriculture. Alamosa, Colo.: Alamosa First National Bank. Promotional booklet, no date.

"Sangre de Cristo National Heritage Area." National Park Service website. http://www.nps.gov/grsa/parknews/sangre-de-cristo-nha.htm (accessed October 17, 2010).

Shellabarger, R. W. "Papers 1901–1941." Unpublished papers, Western History Collection, Denver Public Library, Denver, Colo., April 1949.

Smith, Brendan. "Great Sand Dunes National Park and Preserve/Baca Ranch Purchase," September 1, 2005. Red Lodge Clearinghouse website. http://www.rlch.org/stories/great-sand-dunes-national-park-and-preservebaca-ranch-purchase (accessed December 8, 2011).

Spude, Robert L. Regional Historian, Cultural Resources Management, National Park Service Intermountain Region. E-mail correspondence July 11, 2012.

Trimble, Stephen. "A Sketch on the History of the Establishment of Great Sand Dunes NM." Letter to Thea K. Nordling, Chief Interpreter, Great Sand Dunes National Monument, August 20, 1975. Copy in Great Sand Dunes National Park archives.

"Trujillo Homestead." National Park Service website. http://nps.gov/nr/feature/hispanic/2004/Trujillo.htm (accessed January 7, 2012).

Truman, Harry S. "Proclamation 2681—Redefining the Area of Great Sand Dunes National Monument, Colorado." March 12, 1946. Available at Gerhard Peters and John T. Woolley, *The American Presidency Project*, http://www.presidency.ucsb.edu/ws/?pid=87055 (accessed October 18, 2011).

Valdez, Andrew. "The Role of Streams in the Development of the Great Sand Dunes and Their Connection with the Hydrologic Cycle." Great Sand Dunes National Park and Preserve website. http://www.nps.gov/grsa/naturescience/upload/ Trp2029.pdf (accessed August 23, 2011).

"Water Facts." Citizens for San Luis Valley Water. Copy of press release, n.d., n.p. Located in AWDI newspaper clippings file, Colorado Room, Nielsen Library, Adams State College, Alamosa, Colo.

Williams, Gerald W. "National Monuments and the Forest Service." National Park Service website. http://www.nps.gov/history/history/online_books/fs/monuments.htm (accessed July 11, 2011).

INTERVIEWS

Bunch, Fred. Telephone interview, March 15, 1996.
———. Telephone interview, November 4, 1996.
———. Interview, January 30, 1997.
———. Telephone interview, March 7, 1997.
———. Telephone interview, August 20, 1997.
———. Interview, August 26, 1997.
———. Interview, May 9, 2009.
———. Interview, October 12, 2009.
———. Interview, October 13, 2009.
———. Interview, May 14, 2010.
———. Interview, June 2, 2011.

Burget, Mark. Telephone interview, November 19, 2009.

Canaly, Christine. Telephone interview, December 8, 2011.

Chaney, Steve. Telephone interview, September 22, 2011.

Curtis, Ralph. Telephone interview, March 6, 1997.

Gardner, Harvey. Telephone interview, October 21, 1996.

Hatcher, Christopher. Telephone interview, November 22, 2011.

Hesse, Michael. Telephone interview, October 7, 2011.
Koshak, John. Telephone interview, October 22, 1996.
McInnis, Scott. Interview, July 7, 2011.
Peck, Wayne. Telephone interview, November 18, 2011.
Rio Grande Water Conservation District. Telephone interview, September 26, 2011.
Robbins, David. Telephone interview, December 7, 2011.
Robertson, Paul. Interview, October 14, 2009.
Smith, Erin. Telephone interview, September 30, 2011.
Valdez, Andrew. Interview, June 3, 2011.
Zimmerman, Robert. Telephone interview, August 23, 2011.
———. Telephone interview, December 6, 2011.

INDEX

Page numbers in *italics* indicate illustrations.

acre-feet, 7, 100, 136, 137, 140, 152
Adams, George, 87–88, 141, 165
Adams, William Herbert, 109, 112
Adams State College, 40, 41, 93, 106, 109, 144, 170
Adee, Niel G., 164, 165, 240–41n19
aeolian, 10, 14
aeolian system, 147, 199; diversity of resources in, 160; ensuring protection for, 156, 161, 162, 168, 169, 173, 195; Medano-Zapata and, 163; private land and, 161; resource management strategy for, 150; Valdez briefing on, 162
Alamosa, Colo., 40, *41*, 86, 111, 169–70; Alamosa National Wildlife Refuge and, 137; chamber of commerce, 104, 106, 125; irrigation network and, 96; national monument and, 119; PEO chapter of, 109; railroad and, 95; Salazar born in, 171; "Singing Sands of," 203; Spanish and bison near, 36; Toll and, 112; wood gathering, 91
Alamosa County, 119, 157, 181; roads in, 111
Alamosa Daily Courier, 121
Alamosa Journal, 106, 111, 112
Alamosa National Wildlife Refuge, 137, 138, 178
Albright, Horace M.: Elizabeth Spencer and, 115, 117–18; Toll and, 110, 111, 114, 161
Allard, Wayne, 170–74, *173*, 178, 180, 187, 188; Department of Justice waiver, 194; federal water rights and, 182–84; national park dedication and, 195; The Nature Conservancy and, 189, 190; Senate Appropriations Subcommittee and, 194, 196; Senate bill introduction and, 182; Senate bill passage and, 184; Summit in the Sand and, 173–74; Yale University and, 191
Altithermal era, 24
American Indians. *See* Native Americans
American Water Development Incorporated (AWDI), 4, 8, 160, 161, 162, 178; Baca Grant and, 140, 142; Boyce and, 151–53; claim to groundwater and, 143; Colorado Supreme Court and, 150, 157; court trial and, 144; details of water export plan, 140, 142; lawsuits and, 193–94; mineral rights and, 189–90; opposition to, 142; scientific data and, 145, 155; Stockman's Water Company and, 152
Ancestral Puebloans, 28, 29
Andrews, Sarah, 146
anthropogenic, 5, 22
antidunes, 4
Antiquities Act of 1906: cultural artifacts and, 104–105; Hoover and, 118; national monuments and, 159
Anza, Don Juan Bautista de, 40, 44, 211, 220n28
Apache Indians, 29, 36, 44, 49, 163; culturally peeled ponderosa pine and, 4, 31

271

Apodaca, Apollina, 42
aquifers, 100, 101, 132, 147, 150, 151, 156; capillary rise and, 17; confined and unconfined, 99, 136, 142, 203, 208; estimated size of, 6–7, 213n2
Arapaho Indians, 29
Arizona-Colorado Land & Cattle Company, 142
Arkansas River, 44, 50, 52, 153
arrastra, 42
artemisia, 53, 64–67
artesian well, 8, 72, 96, *99*, 100, 138; definition of, 98
AZL Resources, Inc., 142

Babbitt, Bruce, 171–74, *173*, 183, 189
Baca Grant, 69, 80, 87–89, 165, 211; AWDI and, 140–42, 145; history of, 140–42, 236n19. *See also* Baca Ranch
Baca National Wildlife Refuge, 178, 189, 195, 196, *197*, *200*, 204
Baca Ranch, 156, 160, 161, 168, 169, 172, 184, 191; AWDI and, 140, 142, 150–53; as biological hotspot, 151, 163; Boyce and, 151–53, 155, 185, 190, 193; Farallon Capital Management and, 151, 152, 160, 185, 187, 189, 190–94; Joel Hefley and, 177, 181, 186, 188; Lexam Explorations and, 189–90, 204; The Nature Conservancy purchase of, 174, 189–96; Saguache County and, 151, 171, 174, 178, 180. *See also* Baca Grant
Beale, Edward F., 63, 224n37
Bean, Glen, 122, 126
Beckwith, E. G., 53, 65–67, 72, 74
Bedford, Charles, 194
Benton, Thomas Hart, 60
Big Spring Creek, 112, *207*
bison, 18, *23*, 56, 59, 149, 204, *207*, 210; Medano-Zapata Ranch and, 163, 165–68, 197, 203; Paleoindian hunters and, 20, 22–24, 201; Spanish and, 36, 45, 46, 84, 218n8
Bison antiquus, 22, *23*, 24. *See also* bison
Black Canyon of the Gunnison, 170, 171, 184, 195, 199
Blackmore, William, 69, 70, 72, 79
Blanca Peak, 15, 29. *See also* Mount Blanca; Sierra Blanca
Blanca Wildlife Habitat Area, 138
Blenden, Mike, 178, 195

Bliss, Edward, 70, 72
blue clay layer, 99, 208
boom and bust, 5, 81, 87
boosters, 5, 13, 51, 52, 62, 67, 104, 108; Gilpin and, 68, 72, 78, 81, 94, 97, 106; Richardson drawing and, 73–74, 77; Sahwatch Lake and, 70–72, 100, 136
Boyce, Gary, 182, 191, 193; Baca Ranch purchase and, 151–52; details of water plan, 152; Farallon Capital Management and, 151, 185, 190; local opposition to water plan, 152–55. *See also* Stockman's Water Company
Brownlee, J. L., 103
Bunch, Fred, 43, 158, 173; Big Spring Creek and, 207; NAGPRA and, 209–10; resource management and, 9, 145–46, 204, 206; Well Y and, 208
Burget, Mark, 167, 172, 174, 192, 195
Bush, George H. W., 150
Bush, George W., 181, 190

Cabeza de Baca, Luis Maria, 69, 140–41
Cabeza de Vaca, Álvar Núñez, 140
Cabeza de Vaca Land and Cattle Company, 152, 160, 174, 193
Campbell, Ben Nighthorse, 169, *173*, 181, 190; Black Canyon of the Gunnison and, 170, 171, 184, 195, 199; Summit in the Sand and, 172–74
Canaly, Christine, 153, 154, 190–91, 204
capillary rise, 17, 100, 144
Carhart, Arthur, 42
Carrico, Jim, 204
Carrico, Lisa, 204
Carson, Christopher "Kit," 59
Castle Creek, 53, 157, 158
center-pivot sprinkler, *134*, 135–36, 211
Chaney, Steve, 170, 174, 189, 195, 203; Baca mineral rights and, 190; "extended road show" and, 175; monument to park and, 168, 192; Valdez briefing to, 162; Wellman and, 161–62; Yale University and, 194
Chenoweth, John Edgar, 126
Cheyenne Indians, 29
Citizens for Colorado's Water, 154
Citizens for San Luis Valley Water, 143, 153, 154, 190
Clarke, Bert, 122–23
Clinton, Bill, 188

Clinton Administration, 171, 185, 188
Closed Basin, 101, *135*, 136–38, 140, 152; hydrology of, 14, 77, 100; Sahwatch Lake and, 70, *71*, 72, 100, 136. *See also* Closed Basin Project
Closed Basin Project, 136–40, 150–53, 180. *See also* Closed Basin
Clovis people, 19–22, 24–26, 29, 30, 163, 208
Cochetopa Pass, 60, 64, 68
Cody Complex, 24
Collins, Alfred, 141
Colorado Bowhunters Society, 178
Colorado Central Magazine, 179
Colorado Division of Wildlife, 143, 178
Colorado Environmental Coalition, 179
Colorado General Assembly, 133, 158
Colorado House of Representatives, 109, 182
Colorado Mountain Club, 42, 179
Colorado State Land Board, 153–54, 192
Colorado State Senate, 109, 111
Colorado State University, 8, 22, 159
Colorado Stockgrowers Association, 82
Colorado Supreme Court, 144, 150, 157
Comanche Indians, 29, 36, 44, 47, 49, 63
Conejos Grant, 46, 48
Conejos River, 36, 44, 47, 55, 68
Conway, Sean, 182, 188
Cook, John, 162
Corlett, George, 109, 110
Costigan, Edward, 115
Crestone, Colo., 87–89, 162, 180, 189
Crestone Peak, 46, 152
Cuerno Verde (Green Horn), 44
Curtis, Ralph, 153, 183

Danielson, Jeris, 151–53
Deadman Gulch, 79
Death Valley National Monument, 111, 178
DeGette, Diana, 187
del Norte, Rio Grande. *See* Rio Grande River
Del Norte, Colo., *7*, 48, 64, 84, 104, 151; irrigation and, 95–97; PEO chapter of, 109; Rio Grande River and, 14, 140
Denver and Rio Grande Railroad, 90, 95
Denver Post, 102, 112, 170, 189
Dickey, Valentine B., 164
Dickey, William W., 164
Doering, S. E., 103, 105

Dollar Lake, 27, 28, 83. *See also* San Luis Lake
Dorris, George, 86
Duncan, Colo., 88–90
Duncan, John, 88
dune types, *148*, 149
Durkee, Eugene, 164
Durkee, William Wells, 164–65

Eiseley, Loren, 131
Eisenhower, Dwight D., 127
Elephant Butte Reservoir, 133, 140
El Pomar Foundation, 167
Endlich, F. M., 12–13
Entz, Lewis, 154
environmental history, 3, 4, 66
escape dunes, 16
evapotranspiration, 137
Everhart, Ron, 175

Farallon Capital Management: Baca Ranch purchase and, 151–52, 160; Baca Ranch sale and, 185, 187, 189–94
Federal Reserved Water Right, 143, 144
Fewkes, J. Walter, 105–106
Fiege, Mark, 206
flour gold, 89–90, 116, 202
Flower, Henry C., 165
Folsom people, 9, *22*, 24, 30, 163, 210
Ford, Gerald, 129
Fort Garland, 63, 112, 224n36
Fort Massachusetts, 63, 64, 224n36
Foshay, Wilbur, 125–26
Fowler, Jacob, 13, 59, 84
Franke, Paul, 120
Free Timber Act of 1878, 91
Frémont, John Charles, 60–63, 68, 88, 210
Friends of the Dunes, 143
fur trappers, 5, 13, 51, 58–60, 67, 75, 80, 84

Garnett, Stanley L., 193
Gates Family Foundation, 167
George, Russell, 195
Ghost Forest, 4, 16, *17*, 92
Gilbert, Paul, 42, 102–104, 111–12
Gilpin, William, 50, 79, 80, 100; Baca Grant and, 69, 87, 88, 141; boosters and, 68, 70, 72, 78, 81, 106, 136; Gilpin's Garden and, 72, 78, 94–95, 97, 101, 132, 145; tribe of, 68, 72, 78
Glenn, Hugh, 58
Glickman, Dan, 188

gold mining, 67–70, 78, 79, 163, 189; Great Sand Dunes and, 3, 89, 90, 94, 101, 108, 116–17, 125, 130, 202; La Caverna del Oro, 42, *43*, 103; Sangre de Cristos and, 86–90; Spanish and, 39, 40, 42, 44, 87, 131
Gomez, Antonio Matias, 46, 47, 74, 77
greasewood, 62, 66, 92, 101, 137, 211. *See also* phreatophytes
Great American Desert, 59, 60, 62, 67, 147; early descriptions of, 50, 55–58, 63, 64; Gilpin and, 68, 70, 72, 78, 80, 101, 132, 145; Long and, 51, 56; Pike and, 51, 55–57, 68
Great Outdoors Colorado (GOCO), 167, 192
Great Sand Dunes: bison and, 20, 22, *23*, 24, 201, *207*; cattle and, 28, 83, 84, 94, 113, 122–23, 126–27, 129; early newspaper descriptions of, 102–108; geologic formation of, 10–17, 215n17; gold mining and, 3, 89, 90, 94, 101, 108, 116–17, 125, 130, 202; hydrology of, 4–6, 17, 143–45, 147, 148, 150, 162, 181, 208; Native American artifacts and, 103–105, 113, 115, 209; Paleoindians and, 20, 22, *23*, 24, 26, 45, 105, 201, 207; Pike and, 50–58; sand transport and recycling at, 15–17, 132, 143; skiing at, 113, 115–16; Toll and, 110–15, 119; vegetation on, 16, 53, 55, 65–67; web-footed mustangs and, 4, 84; wildlife of, 17–18. *See also* Great Sand Dunes National Monument; Great Sand Dunes National Park and Preserve
Great Sand Dunes Country Club and Inn, 166
Great Sand Dunes National Monument: atomic testing at, 124; cattle trespassing at, 122–23, 126–27, 129; differences from national park, 105, 158–60; early access roads to, 112, 113, *120*, 121, 123, 125–27, 129; early map of, *125*; early years of, 119–29; entrance station, 121–23, *124*; establishment of, 105–19; headquarters of, 121, 122, *124*; Mission 66 and, 127–29; PEO Sisterhood campaign for, 108–10, 114–19, 121; Pinyon Flats campground, *128*, 129; resource management strategy and, 146–50, 162; U.S. Army training at, 122, *123*; Visitor Center, *128*, 129; wood collecting and, 93–94. *See also* Great Sand Dunes; Great Sand Dunes National Park and Preserve

Great Sand Dunes National Park Advisory Council, 180, 194, 203
Great Sand Dunes National Park and Preserve: economic impact of, 159–60, 169; General Management Plan of, 194, 198, 203; hunting in, 169, 178, 180; initial concept of, 158; initial idea for preserve, 178; justification for creating, 162, 168–70; legislative process for creating, 170–98; management of, 198–201; as new paradigm for conservation, 199–201; noise levels and, 204; official dedication of, 195–96; Resource Management Division and, 197, 209; Summit in the Sand at, 172–75, *174*, 197; U.S. quarter depicting, 205. *See also* Great Sand Dunes; Great Sand Dunes National Monument
Great Sand Dunes National Park and Preserve Act of 2000 (H.R. 4095), 180, 185–89
Great Sand Dunes National Park and Preserve Act of 2000 (S. 2547), 182; water rights and, 182–84
Great Sand Dunes Tiger Beetle (*Cicindela theatina*), 18
Greenhorn Mountain, 44
groundwater, 5, 8, 16, 130, 131, 168, 169, 172, 191, 203; AWDI and, 4, 8, 140, 142–45, 150, 160; Boyce and, 151–53; Closed Basin and, 72, 77, 100, 137; dune stabilization and, 17, 144; Joel Hefley and, 181, 182, 185; nontributary, 143; pumping of, 5, 98–101, 104, 131, 132, 134, 136, 147, 149; tributary, 143
Gunnison, John Williams, 62, 64, 65

Hamel, A. G., 103, 105
Hardy, Guy, 110, 114–15, 117, 118
Hastert, Dennis, 184–87
Hastings, Frank, 80
Hatcher, Christopher, 185–86
Hayden, Ferdinand V., 11, *12*, 13, 69, 118, 164; Closed Basin and, 70, 72
Headlee, A. Elmer, 109, 110, 112
Heap, Gwinn Harris, 63–65, 224n37
Hefley, Joel, 176–77, 181–82, 184–88, 201
Hefley, Lynn, 182

Henry, Theodore C., 96
Herard, Jean François (Frank), 79–80, 83
Herard, Ulysses Virgil (Ulus), 4, 80, 83–84, 85, 94
Herard Homestead, 4, 80, 83–85, 125
Hesse, Mike, 170
Hewett, Edgar L., 28
Hill, Nathaniel P., 69, 72
Holocene Epoch, 21
Homestead Act (1862), 95
Hooper, Colo., 91, 96, 100, 154
Hoover, Herbert, 114, 118, 129, 161, 173
Horn, Elisha P., 42
Hornick, Peter, 152, 193, 194
Hosford, Richard W., 165
Hotel Liberty, *89*
Hull, Edward, 164
Hutchinson, Art, 203, 204
hydrology: AWDI and, 143–45, 155; of Closed Basin, 14, 72, 77, 100, 138; of Great Sand Dunes, 5, 6, 17, 147, 150, 161, 162, 208; hydrologic regime, 181; of San Luis Valley, 99, 103, 130, 149, 178; surge flow and, 4; tritium and, 148

Indian Spring (Big Spring), 26, 28, 46, 112, 207, *209*, 210
Indians. *See* Native Americans
interdunal ponds, 9, 28, 53; disappearance of, 5, 8; discovery of, 102–104, 111; Native American artifacts at, 103–105; Richardson's lake and, 74

Jackson, William Henry, *27*, 74, *75*, *76*, 211
Jicarilla Apache. *See* Apache Indians
Johnson, Edwin C., 121, 125–26
Johnson, Ross, 13–14
Jornada del Muerto (Journey of Death), 124

Kellogg, Edwin H., 74–75, 77
Kern, Edward, 62
Kern, Richard, 61–62, 65, 223n31
King, Glen, 119
Kiowa Indians, 29
Kit Carson Peak, 141, 152, 196
Kuenhold, O. John, 203

La Caverna del Oro (Cave of Gold), 42, 43, 103
La Garita Caldera, 10
Laguna Grande, 46, 75, *76*, 77

Lake Alamosa, 11, 13, 14
Lake of the Dead, 26, *27*. *See also* Sip'ophe
La Mina de los Tres Pasos (Three Steps Mine), 42
Lamm, Richard, 142
Land and Water Conservation Fund, 190, 191
Latham, Hiram, 82
Leshy, John, 183
Levin, Richard Charles, 191
Lew, Jacob, 188
Lexam Explorations, Inc., 189–90, 204
Liberty, Colo., 89, 90
Linger, Bob, 240–41n19
Linger, George W., 165
Linger Ranch, 112. *See also* Medano-Zapata Ranch
Little Ice Age, 45
Long, Stephen H., 51, 56, *57*
Los Alamos, N.Mex., 124
Lost River, 107, *108*. *See also* Medano Creek
Lowther, George, *43*
Luther Bean Museum, 40, 41, 93

Madole, Richard, 14
main dunefield, 94, 96, 112, 145, 198, 203; aeolian system and, *15*, 146, 147, 156, 162, 172, 197; capillary rise and, 17; cattle and, 123; interdunal ponds and, 5, 8, 28, 74; Medano Creek and, 4, 116; size of, 107, 160, 172, 197, 213n1; Toll and, 112; types of dunes in, *148*, 149; vegetation on, 16, 53. *See also* Great Sand Dunes
Mainella, Fran, 195
mammoth, Columbian, 20, 21, 208
Manhattan Project, 124
Manifest Destiny, 5, 60
Martin, Paul, 21
Martin, Steve P., 195
Master of Them All (tractor), *93*
McClure, Bill, 174–75
McDade, Joseph M., 177
McInnis, Scott: Baca Ranch and, 189–90, 192, 194; Great Sand Dunes National Park legislation and, 170–71, 178–80, 188, 195; Joel Hefley and, 177, 181–82, 184–87; race with Wellman, 157, *158*; Summit in the Sand and, *173*, 174–75; Zimmerman and, 157, 160, 162, 173, 196

McKee, Ed, 147
Mears, Otto, 80
Medano Creek, 198, *200*; Ghost Forest and, 16, 92; gold mining and, 3, 90, 116–17, 125; hydrology of, 3, 16, 17, 96, 103, 147, 208; as Lost River, 107, *108*; Pike and, 53, 157; Richard Kern and, 61; Richardson's lake and, 74; surge flow and, 4, 129; Toll and, 112; tritium and, 148; Ulus Herard and, 4, 84, 125; U.S. Forest Service and, 93; U.S. Mint and, 205
Medano Park, 80, 83–85
Medano Pass, *15*, 197, *200*; Beckwith and, 65; Duncan and, 88; Frémont and, 60; Herard homestead and, 4, 80; Pike and, 3, 52, 222n8
Medano Ranch, *166*, *167*, 206; bison and, 203; Dickey Brothers and, 164; The Nature Conservancy and, 166–68, 194, 197; Teofilo Trujillo and, 83; Toll and, 112. *See also* Medano-Zapata Ranch
Medano Springs, 46, 83, 84, 164
Medano Springs Land and Cattle Company, 164–65
Medano-Zapata (Zapato) Grant, 46, 74–77
Medano-Zapata Ranch, 160, 161, *166*, *167*, 206; as biological hotspot, 151, 163; history of, 163–67; The Nature Conservancy purchase of, 166–69; wildlife of, 163. *See also* Linger Ranch; Medano Ranch; Zapata Ranch
megafauna, 21, 22, 25, 30, 31
Mesa Verde National Park, 105, 119, 120, 126
Mexican-American War, 69, 143
Mining Law of 1872, 86
Miocene Epoch, 11, 13
Mission 66, 127–29
Moffat, Colo., 96, 100
Monte Vista, Colo., 96, 108, 109, 121, 151
Monte Vista National Wildlife Refuge, 178
Montville, 80, 119
Mosca, Colo., 91, 92, 96, 100, 111, 123, 129
Mosca Pass, 13, *15*, 89, 119, 123, *200*; Frémont and, 60; Jackson and, 74, *75*, *76*; Pike and, 222n8; Richardson and, 73; Robidoux Pass and, 80; Toll and, 112

Mount Blanca, 15, 80, 162. *See also* Blanca Peak; Sierra Blanca
Mount Herard, 62, 205, 207
Music Pass, *15*, *200*

National Audubon Society, 179
National Environmental Policy Act (NEPA), 204
National Parks and Conservation Association, 179
National Park Service (NPS), 6, 199, 204, 206, 207; AWDI and, 4, 143, 145, 150, 155; Closed Basin and, 138–40; Great Sand Dunes National Monument and, 93, 106, 110, 115, 118–19, 121, 126, 146, 159–60; Great Sand Dunes National Park and Preserve and, 3, 8, 158, 161, 168, 176–78, 186, 188–89, 194–96, *200*; groundwater and, 17, 144, 183, 203; interdunal ponds and, 5, 8, 9, 111; Land and Water Conservation Fund and, 190; Mission 66 and, 127, 129; Native American Graves Protection and Repatriations Act (NAGPRA) and, 209–10; The Nature Conservancy and, 163, 169, 194, 195, 197, 199, 203; Organic Act (1916) and, 176; Toll and, 110, 111, 114, 119, 161; water rights and, 133, 143, 169, 183, 203
National Register of Historic Places, 122, 168, 206
National Wildlife Federation, 179
Native American Graves Protection and Repatriations Act (NAGPRA), 209–10
Native Americans, 5, 6, 35, 42, 46, 50, 67; altering landscapes and, 30–31, 49, 53, 201–202; cultural artifacts and, 6, 19, 104–105; European contact and, 34–36, 45; Great Sand Dunes and, 26, 108, 209; Native American Graves Protection and Repatriations Act (NAGPRA) and, 209–10; San Luis Valley and, 29, 37, 44, 45, 49, 58, 80, 131, 141, 163, 202. *See also names of specific tribes*
Nature Conservancy, The (TNC), 172, 199; Baca Ranch and, 151–52, 160, 163, 174, 189–96; Medano-Zapata Ranch and, 152, 163, 166–69, 194, 197, 203, 206
Navajo Indians, 29, 32, 36, 44, 49, 141, 163

Needles, The, 151
Newhall Land and Farming, 142, 189
"No Dam Water Project," 152
Norton, Gale, 190, 194–96

Oakes, Daniel C., 74–75, 77
Ogburn, Robert, 144–45, 150, 193
Old Spanish Trail, 44, 69
Oñate, Don Juan de, 32, *33*, 34, 36–37, 84
optically stimulated luminescence (OSL), 14, 149
Ota, Hisayoshi (Hisa), 165–67
Otaka International, Inc., 165
Owens, Bill, 181
Owens Valley, Calif., 142

Packard Foundation, 192
Paleoindians, 19, 22–26, 45, 105, 201, 207
Paleolithic hunters, 5, 6, 9, 25, 26, 30, 131
Payment in Lieu of Taxes (PILT), 178
Peck, Allen, 104, 105
Penry, Josh, 179, 188
PEO Sisterhood, 108–10, 112, 114–17, 119, 121, 129, 161. *See also* Spencer, Elizabeth
Phipps, Lawrence C., 110
phreatophytes, 137, 138. *See also* greasewood; rabbitbrush
Pike, Zebulon Montgomery, *51*, *54*, 58, 74, 112, 129, 157, 173, 210; Beckwith and, 53, 65–66; description of Great Sand Dunes and, 3, 53, 66, 150; Great American Desert and, 51, 55–57, 66, 68, 150; and his "dam'd set of rascals," 50, 163, 221n3; Medano Pass and, 52, 222n8; stockade and, 55, 223n25; Red River and, 50, 52, 222n6; Wilkinson and, 50, 221n1, 221n2
Pikes Peak, 44, 52, 67
piñon-juniper, 18, 92, 93, 146, 147
piñon pine (*Pinus edulis*), 18, 62, 74, 146, 147; climate and, 66; wood gathering and, 91–94
Pinyon Flats campground, 60, *128*, 129, 198, *200*, 210
Pleistocene Epoch, 11, 13, 14, 15, 20, 21
Poncha Pass, 44, 80, 83, 142, 211
ponderosa pine (*Pinus ponderosa*), 124, 147; climate and, 66; Ghost Forest and, 4, 15; Native American peeling and use of, 4,

31, 92, 201; wood gathering and, 92. *See also* Ghost Forest
Presidential Proclamation No. 1994, 118
Preuss, Charles, 60
prior appropriation doctrine, 97–98
Public Law 106–530, 188. *See also* Great Sand Dunes National Park and Preserve Act of 2000 (H.R. 4095)
Pueblo, Colo., 52, 104
Pueblo Chieftain, 90, 170, 179, 191
Pueblo Indians, 30, 31, 35, 37–40, 47, 163; Acoma, 37; Taos, 28, 29, 38; Tewa, 26–29, 202, 205, 208, 211
Pueblo Revolt, 38, 42

rabbitbrush, 53, 67, 92, 101, 137. *See also* phreatophytes
railroads, 5, 91, 96, 99, 177; Denver and Rio Grande, 90, 95; Rio Grande and Sangre de Cristo, 87–88; surveys for, 52, 60, 61, 63, 64, 131; Union Pacific, 69, 82, 95
Red River, 50, 52, 222n6. *See also* Pike, Zebulon
reentrant, 13, 15
Refuge Revenue Sharing Act (1935), 179
reversing winds, 16, 149
Richardson, Charles Samuel, 6, 72–75, 77, 226n66
rift fill, 11
rifting, 10, 11
Rines, Howard S., 122
Rio Grande Canal, 96–98
Rio Grande Compact, 133, 136, 140, 150, 172
Rio Grande County, 112, 151
Rio Grande National Forest, 91, 178, 189, 196
Rio Grande Rift, 11
Rio Grande River, *7*, 29, 47, *135*; allocation of resources from, 96–98, 100, 132, 133, 151, 155; ancestral, 11, 14; Closed Basin and, 70, 100, 136–37, 140; Lowther and, 43; Pike and, 53, *54*; San Luis Valley and, 7, 13, 58, 59, 64, 67, 80, 81; Spanish and, 32, 34, 35, 39, 131
Rio Grande Water Conservation District, 178, 183, 191; AWDI and, 142–43; Boyce and, 153–54; establishment of, 133
riparian water rights, 98

Rivera, Juan Maria de, 43
Robbins, David, 177–78, 180, 183, 238n41
Robidoux, Antoine, 80
Robidoux Pass. *See under* Mosca Pass
Rocky Mountain Bighorn Society, 178
Rocky Mountain Bison, Inc., 165, 167
Rocky Mountain Elk Foundation, 178
Rocky Mountain News, 104, 184
Rocky Mountains Cooperative Ecosystem Studies Unit (RMCESU), 8
Ruckelshaus, William, 142

sabkha, *15*, 146, 172, 184, 199; aeolian system and, 147, 150, 156, 160, 162; Closed Basin and, 14, 72; Medano-Zapata Ranch and, 163, 167, 169, 197; Soda City and, 149
sagebrush, 24, 62, 64, 66, 95, 196
Saguache, Colo., 95
Saguache County, 151, 171, 174, 178–81, 193
Sahara Desert, 55, 104, 107
Sahwatch Lake, 70, *71*, 72, 100, 136. *See also* Closed Basin
Salazar, Ken, 180, 186, 195; Summit in the Sand and, *173*; water rights and, 171–72, 184
saltation, 15
Sand Creek, 88, 96, *200*, 208; AWDI and, 143–44; interdunal ponds and, 103; sand recycling and, 16, 17, 147; tritium and, 148
Sand Dunes Summit. *See* Summit in the Sand
sand provinces, 146
sand recycling, 16, 17, 132, 143. *See also* Medano Creek; Sand Creek
sand sheet, 14, *15*, 16, 146, 184, 196, 198, 199, 207, 208; aeolian system and, 147, 150, 156, 160, 162; Baca Grant and, 141; Medano Ranch and, 163, 167, 169, 197; vegetation and, 149
Sandy Place Lake, 26, 28. *See also* Sip'ophe
Sangre de Cristo Grant, 46, 48, 69–70
Sangre de Cristo Mountains, *61*, *62*, *76*, 81, *135*, *167*, *173*, *197*, *200*; cattle and sheep in, 83; fire and, 87, 91; Frémont and, 60–62; fur trapping in, 59–60; Great Sand Dunes and, 10, 13, *15*, 16, 180, 206; Heap and, 63; Herard homestead and, 79, 80, 125; hunting in, 84, 178; legend of naming, 39–42, 211; mining and, 42, 44, 78, 86–90, 125; Pike and, 52–53; ponderosa pine and, 16, 31; Richardson and, *73*; Spanish and, 36, 42, 44, 46; uplift of, 11; watershed of, 3, 70, 107, 146, 147, 150, 156, 160, 162, 172, 196, 198, 199; as White Mountains, 52–53; Wilderness Act and, 129; wood collecting in, 78, 90–94
San Isabel National Forest, 91, 103, 104, 106
San Juan Mountains, 64, 69, 210; Frémont and, 60–61; Great Sand Dunes and, 10–11, 13, 14; mining in, 46, 77, 78, 86, 90, 95; Richardson and, *73*, 77; timber and, 78, 90, 91
San Luis, Colo., *7*, 47, 48, 63, 68, 86, 95
San Luis Hills, 11
San Luis Lake, 14, 24, *27*, 40, *76*, *135*, 138, *200*, 211; Laguna Grande and, 46, 74–77; Richardson and, 74, 77; Sahwatch Lake and, 70, 71; Sip'ophe and, 26–29
San Luis People's Ditch, 48
San Luis Valley, *7*, *135*; Anglo-American settlement of, 42, 80–86, 91, 94, 95, 97; aquifers in, 6, 99–101, 136, 142–43, 150, 152, 181, 182, 213n2; boosters and, 67–72, 78; climate of, 8, 11, 14, 20, 24, 37, 45, 77, 164, 208; cultural values and, 6, 69, 101, 201–202; Frémont and, 60–62; fur trapping in, 58–60; Gilpin and, 68–72, 78; Great American Desert and, 51, 55, 58, 63, 64; Great Sand Dunes and, 10–16, 199; Great Sand National Monument and, 103–30, 160; Great Sand Dunes National Park and Preserve and, 158, 160, 169, 170, 179, 180, 184, 185, 190–202; Hispanic settlement of, 45–49, 68, 81–83; hydrology of, 72, 75, 77, 99–101, 103, 131–56; irrigation in, 48, 49, 78, 95–101, 132–40; livestock in, 81–86, 206; Native Americans and, 29–31, 34, 37, 42, 44, 58; Paleoindians and, 19–26; PEO and, 108–10, 114–19, 121, 129, 161; Pike and, 52–55, 66; Richardson and, 72–75, 77; size of, 6, 10, 213n1; Spanish in, 32–45, 84; Spanish descriptions of, 5, 36–37; Toll and, 111–14; water development plans

in, 4, 140, 142–45, 150–55, 181, 182, 186, 187
San Luis Valley Ecosystem Council, 190, 204
Santa Fe, N.Mex., 34, 39, 44, 55, 58, 95
Santa Fe Formation, 11, 13
Schaafsma, Harold, 126–27
Schaffer, Bob, 188
Schiel, James, 64, 65, 72
Seianyedi, 29. *See also* Apache Indians
Shockey, Howard, 42
Short Creek, 89
Shriver, Karla, 154
Sierra Blanca, *12, 15, 27*, 28, 36, 63, 91, 164. *See also* Blanca Peak; Mount Blanca
"Singing Sands of Alamosa," 203
Sip'ophe, 26–29
Sisnaajinií, 29. *See also* Blanca Peak
Smith, Erin, 169–70
Snyder, Mike, 204
Soda City, 149
Sondaya, Jose Luis Baca de, 46, 47, 74, 77
Sowapophe-uvehe, 3
Sowers, Ted, 122
Spanish explorers, 5, 6, 9, 80; descriptions of San Luis Valley, 36, 37, 39, 163; ecological impacts of, 34, 35, 45, 49; gold mining and, 39, 40, 42, 44; Spanish oxcart legend, 42, 89. *See also specific explorer names*
Spanish Ordenanzas, 47
Spencer, Elizabeth, 108, 109, 114–15, 117. *See also* PEO Sisterhood
Spencer, Frank, 106, 108, 112
Stanford, Dennis, 10, 20
Steamtown USA, 177
Stegner, Wallace, 6, 68, 157
Stewart, Dion, 144, 170, 237n30
Stewart, Malcolm, Jr., 165
Stewart, Malcom, Sr., 165
Stewart's Cattle Guard, 22, 163, 206
Stockman's Water Company, 8, 154, 163, 178; Boyce and, 151–52, 155, 190, 193
Strong, Hanne, 145
Strong, Maurice, 142, 145
Subcommittee on National Parks, Forests and Public Lands, 177, 181
subirrigation, 100
Summit in the Sand, 172–75, 197
surge flow, 4, 129. *See also* Medano Creek
Sylvester, Loren B., 165

Taos, N.Mex., 38, 58, 59, 80, 82
Taos Pueblo Indians, 28, 29, 38. *See also* Pueblo Indians
Taylor, Edward T., 121
Taylor, Gary, 168
Tela We-a-gat, 30
Tewa Pueblo Indians, 26–29, 202, 205, 208, 211. *See also* Pueblo Indians
Toll, Roger, 110–14, 115, 119, 161
Toll Report, 111–14, 115, 161, 232n49
Torres, Francisco, 39–42, 44, 211
Treaty of Guadalupe Hidalgo (1848), 60, 141, 143
Trujillo, Pedro, 83, 86, 165
Trujillo, Teofilo, 83, 86, 165
Trujillo Homestead, 83, 86, 206
Truman, Harry, 125

Udall, Mark, 186
U.S. Army, 47, 53, 59, 63, 122, *123*
U.S. Bureau of Land Management, 133
U.S. Bureau of Reclamation, 133, 136, 138, 159, 180
U.S. Department of Agriculture, 106, 136, 159, 188
U.S. Department of the Interior, 98, 106, 119, 189, 190, 192, 203; Allard and, 183, 194
U.S. Department of Justice, 194
U.S. Fish and Wildlife Service (USFWS), 133, 159, 178, 179, 194, 195, 204
U.S. Forest Service (USFS), 112, 133, 159; Baca Ranch and, 194; cultural artifacts and, 103–105; Great Sand Dunes and, 93, 106, 119, 156, 160; interdunal ponds and, 102–105, 111; La Caverna del Oro and, 42, 103; National Park Service and, 146, 169, 178, 188–89
U.S. Geological Survey (USGS), 11, 14, 146, 172
U.S. House of Representatives, 115, 170, 180, 184–87, 194; Appropriations Committee, 121; Natural Resources Committee, 177, 185, 186; Rules Committee, 184
U.S. Senate, 115, 171, 181–85, 187; Appropriations Committee, 190, 196; Appropriations Subcommittee on the Interior, 194
Ute Indians, 46, 47, 49, 63, 163, 217n55; culturally peeled ponderosa pine and, 4, 31, 217n60; San Luis Valley and, 3,

Ute Indians (*continued*):
29–30, 44; Sowapophe-uvehe and, 3; Spanish and, 34, 36, 37, 39, 40

Vaca Partners, 151–52, 191, 193–95
Valdez, Andrew, 145, 162, 172
Valley Courier, 168, 171, 192
Vargas, Diego de, 38–39
Volcanic Mining Company, 116–17

web-footed mustangs, 4, 84. *See also* Herard, Ulysses Virgil (Ulus)
Wellman, Bill: AWDI and, 144; Chaney and, 161–62; environmental history project and, 6, 8, 210; race with McInnis, 157, *158*, 189; Resource Management Strategy and, 145–46, 150
Well Y, 208
wetlands, 181, 196; at Great Sand Dunes, 204, 209; in San Luis Valley, 11, 18, 131, 147, 149, 150, 163
Wet Mountains, 63
Wet Mountain Valley, 42, 52, 63
White, Richard, 3
White Mountains, 52–53. *See also* Sangre de Cristo Mountains

Wilbur, Ray Lyman, 114, 161
Wilderness Act of 1964, 129
Wilderness Society, The, 179
Wilkinson, James, 50, 221nn1–2
Williams, Bill, 60, 65
Winner, Herb, 93
Wirth, Conrad L., 126
Wirth, Tim, 150

Yale University, 151–52, 191, 194
Younger Dryas, 21, 216n34

Zaldivar, Juan de, 34, 37
Zaldivar, Vicente de, 34
Zapata Creek, 164
Zapata Falls, 91
Zapata Ranch, 46, 83, 84, 86, 160, 161, 206; as biological hotspot, 151; history of, 163–68; The Nature Conservancy and, 166–69; Summit in the Sand and, 172–73; wildlife of, 163. *See also* Linger Ranch; Medano-Zapata Ranch
Zapata Springs, 46
Ziegler, George, 122–23
Zimmerman, Bob, 157–58, 160, 162, 173, 196
Zybach, Frank, 134–35

www.ingramcontent.com/pod-product-compliance
Lightning Source LLC
Chambersburg PA
CBHW020831160426
43192CB00007B/612